© Alan Morton

About the Author

Alice Hogge was educated at the University of St. Andrews, Scotland. She lives in London. This is her first book.

GOD'S SECRET AGENTS

Queen Elizabeth's Forbidden Priests and the Hatching of the Gunpowder Plot

ALICE HOGGE

NEW YORK • LONDON • TORONTO • SYDNEY

To Nicholas Fordham

HARPER PERENNIAL

A hardcover edition of this book was published in 2005 by HarperCollins Publishers.

FIRST HARPER PERENNIAL EDITION PUBLISHED 2006.

Library of Congress Cataloging-in-Publication Data is available upon request.

ISBN-10: 0-06-054228-4 (pbk.)
ISBN-13: 978-0-06-054228-3 (pbk.)

06 07 08 09 10 ❖/RRD 10 9 8 7 6 5 4 3 2 1

Contents

Acknowledgements

I should like to thank the following people for the advice, assistance and encouragement they gave me while I was writing this book: Father Thomas M. McCoog S.J., Father Geoffrey Holt S.J., Brother Hotkinson S.J., and their colleagues at the Jesuit Archives, London; the staff of the British Library; Mrs Joan Bond and all at the Catholic Central Library; the staff of the Public Record Office, Kew; the staff of the Oxfordshire Record Office; Professor John Guy; Mr Michael Hodgetts and Mr Julian Foord at Harvington Hall; Mr John Jarmen, Ian and Alan at Baddesley Clinton; Mrs Angela Sills and family; Yeoman Warder Alan Fiddis and his colleagues at the Tower of London; Mrs Jan Graffius at Stonyhurst College; Teresa Squires at Oxburgh Hall; Amanda Troop at Sawston Hall; Downside College; Lady Camoys; Mr Gerard Kilroy; Father Nicholas King S.J.; Father Michael O' Halloran S.J.; Professor V.A. McClelland and *Recusant History*; Philip Punwar; Nigel Barnes, Alan Morton, Lucie Dodds, and all at Models 1; Mr John Hartley; Philip Hogge; Simon and Suzie Fordham; Emma Höglund; Gavin and Ann Marie Hogge; Fred and Kay Hogge; Dr Corinna Peniston-Bird; Andrew Green; Richard Baron; Hywel Morgan; Alex Baillie-Hamilton; Outlook; Robert and Fiona Maida; Wanda Whiteley; Corinna Arnold.

Especial thanks go to Jane Bradish Ellames; to Amanda Russell; to my agent Stephanie Cabot, and all at William Morris; to all at Harper Collins, in particular Arabella Pike, Terry Karten and Kate Hyde; to Joyce Hogge and Dan Goode.

And if without the help of those above I would never have completed this book, then without the help of Nicholas Fordham I should never have begun it. To him I express my deepest gratitude.

Illustrations

God's Secret Agents

Chapter One

'. . . as the waves of the sea, without stay, do one rise and overtake another, so the Pope and his . . . ministers be never at rest, but as fast as one enterprise faileth they take another in hand . . . hoping at last to prevail.'

Sir Walter Mildmay MP, October 1586

ARMADA YEAR, 1588, swept in on a flood tide of historical prophecies and dire predictions. For the numerologists, who divided the Christian calendar into vast, looping cycles of time, constructed in multiples of seven and ten and based on the Revelation of St John and the bloodier parts of the Book of Isaiah, the year offered nothing less than the opening of the Seventh Seal, the overthrow of Antichrist and the sounding of the trumpets for the Last Judgement.[1]

For the fifteenth century mathematician Regiomontanus, although he had not been quite so specific about the year's unfolding, still the promise of a solar eclipse in February, and not one but two lunar eclipses in March and August had not, he had thought, augured well.* Regiomontanus had recorded his findings in Latin verse, concluding: 'If, this year, total catastrophe does not befall, if land and sea do not collapse in total ruin, yet will the whole world suffer in upheavals, empires will dwindle and from everywhere

* Regiomontanus's real name was Johan Muller, from Königsberg, now Kaliningrad, in Lithuania. He supplied Christopher Columbus with astronomical tables on his voyage across the Atlantic.

will be great lamentation.' As the year began, in Prague the Holy Roman Emperor Rudolf II, himself a keen astrologer, scanned the heavens for signs that his was not the empire to which Regiomontanus referred. He could discover little more than that the weather that year would be unseasonably bad.[2]

The printers of Amsterdam rang in the year with a special edition of their annual almanac, detailing in lurid prose the coming disasters: tempests and floods, midsummer snowstorms, darkness at midday, rain clouds of blood, monstrous births, and strange convulsions of the earth. On a more positive note, they suggested that things would calm down a bit after August and that late autumn might even be lucky for some, but this was not a January horoscope many read with pleasure.

In Spain and Portugal the sailors assembling along the western seaboard talked of little else, no matter that their King, His Most Catholic Majesty Philip II of Spain, regarded all attempts to divine the future as impious. In Lisbon a fortune-teller was arrested for 'making false and discouraging predictions', but the arrest came too late: the year had already begun with a flurry of naval desertions. In the Basque ports Philip's recruiting drive slowed and halted 'because of many strange and frightening portents that are rumoured'.[3]

In Rome it was brought to the attention of Pope Sixtus V that a recent earth tremor in England had just disgorged an ancient marble slab, concealed for centuries beneath the crypt of Glastonbury Abbey, on which were written in letters of fire the opening words of Regiomontanus' prediction. It was felt by the papal agent who delivered this report that the mathematician could not, therefore, be the original author of the verses and that the prophecy could stem from one source only: from the magician Merlin. It was the first hint that God might be on the side of the English.[4]

But in England no one mentioned Merlin's intervention in international affairs and the English almanacs that year were strangely muted affairs, proffering the general observation that 'Here and in the quarters following might be noted . . . many strange events to happen which purposely are omitted in good consideration.' With their fellow printers in Amsterdam working round the clock to meet the public's demand for gruesome predictions, it seems odd the English press were grown so coy, particularly when the editor of Holinshed's Chronicles had written the year before that Regiomontanus' prophecy was 'rife in every man's mouth'. But it was not in the Government's interest that England should be flooded with stories of death and destruction, for it was all too likely that any day now it would be visited by the real thing.*[5]

For some four years now England and Spain had been at war: an undeclared phoney war, fought at third hand, on the battlefields of the Low Countries and up and down the Spanish Main, by mercenaries and privateers, most notably the 'merry, careful' Francis Drake. Drake's raid on the port of Cadiz in April 1587 had cost Spain some thirty ships and had bought England a twelvemonth reprieve. But all this did was to postpone the inevitable until the fateful year 1588, because the Spanish *were* coming, with the mightiest fleet that had ever been amassed. Sixty-five galleons like floating castles, many-oared galleys, cargo-carrying *urcas*, nimble *pataches* and *zabras*, all these had been assembling in the west-coast ports of the Iberian peninsula since 1586. Together they could hold some thirty thousand men, numerous cavalry horses and pack animals and all the many carefully counted barrels of food and water needed to sustain a force of such size. The normally tight-fisted Pope Sixtus – 'When it comes to getting money out of him, it is like squeezing his life blood,' wrote the Spanish ambassador Olivares to King Philip – had swallowed his respect for the English Queen, Elizabeth I, and signed a treaty with Spain, promising a million gold ducats (about £250,000) to Philip should

* At the beginning of 1588 Dr John Harvey was commissioned by the Privy Council to write an academic pamphlet denouncing the accuracy of prophecies in general and Regiomontanus' in particular.

he manage to conquer England, so long as the next English ruler, a position on which Philip had designs himself, returned the country to Catholicism.* The Duke of Parma, commander-in-chief of the Spanish Netherlands, was only waiting for the signal to embark his army in a flotilla of flat-bottomed landing-craft, cross the English Channel and sail his forces up the Thames estuary. And all across Europe, rulers and ruled alike stopped what they were doing to watch and wait for the outcome to this clash between the forces of good and evil. This was ideological warfare of a type never before fought in Europe, transcending national boundaries and old-fashioned disputes about landownership. And while the opposing ideologies were, inevitably, somewhat tarnished by political and personal self-interest, nonetheless, in its purest distillation, this war was billed as the deciding round in the conflict between Catholicism and Protestantism, the final answer to a question that had paralysed sixteenth century Christian Europe, the question of what you could and could not believe. Though few could afford to be combatants, no one could afford to be neutral. But the result, it seemed, was a foregone conclusion.[6]

Ranged against England were the combined forces of Spain, Portugal, the Italian States, and the Spanish Netherlands, with France, though as yet undecided, also likely to join the Catholic crusade. The English troops, in comparison with Spain's professional, battle-hardened soldiers, were an ill-trained rabble of amateur militiamen, drafted into service at county-wide musters and required to pay for their own gunpowder for the duration of combat. The officers were no better. Most refused to take orders from anyone lower in social standing than themselves. Had it not been for the remodelling of the Queen's Navy by Drake's fellow sea-dog, John Hawkins, the outcome of the Armada conflict might well have been very different. Still, Hawkins's fleet of twenty-three warships and eighteen smaller pinnaces was heavily outnumbered

* Sixtus had said of Elizabeth 'were she only a Catholic, she would be without her match, and we would esteem her highly'.

by the massive, some hundred-and-thirty-strong Spanish and Portuguese Navy. And his belief that the success of Elizabeth's ships lay in long-range gunnery rather than traditional short-range grappling was not helped by those English gunmakers still busily selling cannons to the Spanish as late as 1587.[7]

So at the beginning of 1588 the odds on the Deity being a Spaniard were temptingly short. 'Pray to God', wrote one member of the Armada force, 'that in England he doth give me a house of some very rich merchant, where I may place my ensign.' Indeed, for all those about to embark with the Armada, England was a place of lucrative spoils and members of the fleet were delighted by how easy it was to obtain credit on the eve of sailing. Many spent their money on fine clothes for the occasion and one returning Englishman reported that 'the soldiers and gentlemen that come on this voyage are very richly appointed'. If the hard-headed bankers of Europe were putting their money on an easy victory for Spain, it was small wonder that in December 1587 a false rumour that the Spanish were coming sent the population of England's coastal towns flying inland for protection.[8]

As May 1588 gave way to a blustery June, a cargo ship under Captain Hans Limburger made its way slowly north from Cadiz, bound for the Hanseatic port of Hamburg. At Cape Espichel, just south of Lisbon, Limburger saw a sight that stunned him. Although Spain's preparations were no secret to anyone in Europe, nothing had prepared the German captain for this. For one whole day Limburger's ship beat slowly past the assembled Spanish fleet, 'the ocean groaning under the weight of them'. At Plymouth Limburger was able to give the port authorities the confirmation they needed: the Armada was under sail and on its way.[9]

Now, after many months of uncertainty, the orders were finally given for all army officers to remain on call and for all troops to be ready to move at an hour's notice. The better-trained soldiers were positioned near the most likely landing sites, to attack the invasion force while it was disembarking and at its most vulnerable. Barriers of logs and chains were brought in to seal off main roads and all routes into towns and cities. Militia groups were instructed in the scorched-earth policy they were to employ should the Spanish once get a foothold on land. Strategic points such as bridges and fording places were put under guard and instructions were given that in the coastal towns and villages no one was allowed to leave once the warning beacons had been lit, under pain of death. Then the nation waited.[10]

For now the bad weather Emperor Rudolf had seen written in the night skies and that had blown and sobbed its way through Europe for the better part of the year began to play its part in the conflict, breathing new life into the spectres of Regiomontanus' prophecy. By the end of June the Spanish fleet was still holed up in the port of Corunna as storms swept the Iberian coastline. A month later it was the English Navy's turn to suffer the high winds and heavy seas as it carried out its daily patrol of the western reaches of the English Channel. 'I know not what weather you have had there,' wrote Admiral Lord Howard of Effingham, commander of the fleet, to Sir Francis Walsingham, Principal Secretary of State, at court, 'but there never was any such summer seen here on the sea.' With the waiting came the whispering. 'There has been a rumour at Court, which has spread all over London,' reported Philip II's eyes and ears in the English capital, 'that the Spaniards have orders from their King to slaughter all English people, men and women, over the age of seven years.'[11]

Finally on Friday, 29 July the Armada was sighted off the Cornish coastline, and for Howard, Hawkins, and Drake, and the men of the English Navy, battle commenced. Throughout the following week, between contrary winds and dead calms, the smaller, more mobile English vessels harried their larger, cumbersome Spanish counterparts the length of the Channel, trying at every turn to disrupt the tightly packed crescent formation adopted by the Armada fleet. Shrouded in a heavy pall of gunsmoke it was hard enough for those in the thick of each encounter to know what was going on about them, but for those onshore and far inland the desperate clawing into wind to gain the advantage, the agonizing and hypnotic slowness of the combatants closing on each other, the silence broken by the roar of gunfire, was all a distant, disconnected dream. In mainland Europe rumour had the Armada safely landed in a defeated, humbled England, with a captive Queen Elizabeth on her way to Rome, to appear, barefoot and penitent, before Pope Sixtus. In England they kept on waiting.[12]

On Saturday, 6 August at 5 p.m. the Spanish fleet dropped anchor off Calais to make contact with the Duke of Parma. About midnight on the following day the tide turned, bringing with it, blazing out of the darkness, English fireships, packed with explosives, and in panic the Spaniards cut their cables and fled. The last chapter in the Armada's story had begun. 'From this piece of industry,' wrote one Spanish officer, 'they dislodged us with eight vessels, an exploit which with [our] one hundred and thirty they had not been able to do nor dared to attempt.' What the fireships started, the storm-force winds now continued, sweeping the scattered Spanish fleet first towards the shoals of the Zeeland banks and then helplessly northwards up the English coast. For another Spanish officer this had become 'the most fearful day in the world'. The Duke of Medina Sidonia, the Armada's ill-fated and much maligned commander, now placed himself squarely in the hands of 'God and His Blessed Mother to bring him to a port of safety'.*[13]

The report of the Armada's inadvertent flight north reached Queen Elizabeth as she was addressing her troops at Tilbury camp on 18 August, more than a week after the event. But even this good news did not come rumour-free, for now the Duke of Parma and his army were said to be on their way across the Channel. It was not until the end of August that the Dean of St Paul's was ordered to announce officially that the Armada had been defeated

* His flagship, the *San Martin*, reached the Spanish port of Santander on 21 September, with 180 of her crew dead from disease and starvation, another forty killed in battle and the rest so weak the ship had to be towed into dock. Only some seventy ships of the fleet returned.

and Philip II's agent in London was able to write home to Spain on 7 September that 'the Lords of the Council went to St Paul's to give thanks to God for having rescued the realm from its recent danger'. Just three days later, though, another alarm was spread that the Armada was on its way back. By the beginning of November, after ten weeks of continued uncertainty, the public's nerves were frayed to unravelling point. Parliament, which was to have met on 12 November, was prorogued until February 'as it was seen that both people and nobles were weary of so much trouble', wrote Marco Antonio Micea, a Genoese resident in London. 'We are in such alarm and terror here that there is no sign of rejoicing amongst the Councillors at the victories they have gained. They look rather like men who have a heavy burden to bear.' Even Elizabeth, who was not normally chary when it came to her own personal safety, was persuaded by her Council to stay away from St Paul's 'for fear that a harquebuss might be fired at her'. Micea noted that the fifty-five-year-old Queen looked 'much aged and spent'. Perhaps the Spanish fleet had made for Scotland and had succeeded in persuading King James VI to avenge his mother, Mary, Queen of Scots' execution of the year before; or perhaps they had rounded Scotland and were now in Ireland, stirring up trouble among the rebels there. In response to this new fear the Queen 'sent Sir Thomas Perrot to raise 2,000 men in Wales, and take them over [to Ireland] with all speed'. The extent of Elizabeth's anxiety may be measured by her willingness to throw yet more money at the conflict.[14]

But if in England no one could quite believe they had won, on the Continent no one could believe that the Spanish had lost. The French ambassador in London had spent his summer merrily reporting stories of heavy English casualties, so when the English ambassador in Paris, Sir Edward Stafford, produced 400 pamphlets giving the English version of events, he was met with frank incredulity. 'The English ambassador here had some fancy news printed stating that the English had been victorious,' wrote Don Bernadino Mendoza, the Spanish ambassador in Paris, 'but the people would not allow it to be sold, as they say it is all lies. One of the ambassador's secretaries began to read in the palace a [report] which he said had been sent from England, but the people were so enraged that he was obliged to fly for his life.' Only Pope Sixtus remained unimpressed by the European rumour-mill, refusing to loosen his grip on the million gold ducats he had promised Spain until he had better proof of Spanish success.[15]

On 24 November England at last felt confident enough to celebrate its victory and 'a solemn procession . . . was held to give thanks to God for the scattering of the Spanish fleet'. Through winter streets hung with blue cloth, Elizabeth 'was carried in a gilded chair . . . drawn by two grey horses royally caparisoned . . . to the great cathedral of St Paul's', from the battlements of which eleven captured Armada banners streamed out above the city. Here, Elizabeth read out a prayer she had composed specially for the occasion. The mood was one of relief, but also, more pertinently, of sober thanksgiving. In the words of the medal struck to celebrate England's victory: 'God blew and they were scattered'. The victory was His.[16]

And in giving this victory to God, Elizabeth extended a process begun by her father, Henry VIII, many years before: the process of nationalizing England's state religion.* It was an inadvertent process, born out of fear: Henry's fear of plunging his country back into yet another round of civil war if he failed to produce a male heir. And it was a process concluded in fear: the fear of standing alone and vulnerable against the richest and most powerful nation in Europe. The God who won the Armada could not be a Protestant God, at least not in the way Protestantism was understood throughout the rest of Europe. That God had demonstrably

* In the context of this book the term 'nationalism' is not intended to convey any preoccupation with English ethnic identity; rather, it is intended to convey the growing sense of English sovereignty at this period, in response to the fragmentation of Christendom and England's loss of its European territories.

failed to help the Calvinists in the Low Countries; to protect the Huguenots of France from the St Bartholomew massacre of 1572; to save William of Orange from assassination in 1584. Neither could He be a Catholic. Not when the Armada carried at its head a consecrated standard bearing the rallying cry 'Arise, oh Lord, and avenge thy cause'. Not when each ship had been provided with its own priest and each member of the expedition had received absolution in advance for the blood he would shed.* Therefore, He could only be an Anglican God, that compromise and most English of Gods, who continued to frustrate both Catholics and Protestants in almost equal measure.† So the subtle propaganda ran. In a Europe ravaged by wars of religion, only the English had chosen correctly.[17]

And how lucky it was that God should be an Anglican, an Englishman, for no one believed for a moment that England's conflict with Catholic Spain was over. Indeed, at the beginning of November the Venetian ambassador to Spain reported home to the Doge that 'In spite of everything, His Majesty shows himself determined to carry on the war.' So though the coach that bore Elizabeth to St Paul's Cathedral 'was open in front and on both sides' so that she might better be seen by crowds of cheering Londoners, yet her Government was taking no risks. An order had been given 'that in every household along the route no one should be allowed to look out from the windows while she was passing, unless the householder was prepared to stake his life and entire

* The Duke of Medina Sidonia's sailing orders to his fleet confirmed that 'the principal foundation and cause, that have moved the King his Majesty to make and continue this journey, hath been, and is, to serve God; and to return unto his church a great many of contrite souls, that are oppressed by the heretics, enemies to our holy catholic faith'.

† The terms Catholic, Protestant and Anglican should be applied with a certain linguistic caution. Broadly, the word 'Catholic' was by now recognized to refer only to those followers of the Roman Church. The word 'Protestant' referred initially only to German church reformers, though by the late sixteenth century it had acquired a wider reference and was in use in England to describe followers of the State Church. English Protestantism, though, was a very different animal from its European counterparts. The word 'Anglican' seems first to have been used as a derogatory term by King James VI of Scotland in 1598. It was not until much later that it became synonymous with the Church of England and English Protestantism.

fortune on his trustworthiness.' This was an England gripped in the jaws of fear and suspicion.[18]

On Friday, 28 October 1588 a sailing ship beat its way slowly up the Norfolk coast. On board, its passengers scanned the shoreline for a suitable landing place. Having spotted what looked like a safe point between Happisburgh and Bacton, some miles to the south of Cromer, they ordered the crew to drop anchor until nightfall. As darkness fell the ship's boat was launched and headed into shore. When it returned to the ship, it left standing on the beach two young Englishmen of whom the English Government had every reason to feel fearful and suspicious. The pair were Catholic priests belonging to the Society of Jesus, and their intention was to succeed where the Spanish Armada had failed: to return England to the Catholic Church. If the Armada was the latest in a progression of Franco-Spanish 'Enterprises' sanctioned by the Pope and designed to restore English Catholicism through force of arms, then these two young men represented Rome's second line of spiritual attack: force of argument.*[19]

Argument had always been the Christian Church's best weapon against heresy, chiefly because most heretical behaviour was thought to be a consequence of ignorance, poor judgement, or an imperfect understanding of the teachings of Christ.† Such heretics

* In 1585 Philip II had offered himself to Pope Sixtus V as the sword of the Catholic Church in the fight to reconcile England with Rome. Previously, French Catholics, led by the powerful Guise family, had planned a series of invasions, with the intention of deposing Elizabeth and replacing her with the half-French Mary, Queen of Scots (Mary's mother was Mary of Guise). Mary was Elizabeth's de facto heir apparent, although Elizabeth never acknowledged her as such.

† Heretics, as defined by St Thomas (II-II: 11: 1) were those who, having professed the faith of Christ, then proceeded to corrupt it from within. They differed from Infidels who refused to believe in Christ at all, and from Apostates who renounced Christianity for another faith, or no faith altogether. The word heresy derives from the Greek word for choice.

were not sinful therefore, merely misguided, and required little more than clear reasoning to make them see the error of their ways. Of course there were other heretics who wilfully rejected Christ's doctrines – out of pride, or a lust for power perhaps. They *were* sinners and merited the full weight of the Church's wrath, which, ever since the eleventh century, had usually meant burning at the stake. There was a further subset of heresy still: schism, the rejection of papal supremacy. For the rebellious schismatic (who might uphold all Christ's other teachings but this one) as for the misguided heretic, argument was deemed the best form of correction. So while Rome supported the invasion of England and the deposition of Elizabeth – her wilful heresy had imperilled the souls of her countrymen and God would forgive the use of force against her – it also dispatched its army of arguers. It was belief in the divine purpose of their argument that filled the two Englishmen now standing in the dark of a Norfolk beach, straining to hear above the noise of the waves on the shingle any sound to suggest their landing had been observed.*[20]

Neither man had been long enough in Rome to forget the seeping chill of late October English rain. The bad weather that had so hampered the Armada had not abated and the year was ending with as cold a spell as it had begun. But the rain and the cold were the least of the two men's worries as they now tried to put as much distance between them and the coast before dawn. In the dark it was impossible to pick a path that did not lead them up to a house instead of out into open fields. Twice, three times, a dog barked as they neared one of the fishermen's cottages flanking the beach and hastily they retraced their steps. Finally, they headed into a nearby wood to take cover until first light. There, in whispers, they decided it would be safer to separate and each make his own way to London; that way, if one of them should be caught, the other still had a chance of reaching the capital undetected. As soon as it was light enough to see, the older of the two men, twenty-seven-year-

* Every detail of this landing is taken from an account of it written up afterwards by one of the two Jesuits, Father John Gerard.

old Edward Oldcorne, son of a Yorkshire bricklayer, made his way northwards out of the wood towards the town of Mundesley. On the road he fell in with a party of sailors, demobbed and returning home after the defeat of the Armada, and in their company and with the cover they unwittingly afforded him, he made his way to London.

Meanwhile, his companion was leaving the wood by a different path. John Gerard was twenty-four. He was born on 4 October 1564, the son of the prominent Lancashire landowner Sir Thomas Gerard, a former county sheriff.* At the age of five John Gerard was removed from his parents' care when it was discovered his father was involved in a scheme to rescue the newly imprisoned Mary, Queen of Scots from Tutbury Castle in Staffordshire and restore her to the Scottish throne. Sir Thomas Gerard was arrested and held in the Tower of London until 1573. On his release he collected his eight-year-old son from the family of strangers on whom he had been forcibly billeted and returned with him to Lancashire, to Bryn Hall, the Gerards' estate. Whatever the effect on John Gerard of being ripped from his home at so young an age, these events did little to persuade him to conform to the State Church. At the age of just twelve he was sent down from Exeter College, Oxford for refusing to attend a Protestant Easter service.[21]

In the summer of 1577 Gerard applied for, and received, a licence to travel abroad to study. For the next three years he attended lectures at Dr William Allen's English College, first at the University of Douai in the Spanish Netherlands, then, when the college was expelled from there, at the University of Reims. Allen, an exiled Oxford academic, had opened the English College as a training school for those English boys still wishing to enter the Catholic priesthood, now that this was forbidden to them in their own country. The college also offered a thorough education to any English student unwilling to swear allegiance to the new Church of England, an oath required of all those graduating from the

* The sheriff was the Crown's chief executive officer in a county, in charge of keeping the peace, dispensing justice, and overseeing local elections.

universities of Oxford and Cambridge. At Reims Gerard first came into contact with a member of the Society of Jesus, an English Jesuit named Lovel, and from Reims he travelled to Cleremont, the school of the French Jesuits near Paris, determined to join the Society himself. He was not yet sixteen.[22]

The Society of Jesus was a new religious order, founded in 1540 by a Spanish ex-soldier called Ignatius Loyola, with the specific aim of converting the heathen and reconciling the lapsed. Loyola dreamed of countering the rise of Protestantism and restoring the Catholic Church to its former pre-eminence in Europe. To this end he had brought all his army training to bear on the problem in hand: 'I have never left the army,' he explained, 'I have only been seconded to the service of God.' His Jesuits operated as a tightly knit organization bound by a rigid, even military discipline, and Gerard, who used his time at the college to continue his studies, was quickly impressed by the elite band of priests who taught him. When illness forced him to return to England in the spring of 1583, he spent his convalescence disposing of property and possessions in preparation for a new life among them.[23]

His difficulty came in leaving the country for a second time. To leave England without State permission was a crime according to English law. To leave England to train as a Catholic priest was still worse a crime, the effects of which were often felt by the criminal's family in his absence. Gerard chose to leave the country without a licence. With a party of other Catholics, all heading abroad with intentions similar to his own, he set sail from Gravesend early in November 1583. The weather was against them. After five days at sea, making heavy progress into strong winds, they were forced to put in to Dover. At Dover it was revealed they had a spy in their company when the entire party was arrested by customs officers and sent up to London for questioning. The spy, Thomas Dodwell, reported back to the Privy Council how the group had bribed 'Raindall, the [officer] of Gravesend, [who] receiveth money of passengers, suffering them to pass without searching.'[24]

While his companions were imprisoned, the nineteen-year-old

Gerard (whose cousin Sir Gilbert Gerard was Master of the Rolls and held some sway with the Government) was taken into custody first by his uncle, George Hastings, brother to the Earl of Huntingdon, then by the Bishop of London. Both men set about encouraging Gerard to convert to the Protestant faith. Both men failed. It was a measure of his strength of will that the teenager held out against the arguments of his two more powerful opponents, with the threat of imprisonment, and worse, hanging over his head. But whatever fear Gerard might have felt, he left the Bishop of London's palace for prison still protesting his Catholicism.[25]

John Gerard was committed to the Marshalsea prison in Southwark on 5 March 1584. His year-long imprisonment was spent in the heady company of like-minded rebels against the nationalized Church: laymen and women arrested for their refusal to attend Protestant services and a number of priests awaiting execution. It was an intoxicating education. When at last his friends were able to secure his release from prison, in return for paid guarantees that he would not leave the country, his desire to become a Jesuit burnt fiercer than ever. At the end of May 1586 his chance came. An old friend of his, Anthony Babington, agreed to stand bail for him if he failed to appear before the authorities at the next quarter and John Gerard escaped to France. He was twenty-one.[26]

From France, Gerard travelled south to Italy, where he entered the English College of Rome, the companion school to Dr William Allen's successful Reims institution. By now his general eagerness to become a priest had transformed itself into the specific ambition of becoming a priest on the English mission. Pope Sixtus V granted him dispensation to take his holy orders early, some months short of the statutory age. The Society of Jesus agreed to admit him into their ranks as a novice and let him finish his training as he worked. And at last, on 15 August 1588, John Gerard became a Jesuit priest in the company of Edward Oldcorne. He was ready to return home.[27]

Throughout late August and all of September, as the Armada fleet underwent its grim circumnavigation of the British Isles and rumour ran unchecked through the courts of Europe, Gerard and Oldcorne travelled north towards England, accompanied by two other priests. 'Passing through Switzerland,' Gerard wrote, 'we stayed a night at Basle and decided to see the old Catholic buildings of the town: the Lutherans usually leave them intact but the Calvinists destroy them.' At Reims, he noted, 'we passed incognito': the seminary city was full of spies. In Paris a prisoner in one of the city gaols calling himself Jacques Colerdin learned of their arrival. On 1 October Colerdin was able to scribble a letter to Sir Francis Walsingham in London, telling him that 'There be 8 Priests over from Rome, whereof John Gerard . . . will be in England within five days.' Colerdin, who described himself to the Archbishop of Paris as 'an English priest and Bachelor in Theology' in a petition he wrote seeking his release, was a Government informer.* His real name was Gilbert Gifford. He was, indeed, a Catholic priest, but since his arrest as an alleged accomplice in the Babington Plot he had found it prudent to switch sides in the religious conflict.† Now he was well placed to point out his fellow seminarians to the English authorities.[28]

From Paris, Gerard and Oldcorne continued on to Eu, some miles north of Dieppe, in preparation for crossing the Channel. But here they received unwelcome news from England. 'The Spanish Fleet', wrote Gerard, 'had exasperated the people against the Catholics; everywhere a hunt was being organised for Catholics and their houses searched; in every village and along all the roads and lanes very close watches were kept to catch them.' Clearly conditions at home were far from ideal for them to attempt a landing in secret and for the next few weeks the pair were forced to kick their heels on the French coast, while their superiors back

* He was arrested in a Paris brothel in 1587, though the precise nature of his crime remains unclear. He died in prison in 1590.

† The 1586 Babington Plot was the last in a series of Catholic attempts to assassinate Elizabeth and free Mary, Queen of Scots in advance of a foreign invasion; it led to Mary's execution. Anthony Babington, Gerard's friend, appears to have been a man of more devotion than sense and how much the plot was the work of *agents provocateurs* remains ambiguous. Gilbert Gifford's role is particularly questionable and most believe he was working for Walsingham from the start. Gerard's father was also arrested for alleged complicity in the plot. He was released in October 1588.

in Rome decided what should be done. At last a letter came through: 'we were free', wrote Gerard, 'either to go ahead with the enterprise or stay back until things in England had quietened down. This was the answer we desired.' Immediately, the two men set about finding a ship.[29]

As John Gerard stood in the shadows of a Norfolk wood choosing the best and safest route to London, he had already committed treason, according to England's latest laws. The act of 1585 'against Jesuits, seminary priests and such other like disobedient persons', one of nine pieces of parliamentary legislation during Elizabeth's reign to seek to redefine treachery in the face of a newly perceived menace, employed bully-boy language to make its point. Any Englishman ordained a Catholic priest since June 1559 would, the act threatened, soon find out 'how dangerous it shall be for them . . . once to put their foot on land within any of her Majesty's dominions'. In returning home, in stepping from his ship's boat onto a Norfolk beach, John Gerard had become a traitor to his country. If caught, he would be punished accordingly. As he left the wood, heading westwards, he was spotted by a group of men walking towards him.*[30]

Gerard takes up the story:

> 'Walking boldly up to them I asked whether they knew anything about a stray hawk; perhaps they had heard its bell tinkling as it was flying around. I wanted them to believe that I had lost my bird and was wandering about the countryside in search of it [then] they would not be surprised because I was a stranger here and unfamiliar with the lanes and countryside; they would

* The coastal counties had borne the brunt of the nation's anxiety during the Armada conflict. Close watches had been kept throughout the summer months for spies, Spanish ships, and anything suspicious; now those same coastal watches were kept busy patrolling the countryside looking for vagabonds. The vagabonds in question were disbanded sailors and militiamen, laid off without their promised pay and in search of food and work. Since the Armada, an order had been issued that any vagabond 'found with any manifest offence tending to stir troubles or rebellion . . . [was] . . . to be executed by martial law'.

> merely think that I had wandered here in my search . . . They told me they had not seen or heard a falcon recently and they seemed sorry that they could not put me on its track. So with a disappointed look I went off as if I were going to search for it in the trees and hedges round about.'[31]

This was his strategy for the rest of that day. Each time he saw someone working in the fields he approached them, asking them the same question: had they seen his hawk? His progress was slow. Occasionally he doubled back on his tracks to make his search more convincing. But gradually he moved inland, away from the sea.

'At the end of the day I was soaked with rain and felt hungry. It had been a rough crossing and I had been able to take practically no food or sleep on board, so I turned for the night into an inn in a village I was passing, thinking that they were less likely to question a man they saw entering an inn.' Inside, he made enquiries about buying a pony and found the people willing to help him. The following morning, Sunday, 30 October, he set off on horseback towards Norwich, no longer in any danger of being taken for a vagrant, but still at the mercy of the county watches. At the village of Worstead he was apprehended.[32]

'They ordered me to dismount, and asked me who I was and where I came from. I told them I was in the service of a certain lord who lived in another county – he did in fact know me well, although these men had not heard of him – and I explained that my falcon had flown away and I had come here to see whether I could recover it.' But this time the watchers refused to release him, insisting he be brought before the constable and the officer of the watch for further questioning. Gerard submitted and was led to the village church where the two men were attending morning service. Now he was faced with a dilemma. 'One of the watchers went in [to the church] and came back with the answer that [the officer] wanted me to come inside where he would see me at the end of the service.' For Gerard it was a sin to enter a Protestant church. So Gerard refused to go in, claiming he was reluctant to leave his horse behind. When the officer at last came out to question him he was clearly angry and suspicious. 'He asked me first where I came from and I named a number of places which I had learned were not far away. Then he asked me my name, employment, home, the reason for my coming, and I gave him the answers that I had given before. Finally on asking whether I was carrying any letters, I invited him to search me.' The officer was unimpressed. He declared 'it was his duty to take me before the Justice of the Peace', and Gerard prepared himself for immediate arrest. Then suddenly the man relented, with the words, 'You've got the look of an honest fellow. Go on then in God's name.' Later, Gerard attributed this stroke of good fortune to providence. Now, he hurriedly set off towards Norwich before the officer could change his mind.[33]

Furnished by a fellow traveller with the name of a suitable inn on the southernmost outskirts of Norwich, Gerard circled the city walls. He avoided the busy London road, which led into the city through the well-guarded St Stephen's Gates, and passed instead over common grazing land to Brazen Doors, a smaller set of gates that opened onto All Saints Green. From there it was a short walk to the inn on Market Hill, at the foot of Norwich Castle.[34]

The inn was busy and Gerard settled himself down to observe. 'I was there only a short time when in walked a man who seemed well known to the people of the house. He greeted me courteously and then sat down by the fire to warm himself. He began talking about some Catholic gentlemen imprisoned in the city and mentioned by name a man, one of whose relatives had been with me in the Marshalsea Prison . . . I listened carefully but said nothing.' When the man left the room, Gerard asked his neighbour who he was. The reply was welcome news to him: 'He is a very good fellow, except for the fact that he is a Papist.' The man was out on bail from the city gaol after a decade in prison for his faith and he was, by common consent, 'a most pig-headed' Catholic.*[35]

* The 'pig-headed' Catholic gentleman was almost certainly the Norfolk landowner Robert Downes, whose Melton estate lay just a mile west of Norwich. He was arrested in 1578 for his refusal to attend the Protestant Church. He lost his estates in Suffolk and Essex

'I kept quiet until the man returned and when the others had gone out I told him that I wanted to have a word with him in some safe place. I had heard he was a Catholic, I said, and was very pleased to hear it because I was one too.' Briefly, Gerard explained how he came to be in Norwich and asked for the man's assistance in getting to London. The man knew of no one travelling to the capital that Gerard could join, and so pass as one of their party, but he did know someone in town who might be able to help him and he left the inn to find this contact.[36]

When he returned to the inn a short while later he asked Gerard to follow him out onto the street and into the thick of the bustling market. While the two men pretended to examine the various goods for sale they were observed from a distance by a third man. Soon he approached the pair and asked them both to come with him. He led the way through the narrow side streets of Norwich towards the city's cathedral. There, in the cavernous nave of the great church, the man questioned Gerard intently before asking him outright whether he was a priest – in which case, said the man, he would offer him all the help he needed.* Gerard asked his name. Then he admitted he was a Jesuit priest sent from Rome.[37]

The printers of Amsterdam, as they compiled their annual almanac for 1588, had predicted that late autumn would be lucky for some. So it had proved for John Gerard. The man in whose hands he had just placed his life was Edward Yelverton, one of the richest Catholics in Norfolk. That same evening Yelverton spirited Gerard out of Norwich and by Monday, 31 October the priest had gone to ground at Yelverton's house at Grimston, 6½ miles north-east of King's Lynn. After forty-eight hours at large Gerard had reached a safe haven. His work, though, had only just begun. And with the events of the summer that work had become harder still.[38]

---

and in 1602 he surrendered most of his life interest in Melton to the Queen. He was still in Norwich gaol in 1598. He died in 1610.

* Catholics believed that though they could not enter a Protestant church where divine services were held, the nave of a cathedral was part of the general precincts of the building and therefore not sacrosanct.

The Spanish Armada had achieved what no amount of religious reforms and parliamentary legislation over the preceding decades had been able to do. It had united a fractured nation behind its unhappy compromise of a nationalized Church in opposition to the Catholic crusade. It had bound Anglicanism and Englishness together seemingly indissolubly. And it had transformed all those who could not bring themselves to accept this new Church of England into potential traitors, fifth columnists willing, in the rhetoric of a royal proclamation issued that July, 'to betray their own natural country and most unnaturally to join with foreign enemies in the spoil and destruction of the same'. To be a Catholic in the year 1588 was to be an *unnatural* Englishman. It was to be worse than that still, as a story told by a fellow Jesuit illustrated. In late August, as many Catholics awaited execution for their alleged treachery, a 'certain lady went to a man of importance asking him to use his influence that the death of one of the condemned might be delayed. The first question was whether the person whose cause she pleaded was guilty of murder. She replied that he had not been condemned for any such thing, but only for the Catholic religion. "Oh dear," said the gentleman, "For his religion! If he had committed murder I should not have hesitated to comply with your request; but as it is a question of religion, I dare not interfere."' To the English in Armada year, even homicide was less of an evil than Catholicism.[39]

And Gerard had come to join a covert mission, the first undertaking of which was to ask English Catholics to stand up and be counted, to demonstrate their faith by refusing to attend the Protestant Church, to *identify* themselves to a Government seeking to eradicate them as traitors. Small wonder then that in an England still reeling from the events of that year, in an England crying out for revenge, this task had acquired Herculean proportions.*[40]

* One eyewitness wrote of the vengeance taken against English Catholics following the Armada's defeat: 'When the danger of the war at sea was over, and the army conscripted upon land dispersed, our rulers turned their weapons from the foe abroad and plunged them into the bowels of their own nation. The hatred stored up against the Spaniards they are wreaking with a sort of bestial fury upon their own fellow citizens and subjects.'

Fourteen years earlier, though, for the first young English priests to embark upon this newly begun mission, their venture must have seemed no less a trial of strength. They may have been returning home to their family and friends, but at the same time they were entering an unfamiliar, hostile world: a beleaguered and suspicious England whose savagery had yet to be put to the test. And they were entering that world in secret, in disguise, in the very manner of the political secret agents their homeland believed them to be. Formidable, too, had been the challenge of establishing the mission at the start, of drawing together the dejected exiles from an Oxford University shattered by religious reforms, of schooling them in martyrdom and of dispatching them home in increasing numbers to face unknown perils.

For his ninth task Hercules had only to steal the Amazon Queen's belt – a gift she herself had been willing to grant him. John Gerard and his fellows were attempting to steal the souls of English men and women back to the Catholic Church, and to do so under the nose of the, perhaps, even more redoubtable English Queen. It was a labour that would come at a high price.

Chapter Two

'If God himself on earth abode would make
He Oxford sure would for his dwelling take.'
(Sixteenth century)

On 10 December 1566, eight years into the reign of Queen Elizabeth I and twenty years prior to John Gerard's secret Norfolk landing, Magdalen College acquired a new tenant for its property at 3 Castle Street, standing in the shadow of Oxford Castle. The tenant's name was Walter Owen. He was a twenty-six-year-old carpenter with a wife and young family. In time, all four sons from this family would join the mission to save English Catholicism. Two would die for it. One, in death, would hold in his hands the life of almost every Catholic involved in it. His name was Nicholas Owen.[1]

Few facts are known about Nicholas Owen's childhood: an approximate date of birth (some time between 1561 and 1564), a joinery apprenticeship (in February 1577) to Oxford's William Conway, and the location of the Owen family's house on Castle Street – little more. Across the road from this six-room, two-storey tenement stood the twelfth century parish church of St Peter-le-Bailey, a 'very old little church and odd'. Four doors to the left of the house was 7 Castle Street, called Billing Hall or the Redcock. Here, in 1298, it was said a clerk had caused the Devil to appear. A few yards to the right of the house were the butchers' shambles, a row of shops in the middle of the newly paved Great Bailey Street. Here, the blood and offal spilt from the freshly killed carcasses coursed over the gravel and into the drainage channel running down the centre of the road. Heaven, hell and the stench of blood: Nicholas Owen was raised within the axis of all three.[2]

Further still to the left, high up on the hill, stood the tall ten-sided keep of Oxford Castle, its walls as thick as a man was tall. Within that keep there stood another, its walls only slightly less thick, and inside that there was the well chamber, with a well shaft so deep you could not fathom the bottom of it. Walls within walls, chambers within chambers: Oxford Castle would provide plenty of inspiration for Nicholas Owen in his later life. Beyond the keep there lay the castle gaol and next to the castle gaol there stood the gallows.[3]

The castle was in a ruinous state by the time of Owen's birth: the seat of Oxford's civic power had long ago shifted to the city's Council Chambers. But beyond the parish of St Peter-le-Bailey, to its west, lay evidence of a more recent power-shift still. Here were the remains of Oseney Abbey, lately Oxford's cathedral. All that was left of it were the church walls, the dovecote and the out-buildings; the rest of the stone had been stripped from the site and carried over to the construction works at Christ Church College. Further to the south a similar process was at work as the Franciscan and Dominican friaries were dismantled piece by piece by city speculators and sold off to make new townhouses.[4]

The landscape of England was being re-drawn. The castles and manor houses of the old feudal aristocracy had shared their domination of the English countryside with the spires and steeples of the abbeys, priories and monasteries. They had stamped their authority on the public consciousness by the sheer scale of their physical presence. But both aristocrats and abbots had found themselves systematically stripped of that authority in Tudor England. And all those who had bowed to the seigniority of Nobility and Church, who had prospered under it or were sheltered by it, were left shivering in the brisk winds of change.

This was the birth of modern England, with a newly re-worked relationship between Parliament and monarch, and an increasing dependence on the unstoppable middle classes. Of course for some it had been an entirely unwanted pregnancy, but for many, many more across the nation, some with, some without a vested interest in the old order, but all of them sharing a strong desire for stability and the certainty of tradition, it would prove a difficult birth: bloody and unutterably painful. And nowhere was this truer than at Oxford.

On 3 February 1530, just over thirty years before Nicholas Owen was born, William Warham, Archbishop of Canterbury, peremptorily thrust Oxford University into the centre of the controversy of the hour. He wrote to the Vice-Chancellor asking him and Oxford's academics to provide a unanimous opinion on the validity of the marriage of Henry VIII to Catherine of Aragon.[5]

The request was as untactful as it was unwelcome: Cardinal Wolsey, the university's wealthiest patron, had been arrested only the year before for failing to provide Henry with the verdict he was looking for. And now it must have seemed to those at Oxford, asked to enter after Wolsey into this most explosive of minefields, that the Cardinal's downfall would soon be followed by their own. So Oxford dragged its heels. When Cambridge, to the same request, came quickly back with the answer Henry wanted, the relief in the fens may have been palpable but the spotlight now shone ever more brightly on the midlands. And still Oxford dragged its heels.[6]

In early April the King could wait no longer. His agents, led by the Bishop of Lincoln, descended on the university, hotly pursued by a strongly worded letter from Henry himself. While the bishop worked on Convocation, persuading them to hand the matter over to the university's theologians, Henry reminded the 'youth' of

Oxford precisely where their loyalties lay. This two-pronged attack produced the desired result. Though the Faculty of Arts grumbled that the Faculty of Theology had no right to speak for the university as a whole, the combination of manipulation and not so veiled threat had won the day. On 8 April Oxford University gave Henry the answer he was looking for: his marriage was invalid. But its tardiness in doing so was neither forgiven nor forgotten.*[7]

In September 1535 Henry's agents were back in Oxford as part of a whirlwind tour of the country in preparation for the dissolution of the monasteries and by now the bloodshed had begun. In July of that year one of Oxford's most illustrious former scholars, Sir Thomas More, was executed on Tower Hill, a victim of the new Treason Act: a piece of legislation that efficiently turned loyalty to the Pope in Rome into treachery to the English State. More's scruples did not trouble Henry's agents, though: Dr Richard Layton, 'a cleric of salacious tastes', and his assistant, John Tregwell, brought with them to Oxford an estate agent's eye for a property and a prospector's nose for gold, neatly masked behind an official mandate to root out opposition to the King's new church. Their report, when it came, hit the university a sickening blow.[8]

The first wave of the dissolution saw all of Oxford's religious houses shut down: the Benedictine-run Canterbury, Durham and Gloucester Colleges, and the Cistercian-run St Bernard's College. Their assets were stripped, their real estate sold off to the highest bidder and their inhabitants turned out onto the streets. Scores of academics and undergraduates now found themselves jobless, homeless and penniless. Hard hit too were the university libraries. College after college was ransacked for its illuminated books (seen as symbols of a despised papist idolatry), which were then carried out in cartloads and destroyed. In New College quadrangle the pages of the scattered medieval manuscripts blew thick as

* When Thomas Cromwell was made Chancellor of Cambridge in 1535, on the execution of Cardinal John Fisher, Oxford graduates saw Government preferment steered past them towards the students of Cambridge.

the autumn leaves, reported Layton. One enterprising student, a Master Greenefeld from Buckinghamshire, gathered them up and used them to make 'Sewells or Blanshers [game-scarers] to keep the Deer within the wood, and thereby to have better cry with his hounds.'[9]

Oxford reeled under this attack. The developing university had drawn the wealth of the monasteries to the city. Those same monasteries had spawned the abbeys and priories, and they, in turn, had financed the building of Oxford's first academic halls. And so the university and the city had grown. Even the newer, secular colleges, which escaped the cull but were bludgeoned into a show of loyalty, were dependent on the Church revenue now being siphoned into the Crown's coffers. The destruction of the monasteries brought academic chaos to Oxford. More pertinently, it also brought festering resentment.[10]

Some fourteen years later, in 1549, royal agents called on Oxford again. This time they were Edward VI's visitors, come to enforce his new Prayer Book, the first real doctrinal step towards Protestantism.* Hot on their heels came the German and Swiss mercenary forces of Lord Grey of Wilton. Let Oxford be in no doubt that this regime meant business. Of the thirteen heads of those Oxford colleges still remaining, the visitors could find only two who supported the Government's religious policy; as the Protestant reformer, Peter Martyr, remarked: 'the Oxford men . . . are still pertinaciously sticking in the mud of popery'. But events soon proved that Oxford men were not the only stick-in-the-muds. July of that year burnt with the heat of a countrywide rebellion. On 12 July the Duke of Somerset wrote to Lord Russell, to lament the 'stir here in Bucks. and Oxfordshire by instigation of sundry priests (keep it to yourself)'. Lord Russell scarcely had time to broadcast the news; just six days later Grey's mercenaries had quickly and ruthlessly put out the fires. The eleven-year-old King Edward noted gleefully in his journal that Lord Grey

* Each college was provided at its foundation with an external 'visitor' – part trouble-shooting ombudsman, part spiritual inquisitor.

'did so abash the rebels, that more than half of them ran their ways, and other [*sic*] that tarried were some slain, some taken and some hanged'. Grey left careful instruction that 'after execution done the heads of every of them . . . to be set up in the highest place for the more terror of the said evil people'. The vicars of Chipping Norton and of Bloxham were hanged from the steeples of their own churches, and Johann Ulmer, a Swiss medical student at Christ Church College, wrote home to his patron, the Zurich reformer Heinrich Bullinger, that 'the Oxfordshire papists are at last reduced to order, many of them having been apprehended, and some gibbeted, and their heads fastened to the walls'.[11]

The purges soon followed. Magdalen College lost its president, Dr Owen Oglethorpe, forced out in favour of a suitable Protestant candidate. Christ Church lost John Clement, a former tutor in the household of Sir Thomas More, who now fled to the university at Louvain in the Spanish Netherlands to join the growing community of Oxford disaffected there. Corpus Christi lost its dean and its president, both arrested and carried off to London, one to the Marshalsea prison for seditious preaching, the other to the Fleet prison for using the old form of service on the preceding Corpus Christi day. These were not the only expulsions and Oxford braced itself for a stormy future.[12]

And then the weather turned. On 6 July 1553 Edward died, and the chill winds of reform swung southerly, blowing before them, back from the Continent, the exiles of Oxford: Oglethorpe back to Magdalen, Clement back to practise medicine in Essex. They passed on the dockside a new generation of Oxford men, John Jewel of Corpus Christi, Christopher Goodman, the Lady Margaret Professor, off into exile in their turn, as Queen Mary immediately rescinded her brother's statutes. The next five years brought calm to a city which basked in the warmth of the sovereign's personal favour. They also brought prosperity. Mary tripled the university's revenue and oversaw the foundation of two new colleges, Trinity and St John's, in place of the ruined Durham and St Bernard's.[13]

At Corpus Christi the ornaments and vestments hidden from sight during Edward's reign were triumphantly returned to the college chapel. Communion tables were quickly removed and replaced with new altars; the '6 psalms in English', the 'great Bible' and the 'book of Communion' – all of which had been demanded by Edward's visitors – were destroyed; and, all around Oxford, people picked up the pieces of lives rudely shattered by statute from London. When ex-Bishops Hugh Latimer and Nicholas Ridley, and the former Archbishop of Canterbury Thomas Cranmer, were sent to Oxford for trial, it was signal proof that Mary never doubted the city's loyalty. That the Government wished to demonstrate in an academic setting the shortcomings of doctrinal heresy and that all the accused were from Cambridge merely underlined the wisdom of the decision. When the three men were burnt in a ditch opposite Balliol College it had little impact on the watching crowd – Latimer's 'candle' of martyrdom found a marked lack of oxygen in Oxford.[14]

Mary was swept into power on a wave of popular support, but she found herself left high and dry on virgin sand as England's first queen regnant since Matilda's ill-fated attempt to assert her claim to her father, Henry I's throne. Matilda had plunged the nation into a nineteen-year civil war; throughout the last bitter summer of Mary's reign Englishmen can have felt only relief that they had got away so lightly this time around. Still, the previous five years had brought more than their fair share of misery. On her accession Mary had commanded her 'loving Subjects, to live together in quiet Sort, and Christian Charity, leaving those new found devilish Terms of Papist or Heretic'. Her first Parliament had seemed to speak for the majority of her people. Then had come marriage to the reviled King Philip of Spain, the burning of Protestants and a disastrous war with France, and now those same people, the stench of the execution pyres fresh in their nostrils, awaited her death with impatience. On 17 November 1558 it finally came and 'all the churches in London did ring and at night [the people] did make bonfires and set tables in the street and did eat and drink and make merry for the new Queen', wrote Machyn in his diary. At twenty-five years old that new young queen, Elizabeth, was very much an unknown quantity though. While London celebrated, Oxford held its breath and waited for what the royal visitors would bring.[15]

On Christmas Day 1558, just weeks into the new reign, Dr Owen Oglethorpe of Magdalen, now Bishop of Carlisle and the officiating divine for the festivities, received a message from Queen Elizabeth asking him not to elevate the consecrated host at High Mass that day. The Spanish ambassador, Count de Feria, reported Oglethorpe's refusal to comply with the request: 'Her Majesty was mistress of his body and life, but not of his conscience'. Elizabeth heard mass that day until the gospel had been read and then, as Oglethorpe prepared to celebrate the transformation of bread to body and wine to blood, she rose and left the royal chapel. To those who watched and waited this was the first public indication of which way the Queen might jump.[16]

If Elizabeth herself had wanted a sign of how the battle lines were forming she need not have looked far. Mary's death had brought with it a flurry of bag packing in Geneva, Zurich, Strasbourg and Frankfurt, the centres of Protestantism, as the exiles from the previous reign prepared to return home. After all, Elizabeth, like her brother Edward, had been educated from childhood in the new religion. Meanwhile, the opposing camp was quick to make its objections felt. At Mary's funeral the Bishop of Winchester praised the dead monarch as a good and loyal daughter of the true Church, referring to Elizabeth, throughout, as 'the other sister'. He laid down his challenge to the new Queen, a challenge peculiar to her sex, in the bluntest of terms: 'How can I, a woman, be head of the church, who, by Scripture, am forbidden to speak in church, except the church shall have a dumb head?' At Elizabeth's coronation the Archbishop of York refused to officiate and only Owen Oglethorpe could be persuaded to perform the ceremony. And in France King Henri II, who had ordered that the arms of England should be quartered with those of Scotland upon the marriage of Mary Stuart to the Dauphin, now encouraged his son and daughter-in-law to style themselves King and Queen of England.[17]

Was Elizabeth's choice of religion ever really in doubt then? She *was* the daughter of the 'concubine Anne Boleyn', the woman for whom Henry VIII had broken with Rome in the first place. Her parents' marriage had never been recognized by the Catholic Church and her own legitimacy of birth had long been a subject for parliamentary enactment. Just days before her mother's execution Thomas Cranmer had annulled her parents' marriage and Elizabeth, at a stroke, was both bastardized and disinherited. Her present claim to the throne rested on her father's will and the Succession Act of 1543, which reinstated her as Henry's heir, and the French, in particular, were quick to cast doubts on Parliament's right to tamper with these sacred laws of inheritance – by Christmas 1558 they 'did not let to say and talk openly that Her Highness is not lawful Queen of England and that they have already sent to Rome to disprove her right', wrote Lord Cobham, Elizabeth's envoy in Paris.[18]

The French had indeed sent to Rome. By the New Year Sir Edward Carne was reporting back from the Holy City that 'the ambassador of the French laboreth the Pope to declare the Queen illegitimate and the Scottish Queen successor to Queen Mary'. That the French chose to object to Elizabeth's claim out of political self-interest rather than religious scruple was not in question: they had raised similar doubts about the legitimacy of Elizabeth's sister Mary when Henry VIII tried to engineer an overly advantageous marriage treaty between her and the Duc d'Orléans. But their challenge to her title underlined Elizabeth's quandary. If she wished to retain papal supremacy in England she would need to throw herself on the mercy of the Pope. Paul IV had intimated that he was quite ready to consider her claim to her title, but could she really stomach the indignity of going cap in hand to Rome, begging to be excused her bastardy? And could she afford to begin her reign from a position of such weakness? Surely England's throne was her birthright and no Pope could grant her dispensation to wear the crown? So 'the wolves of Geneva' packing to return home to England knew from the start that the odds on Elizabeth seeking a national religion, independent of Rome, were short enough for them to stake their lives on. Now they came back in readiness for that outcome.[19]

They did not have long to wait. On 8 May 1559 Elizabeth dissolved the first Parliament of her reign, giving royal assent to those acts from which her new Church would take its shape: the Act of Supremacy, which settled on Elizabeth the title of Supreme Governor of that Church, and the Act of Uniformity, which agreed the doctrine it should follow.

The reactions followed swiftly. 'A leaden mediocrity,' wrote the newly returned Protestant, John Jewel. 'The Papacy was never abolished . . . but rather transferred to the sovereign,' wrote Theodore Béza in Geneva to Heinrich Bullinger in Zurich. From the first, Elizabeth's was a Protestant settlement that failed to please the Protestants. But neither did it please the Catholics. 'Religion here now is simply a question of policy', wrote the Bishop of Aquila from London, 'and in a hundred thousand ways they let us see that they neither love us nor fear us.' John Jewel expressed his surprise that 'the ranks of the papists have fallen almost of their own accord', and Count de Feria wrote sadly home to Spain to explain why: 'The Catholics are in a great majority in the country, and if the leading men in it were not of so small account things would have turned out differently.'* And in London a zealous mob went on the rampage, stripping the capital's churches of their statues and stained-glass windows 'as if it had been the sacking of some hostile city'.[20]

* De Feria believed that the leading Catholics, in both the Commons and the Lords, had failed to put up a convincing fight during the crucial parliamentary debates from which the settlement sprang. However the Catholics were also under-represented in these debates: ten out of the twenty-six bishoprics were empty when Parliament opened on 25 January.

From the start Elizabeth's religious settlement was a compromise. Like all compromises it failed to satisfy anyone and like all compromises it would be subjected to stresses and strains as each dissatisfied party tried, in turn, to wrest back the advantage. But it is the fact that there was need for a compromise that is of significance to this story, because it suggests a country divided into pressure groups of equal fighting weight.

The religious changes of the English Reformation, so decisive and so devastating for a select few in key positions of authority, had filtered slowly through the rest of the country, dependent upon the efficiency and willingness of those officers charged with their enforcement. By 1558 England's religious spectrum was a kaleidoscope of colours ranging all the way from the most Roman of purples to Puritan grey. It is impossible to estimate the precise number of confirmed Catholics and Protestants, together with the number of relative indifferents, in England at Elizabeth's succession. It is equally impossible to arrive at a precise and consistent definition for English Catholicism or English Protestantism at this time: these were not hermetic terms upon which everyone could agree and with which everyone could identify. Indeed it is unlikely that everyone could have told you *what* they were, Catholic or Protestant, if questioned. It is highly probable that in reaching a compromise settlement the Government paid close attention to the predictable response of the powerful and predatory Catholic nations of Europe. But it is certain that such a compromise would not have been necessary had England not been divided, top to bottom, on this matter of religion. The England Elizabeth inherited was definitely not a Protestant country.

For every Londoner in the largely pro-Protestant capital who went on a spree of vandalism, there was someone else in the shires and villages quietly secreting away the statues, crucifixes and church plate for happier times. For anyone in the south of England – and close to the seat of government – prepared to toe the party line if it led to promotion, there was someone else in the north of the country and far from influence, stubbornly doing as he pleased. For anyone whose heart belonged to Geneva and who felt they had been betrayed by the Queen, there was another whose heart belonged to Rome, who was smarting just as badly. And for Elizabeth, whose heart belonged firmly to England, the challenge lay in holding these two opposing forces in their precarious balance long enough to allow civil divisions to heal over and to effect the urgently needed overhaul of the country's economy and the repair of its diplomatic relations with the rest of Europe. Because the twenty-five-year-old Queen can have been in little doubt that, as far as Europe was concerned, she and England were in for a turbulent future.

And for the ordinary man and woman in the street, the challenge lay in working out precisely what was required of them by this latest change to the national religion.

On the face of it these requirements were simple. The new Act of Uniformity demanded each subject's presence at their parish church every Sunday and holy day. Failure to do so, without reasonable excuse, would result in a twelve pence fine for each offence, or the 'censure of the church' and possible excommunication, with the consequent loss of civil rights.

Going to church on a Sunday had long been a tradition inspired by faith but enforced by the ecclesiastical courts and over the centuries the machinery of that enforcement had become powerfully efficient: the long arm of the Church's law reached the full length and breadth of England. It was this machinery, in place, fully operational and re-greased with parliamentary drive in place of holy oil, that Elizabeth's ministers now used to unseat the religion that had devised it: the Catholic Church in England was hoist with its own petard.

There were other requirements too. Subjects were not to speak in a derogatory fashion about the new Prayer Book, nor to cause a clergyman to use any other form of weekly liturgy than the one specified by the Queen's officers. The penalties for this were fines of 100, then 400 marks, and, thereafter, life imprisonment. And they were not to be caught defending the papal supremacy; not unless they were prepared to forfeit first their goods, then their liberty, then their life.

But so long as they kept their weekly appointment at the Queen's new Church and their mouths tightly shut, there was nothing to stop them benefiting from Elizabeth's lenient attitude towards Catholicism. And this was lenience rather than a move towards outright religious tolerance – a lenience born of political realism. There was no question that Elizabeth wanted to eliminate the Catholic Church in England. There were few, if any, sixteenth century monarchs who could afford to tolerate so strong a rival within their own dominions and Elizabeth's crown was more vulnerable than most. But realistically this was not going to happen overnight. So the Oath of Supremacy was tendered to all office holders. Anyone who could not swear 'that the queen is the only supreme governor of this realm . . . as well in all spiritual or ecclesiastical things . . . as temporal' was evicted from that office and denied any further position in the new administration, but elsewhere Elizabeth's behaviour remained conciliatory.[21]

Deleted from the new Prayer Book was thc offensive Edwardian reference to 'the tyranny of the Bishop of Rome'. The new Communion service became a careful amalgam of phrases from successive earlier prayer books. It was a mouthful to say – 'The body of our Lord Jesus Christ, which was given for thee, preserve thy body and soul into everlasting life: and take and eat this in remembrance that Christ died for thee, and feed on him in thine heart by faith, with thanksgiving' – but it remained sufficiently ambiguous to satisfy both those who believed in the real presence in the sacrament and those who denied it. Churchgoers seeking the Virgin Mary were offered a newer and more vital virgin to adore: 'they keep the birthday of queen Elizabeth in the most solemn way on the 7th day of September, which is the eve of the feast of the Mother of God', wrote the Catholic Edward Rishton. And although the churches were stripped of their decoration, they still hung on to 'the organs, the ecclesiastical chants, the crucifix, copes, [and] candles'. Rishton observed: 'The queen retained many of the ancient customs and ceremonies . . . partly for the honour and illustration of this new church, and partly for the sake of persuading her own subjects and foreigners into the belief that she was not far . . . from the Catholic faith.' It convinced the French ambassador. He wrote home, duly impressed, that the English 'were in religion very nigh to them'. And, Rishton added, 'the Queen and her ministers considered themselves most fortunate in that those who clung to . . . [Catholicism] . . . publicly accepted, or by their presence outwardly sanctioned, in some way, the new rites they had prescribed. They did not care so much about the inward belief of these men.' No one, it seemed, was keen to start opening windows into men's souls at the beginning of Elizabeth's reign.[22]

And if the new Anglican Church was built upon the bedrock of compromise, then many who attended it did so in the same spirit. When, in the summer of 1562, a number of prominent Catholics approached the Spanish ambassador, and through him Rome, to ask if they might worship in the Queen's new Church, the answer they received (in the negative) was not considered absolute enough to act upon, so worship there they did. When many local priests became aware of the level of Catholic feeling in their parishes they made adjustments accordingly; so Catholics might have 'Mass said secretly in their own houses by those very priests who in church publicly celebrated the [Protestant] liturgy'. To compromise made sound political sense.[23]

But could it ever make spiritual sense? The notorious sixteenth century Cambridge academic, Dr Andrew Perne of Peterhouse, 'was known to have changed his religion three or four times to suit the change of ruler', but when Perne was asked by a close friend 'to tell her honestly and simply which was the holy religion that would see her safe to heaven', he replied, 'I beg you never to tell anyone what I am going to say . . . If you wish, you can *live* in the religion which the Queen and the whole kingdom profess – you will have a good life, you will have none of the vexations which Catholics have to suffer. But don't *die* in it. Die in faith and communion with the Catholic Church, that is, if you want to save your soul.' Perne never had the chance to heed his own advice: he died suddenly, on the way back to his room after dining with the Archbishop of Canterbury, caught out not only in the wrong faith, but also in the headquarters of that faith, Lambeth Palace itself. But this was the dilemma facing all Englishmen now: how did you square your political survival with your spiritual salvation, if, like vast numbers of your fellow countrymen, you still regarded yourself as Catholic? Happy were those whose conscience and the law agreed. For those others, the future, both in this world and the next, looked much more uncertain.[24]

In November 1561, three years into Elizabeth's reign, the mayor of Oxford had the unpleasant task of informing the Privy Council that 'there were not three houses in [Oxford] that were not filled with papists'. And, added the new Spanish ambassador, Bishop Alvaro de la Quadra, in his regular gossip-filled letter to the Duchess of Parma, 'the Council were far from pleased, and told the Mayor to take care not to say such a thing elsewhere'. But to those with any knowledge of the city's past, this level of defiance will have come as no surprise: Oxford was running true to form. Deep in the cellars below the Mitre Inn on the High Street, at the Swan Inn, the Star Inn and the Catherine Wheel, Oxford's Catholics were meeting in secret and in droves to celebrate their forbidden mass.*[25]

If the city of Oxford was reluctant to embrace the new Church, then its university was proving even more mutinous. In May 1559 the Swiss Protestant Heinrich Bullinger was confidentially advised against sending his son to college at Oxford, for 'it is as yet a den of thieves, and of those who hate the light'. That same

* Mrs Williams of the Swan was more than usually defiant in receiving Catholic priests: her husband was a justice of the peace and a city alderman.

month John Jewel, now Bishop of Salisbury, was noting with some frustration that 'our universities are in a most lamentable condition: there are not above two in Oxford of our sentiments'. And when Elizabeth's visitors arrived at the university that year to enforce the new religious settlement, they were daunted by the strength of Catholic opposition they encountered.[26]

At New College they avoided asking everyone to subscribe to the Oaths of Supremacy and Uniformity for fear of the number of refusals, reported Nicholas Sanders, a fellow of that college. The Bishop of Winchester, the visitor responsible for New College, found similar hostility at his other wards, Trinity, Corpus Christi and Magdalen. Here, too, he declined to look closely. Instead, he and his fellow visitors concentrated their attention on what they saw as the root of the problem: the men in charge. Within two years only one of Oxford's college heads appointed during the previous reign remained in office and with that the Council seemed to be content. Let these new replacements keep their house in order and play the heavy hand. That the sole surviving college to retain its Marian head, New College, was the scene of widespread, Council-led purges throughout the first decade of the reign merely seemed to support the wisdom of the Government's policy.[27]

Then fate stepped in to send the precarious balance of European power reeling. In July 1559 an unlucky tilt at a French court tournament left King Henri II dead, his fifteen-year-old heir, François, in the sway of his zealous cousins the Guises, and his teenage daughter-in-law, Mary Stuart, sufficiently emboldened to have herself heralded with cries of 'Make way for the Queen of England!' A nettled Elizabeth was soon persuaded by her Council to send money to help the Protestant, anti-French rebellion in Scotland and quickly the situation spiralled into open confrontation.[28]

In early 1560, mindful of the need to present a strong show of national unity in times of danger and fearful that the conflict had fallen far too neatly into battle lines of an awkwardly religious nature, Elizabeth sent her visitors back to Oxford. Soon Bishop de la Quadra was reporting home that 'Oxford students . . . [known to be Catholic] . . . have been taken . . . [and imprisoned] . . . in great numbers'. Was this how it was going to be from now on? Each time an enemy threatened was any Englishman not *seen* to be standing foursquare behind the Queen's new church and openly obeying her laws liable to arrest and imprisonment? The detention of six Oxford students the following year, for resisting the mayor's attempts to remove their college crucifix, seemed to confirm this. As Elizabeth braced herself for the return home to Scotland of the newly widowed Mary, it was more the openness of the students' defiance that earned them their prison sentence: after all, the ultra-conservative Elizabeth still kept a crucifix in her own royal chapel.[29]

A pattern was being established, a pattern that those English Catholics arrested for attending mass at the French embassy in February 1560, even as the situation in Scotland worsened, may have been able to spot for themselves. The rationale behind it was simple. Had England's fortunes been entirely separate from those of Europe then Elizabeth and her government could have been content to settle back and let the dismantling of the English Catholic Church be a gradual one, sure in the knowledge that in time the majority of their countrymen would come round to their way of thinking. But England was as entangled with the rest of Europe as religion ever was with politics.* It was a part of the Christian Church, the Church that had bound Europe together. That Church was now divided into factions and while Europe was still known as Christendom, England, like it or not, was integral to that factional struggle. And it was vital to Europe's equilibrium: its fragile diplomatic alliances with France and Spain in turn keeping either of those two nations from ever singly dominating the European stage – a necessity for Europe, but a

* England's sense of growing isolation from the rest of Europe, in spite of these entanglements, features strongly in the State Papers of the time. The Spanish ambassador reported back to Philip II a speech made by Sir William Cecil to the House of Commons in 1563, in which Cecil declared, 'They had no one now to trust but themselves, for the Germans, although they had promised the Queen great things, had done nothing and had broken their word.'

constant irritation to successive generations of ambitious French and Spanish monarchs.[30]

With this the case, conflict was inevitable. For though Elizabeth might have no stomach for religious persecution, still she needed to keep her throne safe from predatory interlopers from across the narrow English Channel. And though England's Catholics might be loyal to England, still they began to find themselves the focus of increasing and unwelcome Government-imposed restrictions every time affairs in mainland Europe took a turn for the worse.

But if this was a pattern that would emerge more clearly as Elizabeth's reign progressed, then Oxford's particular place within that pattern was predictable from the start. And from the start Elizabeth tried to forestall it.

On Saturday, 31 August 1566, 'about 5 or 6 of the clock at night', Queen Elizabeth I rode into Oxford. Her wooing of the city, and its university, had begun.[31]

At the head of the royal procession were the Queen's heralds. Behind them came the Earl of Leicester, in his official role as Chancellor of the university, then the Mayor of Oxford and his party of aldermen, the noblemen of the court, and finally Elizabeth herself. Her

> 'chariot was open on all sides, and on a gilded seat in the height of regal magnificence reposed the Queen. Her head-dress was a marvel of woven gold, and glittered with pearls and other wonderful gems; her gown was of the most brilliant scarlet silk woven with gold, partly concealed by a purple cloak lined with ermine after the manner of a triumphal robe. Beside the chariot rode the royal cursitors, resplendent in coats of cloth of gold, and the marshals, who were kept busy preventing the crowds from pressing too near to the person of the Queen . . . The royal guard,

> magnificent in gold and scarlet, brought up the rear. Of these there were about two hundred . . . and on their shoulders they bore . . . iron clubs like battle-axes.'

Through the north gate they streamed. Down Northgate Street (now Cornmarket), where the scholars who lined the road sank awe-struck to their knees and called out *Vivat Regina Elizabetha*, hearing their cry taken up by the townspeople leaning from the windows and crammed precariously together on the roof-tops above them. To Carfax, where Giles Lawrence, Oxford's Regius Professor, welcomed the Queen with an oration in Greek to which Elizabeth responded warmly in the same tongue, thanking Lawrence for his speech and praising it as the best she had heard in that language, adding coyly 'we would answer you presently, but with this great company we are somewhat abashed'. Lawrence was transfixed.

On down Fish Street (St Aldates) the procession flowed, to Christ Church College, where the gate and walls were festooned with verses in Latin and Greek in admiration of Elizabeth and where, beneath a canopy borne by four Doctors of the university, the Queen was ushered slowly across the quadrangle into the cool and calm of the great cathedral. Here Elizabeth knelt in prayer as Dr Godwin, Christ Church's Dean, gave thanks for her safe arrival in the city. To the sound of cornets the choir sang the *Te Deum* and then wearily Elizabeth slipped away through the gardens in the lengthening dusk, to her lodgings in the east wing, to prepare for this, her latest charm offensive.

It was the Queen's first visit to Oxford. An earlier attempt two years before had been called off at the last moment when plague broke out in the city. But this delay merely ensured that by the time Elizabeth made her dramatic appearance at the north gate anticipation had grown to fever pitch. It also meant that those charged with arranging the visit had left little to chance.

On the Wednesday before the Queen's arrival the Earl of Leicester and Sir William Cecil had ridden the eight miles from the Palace of Woodstock to Oxford, through the sluicing rain of a late summer downpour, to check for themselves that everything was in order. Leicester, as Oxford's Chancellor, was host for the week and with his ambition to marry the Queen still intact at this date – just five years earlier, with his brother-in-law acting as go-between, he had approached the Spanish ambassador and offered to return England to the Catholic Church if Spain backed their wedding, a far cry from his later reincarnation as the scourge of English Catholicism – there was more at stake for him here than mere proprietorial embarrassment should Oxford's hospitality fail to please the Queen. But for Sir William Cecil, Elizabeth's Principal Secretary of State and her chief adviser on all policies relating to Church and foreign affairs, Oxford's performance was a matter for greater concern still.[32]

Each day of the royal visit Elizabeth and her entourage would attend debates and disputations, the art of which formed the basis of every student's education. On the Tuesday a rising young Oxford star, Edmund Campion of St John's College, would triumph in the Natural Philosophy Disputation, proposing 'that the tides are caused by the moon's motion'. Elizabeth, who in later life would be revered as the moon goddess, Cynthia, the 'wide ocean's empress', was delighted with Campion's speech; Cecil and Leicester immediately offered to become his patrons.* But it was indicative of the Government's continued anxiety over the problem of Oxford's religious insubordination that Cecil had provided the students in advance with a list of preferred subjects for these debates. Thursday's Divinity Disputation took as its Council-chosen theme 'Whether subjects may fight against wicked princes?', allowing little scope for awkward theological reasoning.† It would have been a brave – and short-lived – undergraduate who dared to

* She is depicted as such in the *Rainbow Portrait* of *c.*1600 and was the subject of Walter Ralegh's *Book of the Ocean to Cynthia*, in which he describes the anguished nature of his relationship with the Queen.

† Cecil's own choice of suitably non-controversial debating matter for the week ahead was 'Why is ophthalmia catching, but not dropsy or gout?'

denounce Elizabeth's break with Rome as wicked to her face; more embarrassing and more damaging still to the royal party would have been a spirited and unopposed defence of the Catholic faith. Oxford's young students were to be given little opportunity to air their religious grievances.[33]

But Elizabeth favoured the carrot over the stick whenever possible. In addition she held a deep and unshakeable regard for learning and was determined to see Oxford back in the vanguard of European scholarship after so many decades in the wilderness of religious upheaval.* She had a captive audience of some seventeen hundred students – all of whom had elected to remain at the university despite the term being officially over – and if any queen knew how to entrance an audience it was Elizabeth.[34]

So Edmund Campion won his court patronage. George Coriat won half a sovereign. Tobie Matthew of Christ Church won the coveted title of Queen's Scholar, which led to a lifetime of royal preferment and his eventual appointment as Archbishop of York. And all of them won the lavish praise and attention of a queen acutely conscious that her visit needed to serve as a fast-acting panacea for the ills afflicting Oxford. There was banqueting each evening and boisterous theatre in Christ Church's Great Hall, transformed for the occasion into a gleaming, golden 'Roman palace'. And then there was Elizabeth's own speech, given at the church of St Mary the Virgin before the entire university on the final evening of her stay – a speech delivered in faultless, eloquent Latin, a speech in honour of Oxford and of academia, a speech that was welcomed and applauded with unqualified enthusiasm.

As Elizabeth rode out of Oxford the following day, surrounded once again by her glittering procession and by a city liberally hung with verses expressing grief at her departure, she had done much to

* At the start of the sixteenth century Erasmus had placed English learning second only to that found in the Italian universities. Elizabeth's concern over the standard of education in England extended as far as exempting schoolmasters from paying tax.

heal the old wounds left by her father and her brother's brutal and bullish enforcement of religious change. Her leave-taking was as sincere as it was warm: 'Farewell, the worthy University of Oxford; farewell, my good subjects there; farewell, my dear Scholars, and pray God prosper your studies.' Few could have done better under the circumstances. The only problem was it had all taken place several years too late.

Five years before Elizabeth's visit a twenty-nine-year-old Lancastrian, a one-time student of Oriel College and former principal of St Mary's Hall, had left Oxford for Flanders and the Low Countries. There, he was a welcome addition to the exiles of Louvain. And there, just seven years later, at the university town of Douai in the province of Artois, he would rent a 'large . . . and very convenient' house from where he would attempt to turn the ebbing fortunes of English Catholicism. 'We cannot', he would later write, 'wait for better times; we must act now (to make them better).' If the recalcitrant students of Oxford were to be summarily expelled from college whenever Europe threatened and if the men and women of England were to continue compromising their salvation in the name of political survival, then Dr William Allen had found the answer: use the former to educate the latter. It was a simple solution and it would prove devastatingly effective.[35]

Chapter Three

'The very flower of the two universities, Oxford and Cambridge, was carried away, as it were, by a storm, and scattered in foreign lands.'

Edward Rishton, 1585

THE 1560S ENDED with a warning clap of thunder, audible across France and all the way to distant Spain. Rebellion! As the Catholic nations of Europe listened in, England rang to the sounds of revolt.

The uprising was led by the powerful northern earls Percy and Neville, names guaranteed since the Wars of the Roses to strike fear into the heart of any English monarch, let alone one as vulnerable as Elizabeth. Their rebellion marked the last dying gasp of the old feudal order. More than that, it was the angry response of a disgruntled aristocracy, shouldered out of its long-held place in the sun by middle-class *arrivistes* like the Queen's chief minister Sir William Cecil. The Percy/Neville proclamation raged against those 'evil disposed persons, about the queen's majesty, [who] have, by their subtle and crafty dealing to advance themselves . . . abused the queen, disordered the realm, and now, lastly, seek and procure the destruction of the nobility'.[1]

But to rally supporters to their cause the rebels cloaked themselves in the flag of Catholicism. They marched to Durham Cathedral where they tore up the new English Prayer Book and Bible, demanding the restoration of 'the true and catholic religion'.

If this was what it took to spur the slumbering northern counties into action behind them then Percy and Neville were more than happy to make it their campaign slogan – neither man felt any long-standing loyalty to the new Church. Hidden further down the list were their more sought-after demands: the immediate arrest and trial of Cecil and the release from prison of the disgraced Duke of Norfolk.[2]

Elizabeth's response was swift and uncharacteristically brutal. Between 500 and 800 men, all of very little account, were rounded up and executed. Percy and Neville fled the country and the decade closed on a note of queasy anticipation. It did not help that since 1568, Mary Queen of Scots had been living in England as Elizabeth's prisoner. This was the Mary, half Scottish, half French, wholly Catholic, who had claimed Elizabeth's crown as her own some ten years earlier. Mary had lost her French throne on the death of her first husband, her Scottish throne on the murder of her second. Now separated from her third husband, there were many who thought that, as Elizabeth's presumed heir, she was entitled to another throne yet – England's.

Then in February 1570 a new Pope, Pius V, a fanatical firebrand of great zeal but uncertain common sense, took it upon himself to fuel the conflagration further. He issued his bull *Regnans in Excelsis*, excommunicating 'Elizabeth, pretended queen of England', releasing English Catholics from their allegiance to her, and openly encouraging her overthrow, an appalling concept in a world that believed in a monarch's divine right to rule. And the rulers of Europe were duly appalled, particularly as none was at present in the position to make good Pius's threat. Philip II of Spain refused to let the bull be published anywhere in his dominions, openly reassuring Elizabeth that he had no intention of breaking the Anglo-Spanish amity. Privately, he complained that the Pope had 'allowed himself to be carried away by his zeal'. The Holy Roman Emperor Maximilian fired off an angry response to Pius, receiving in return the peevish reply: 'Why she [Elizabeth] makes such a stir about this sentence we cannot quite understand;

for if she thinks so much of our sentence and excommunication, why does she not return to the bosom of the Church, from which she went out? If she thinks it of no consequence, why does she make such a stir about it?'[3]

But Pius had achieved what Protestant Parliamentarians had so far only dreamed of. In showing that a strict adherence to the Catholic faith was now mutually incompatible with loyalty to Elizabeth, he had bound Anglicanism to Englishness more firmly than ever. And he had given to an anxious English nation the cast-iron proof that the more devout the Catholic, the more danger they presented to the realm. The problem for England's Catholics was that as the roots of Elizabeth's new Church began to take hold, the only active Catholics left in the country were, perforce, devout ones. When Edwin Sandys, Bishop of London, opened the Parliamentary session of 1571 with a sermon at Westminster Abbey warning 'This liberty, that men may openly profess diversity of religion, must needs be dangerous', he revealed just how important to the nation's sense of security a solid connection between Church and State had become. He continued, 'One God, one king, one faith, one profession is fit for one Monarchy and Commonwealth. Division weakeneth.'[4]

Paranoia ran rife throughout the 1570s, stalking through the courts of Europe, trailing terror and swift acts of bloody reprisal in its wake. In 1572 some two thousand French Protestants were slaughtered by their Catholic countrymen in Paris on St Bartholomew's Eve, an act that imprinted itself indelibly upon the consciousness of every European Protestant; the French Catholics responsible claimed they had attacked only because they thought they were about to be murdered themselves. Continent-wide, an epidemic of fear and suspicion was spreading. The ideological gulf between Catholicism and Protestantism had reached unparalleled proportions. For the Protestants, the sight of a renewed and invigorated Catholic Church – leaner and keener since the Council of Trent had given it a much needed shake-up – lent substance to the rumours that the Catholics were regrouping for a crusading attack against them.* For the Catholics, meanwhile, the consolidation of the Protestant position only increased the fear that this insidious spread of revolutionary thought would continue, destroying the traditional structure of the civilized world and consigning everyone in it to the fires of hell. Not surprisingly there was little room for compromise. The very words 'Papist' and 'Heretic' carried sufficient emotional charge to unite one side in loathing of the other.[5]

To come of age in the 1570s, like John Gerard in Lancashire and Nicholas Owen in Oxford, was to grow to awareness in the uneasy stillness that heralds a distant but inevitable storm. And picked out brightly against the decade's darkening sky was a series of events, the intervals between which might be counted out like the silence between lightning and thunder to show how fast the storm was approaching.

Early in 1573 a package of letters from the Continent fell into the hands of Bishop Edwin Sandys. Sandys dispatched a party of royal messengers – the ominously named pursuivants – to bring in the intended recipients, and the pursuivants took the well-trodden road to Oxford.†[6]

There, they rounded up a handful of students for questioning but one of the names on their list was missing: Cuthbert Mayne,

* The Council of Trent met in three sessions during the mid-sixteenth century, its purpose to revivify the Roman Church, enabling it to meet the challenge of Protestantism. The Council worked to establish a set of fixed doctrinal definitions for the Catholic faith and to re-order its institutional structure, emphasizing the subordination of the entire Catholic hierarchy to the Pope. Out of the Council of Trent sprang what has been termed the Counter-Reformation, a movement almost as amorphous as the Reformation it opposed, but which can loosely be defined as the attempt at re-conquest of those parts of Christendom lost to the Catholic Church. Rome's army of arguers, as featured in this book, was a component of this movement.

† According to a contemporary Catholic description, 'The pursuivants [were], for the most part, bankrupts and needy fellows, either fled from their trade for debt, and by the queen's badge to get their protection, or some notorious wicked man.'

a West Countryman and member of St John's College, was away visiting relatives. Friends quickly passed word to the student that it would be unwise for him to return to university and soon Mayne found himself boarding a ship off the coast of Cornwall and sailing for Flanders and the English College at Douai.[7]

Over the next few years many more packages arrived in Oxford. Their contents were identical – invitations, from one friend to another, to join the growing fraternity of students overseas – and their summonses were answered by vast numbers of Oxford's disaffected undergraduates. Such was the siren call of Douai.

Then, in 1574, just a year after Mayne's hurried departure to the Continent, four other young Englishmen – one a former fellow of Mayne's old college, St John's – made a second and even more significant Channel crossing. Their names were Lewis Barlow, Martin Nelson, Thomas Metham and Henry Shaw. They were recent graduates of the Douai College and all were newly ordained Catholic priests. Their journey took them from the Low Countries back home again, in secret, to England. Dr William Allen's solution had been put in motion.[8]

The English College of the University of Douai, William Allen's brainchild, was born out of frustration. Allen had departed Oxford in 1561 refusing to swear the Oath of Supremacy required of him by the university authorities, and his flight had taken him as far as the University of Louvain in the Low Countries. There he discovered a flourishing community of English exiles living in two large houses, to which they had given the names Oxford and Cambridge and from which they released a stream of anti-Protestant publications to be smuggled back to England. Allen set to work with a will. When ill health forced him to return home in the summer of 1562, he found among the leaderless English Catholics a religious apathy in stark contrast to the vigour of Louvain.[9]

For the next two and a half years Allen toured England, trying single-handedly, but with isolated success, to communicate a sense of Louvain's vitality to his friends. His dismay at their complacency and their willingness to compromise grew steadily all the while. The Pope's recent ruling that Catholics should not attend Church of England services had been widely ignored. Those 'who believed the faith in their hearts and heard mass at home when they could' were still frequenting their local parish churches, heedless of the dangers of this 'damnable sin of schism', wrote Allen. No matter how they blamed the Government's laws for 'their unlawful acts', England's Catholics were heading for 'the miserable abyss of destruction'. Elizabeth's policy seemed to be working: the old religion was dying by degrees – and not through persecution but through isolation and lack of spiritual guidance. Indeed, it was a measure of the Government's live and let live policy at the time that Allen was permitted to remain so long in England, given his efforts to persuade his friends to break the law. But by the spring of 1565, aware that the Government's patience was not to be tried indefinitely and worn down by the Sisyphean nature of his chosen task, Allen departed for Louvain once more. There the situation at home continued to haunt him. The remedy, though, proved elusive.[10]

Then in the autumn of 1567 Allen travelled to Rome in search of a position as chaplain to the English Hospice there. The opening did not materialize and he set off back to Flanders, accompanied by his friend Dr Jean Vendeville of the University of Douai. Vendeville had just failed to persuade Pope Pius to support his proposal for a crusade against the Turks, but the two friends' conversation over the course of their journey north delivered up an answer to their respective disappointments: couple Vendeville's thwarted missionary zeal with Allen's desire to save England's wavering Catholics. So began the 'oasis in the wilderness of exile'.[11]

Within a few weeks of its opening, on 29 September 1568, Vendeville was writing that the new English College boasted a handful of men 'of great ability and promise'. And from the start the Douai seminary looked very much like being an Oxford affair. Among its first members were John Marshall, former Dean of Christ Church College, Richard Bristow, MA of Christ Church and fellow of Exeter College, and Edward Rishton, MA of Exeter

College. Only one of the new English students, John White, was not an Oxford man.[12]

What news of this reached Oxford? What shape did the rumours take as quickly and quietly they spread about the town? That William Allen had founded a college where exiled scholars 'might live and study together more profitably than apart'? That he was preparing a school of men 'to restore religion when the proper moment should arrive'? That Dr Vendeville saw England as the next great mission?[13]

For many this was welcome news. The Parliament of 1563 had further extended the Oath of Supremacy to all in, or taking, holy orders, to all lawyers, MPs and schoolmasters, and to all university graduates. For good measure, the House of Commons was also insisting on harsher punishments for those refusing to swear to the oath. As Sir William Cecil observed: 'such be the humours of the Commons House, as they think nothing sharp enough against Papists'. A first offence brought with it the penalties of praemunire: loss of lands and life imprisonment. For a second refusal the sentence was death. The new measures brought sharply into focus the choices available to Oxford's students.[14]

Loyalty to Elizabeth carried with it the promise of advancement in a country crying out for new priests for its newest Church. It might also be a path to high office in the service of a queen looking to employ 'men meaner in substance, and younger in years' in her Government, in place of those ambitious aristocrats dismissive of a female ruler and powerful enough to challenge her. Loyalty to Elizabeth was something Elizabeth herself, with her charm, her flirtatiousness and her calculated displays of majesty, was *most* keen to encourage – not surprisingly given the vulnerability of her throne.[15]

Loyalty to your conscience, on the other hand, led to certain ruin: to separation from friends, estrangement from family and crippling poverty – just as the nation's economy began to stabilize. A letter home from one young Englishman who chose conscience over country illustrated the emotional and financial cost of his decision: 'Pray crave my parents' blessing for me, and confer with my mother, and ascertain whether if I should come home, it would turn my father to me.' And, he added desperately: 'my wants are very great. Pray be a means to them [my parents] to help me'. Another letter, this time from the exiled Thomas, Lord Copley, uncle to the Jesuit-poet Robert Southwell, set out the price of conscience clearer still: 'I love my country, friends, and kinsfolk, but I must be content patiently to forbear the comfort of them all, as I am taught by our Saviour himself, rather than to forsake him'. And William Shakespeare, in his play *Richard II* of *c.*1595, would sum up the pains of exile in a couplet:

> Then thus I turn me from my country's light,
> To dwell in solemn shades of endless night.[16]

So why was any Oxford student prepared to make this sacrifice? Of course, for some it must always have been for the sheer excitement of going up against the Establishment. But it was one thing to attend secret mass at the Mitre Inn, to pass on in stolen whispers the latest news from Douai, to argue long into the night in the rarefied, ivoried, once-removed atmosphere of academia – quite another to go over to the other side altogether.

For John Gerard, his reason was that of tradition; perhaps, too, an unspoken need to settle an old score: 'My parents had always been Catholics,' he wrote, 'and on that account had suffered much at the hands of an heretical government.' (Curiously, his was a self-censored family history: his grandfather, Thomas Gerard, had been burnt at the stake at Smithfield in London on 30 July 1540, as a convert to Lutheranism.) In Gerard's fellow Jesuit, Robert Southwell's, case, Catholicism was 'the belief which to all my friends by descent and pedigree is, in manner, hereditary'. But for numerous others – such as Cuthbert Mayne, raised by his uncle, a Protestant parson – the old faith was not *their* old faith. Rather, the 'Old religion [had] renewed its youth' from among the ranks of many families who had already forsaken it. Those students who chose to leave Oxford for Douai, to sacrifice a life of opportunity for one of danger and penury, did so on the basis of ideological certainty.[17]

For some, their certainty sprang from a conviction that Parliament, 'which has not long used to judge causes of faith, or prescribe ecclesiastical laws' (so wrote Lord Copley), had no mandate to tell them what to believe. Others, looking about them at the bloodshed and chaos, the failed harvests and famine that had so blighted England in the preceding decades, saw God's hand at work – their country was being punished for the sin of challenging the established Church. For such students, on the brink of entering this world of bloodshed and chaos for themselves, here was a way of drawing its poison. In Robert Southwell's words, it now became their 'duty . . . by the gentleness of [their] manners, the fire of [their] charity, by innocence of life and an example of all virtues, so to shine upon the world as to lift up the *Res Christiana* that now droops so sadly, and to build up again from the ruins what others by their vices have brought so low'. Still more young undergraduates believed that England had been betrayed by its Government – a Government more concerned with its own immediate survival than with the salvation of the nation. Elizabeth herself may have learned the value of political compromise at a very early age; most Oxford students had never had that need and saw no reason to acquire it now – not with the souls of their countrymen at stake. Later they would be charged with betraying those same countrymen to Spain – their defence would be that the true betrayal had not been theirs, but had come many decades prior to them setting out for Douai.[18]

In a poem of 1581–5 Robert Southwell wrote:

> Then crop the morning Rose while it is fair;
> Our day is short, the evening makes it die.
> Yield God the prime of youth 'ere it impair,
> Lest he the dregs of crooked age deny.[19]

Whatever their motives for escaping to Douai, at William Allen's disposal now was the prime of Oxford youth.

At first Allen did not envisage sending the graduates of his Douai seminary back home to England as missionaries; the impetus for this was Jean Vendeville's and came later. Rather, he thought to prepare them for the happy moment – Elizabeth's death or a foreign invasion – when England would again need Catholic priests. But the syllabus he devised for them was a blueprint training manual for a very specific kind of 'holy war'.[20]

The students would remain at the college for three years. In that time they would learn Greek and Hebrew to augment their existing knowledge of Latin. With these three languages at their disposal they could read the scriptures in their original form, so as 'to save them from being entangled in the sophisms which heretics extract from the properties and meanings of words'. They would study their Bibles with painstaking detail, working through the Old Testament at least twelve times and the New Testament sixteen times. And each week there would be debates in which the students would 'defend in turn not only the Catholic side against the texts of Scripture alleged by the heretics, but also the heretical side against those which Catholics bring forward'. Thus armed, they would 'all know better how to prove our doctrines by argument and to refute the contrary opinions'.

For the advanced students there would be a further course of study: English, the 'vulgar tongue'. 'In this respect', wrote Allen, 'the heretics, however ignorant they may be on other points, have the advantage over many of the more learned Catholics.' The Protestants' use of the Bible in translation gave them an advantage over Allen's priests when preaching to those unschooled in Latin. English classes would correct the inaccuracy and 'unpleasant hesitation' with which many of his trainee missionaries interpreted their scriptures. And William Allen was preparing for a war in which any inaccuracy or hesitation could have devastating consequences.

It was to be a war of words and will in which the sharpest weapons would be the combatant's ability to argue his cause clearly and persuasively, and his unwavering belief in the rightness of that cause. To this latter end it was Allen's 'first and foremost study' to stir up 'in the minds of Catholics, especially of those who are preparing here for the Lord's work, a zealous and just indignation against the heretics' and to set before 'the eyes of the students the . . . utter desolation of all things sacred . . . the chief impieties, blasphemies, absurdities, cheats and trickeries of the English heretics'. 'The result', wrote Allen, 'is that they not only hold the heretics in perfect detestation, but they also marvel and feel sorrow of heart that there should be any found so wicked, simple and reckless of their salvation.'

It was incendiary teaching. And it proved overwhelmingly popular. In December 1575 Allen was summoned to Rome to advise the Pope on the foundation of a second seminary there. By the following year the original Douai College had grown to fill three houses. Swarms of students were 'daily coming, or rather flying to the college', they were among 'the best wits in England' and many were former students of Oxford University.[21]

But not even Douai could escape the decade's disease: paranoia. Throughout the 1570s, as Philip of Spain's army battled to stamp out Protestantism in the Spanish-owned Netherlands, the rumours spread that Allen's students were spies for the Catholic cause. An entry in the Douai Diary of 27 June 1577 reads: 'Dr Bristow admonished us to be more guarded in our behaviour and, as far as possible, to walk less frequently in the streets, because the common people had begun . . . to spread reports and excite murmurs against us.' By August the students were whispering about a coming raid on the college. Finally, in the spring of 1578 the seminary was expelled from the city. The trainee missionaries decamped to Reims, the French university city, where, under the protection of the powerful Guise family, they hoped to continue their studies free from suspicion. It was not to be. By September 1578 Allen was writing to the Governor of Reims, begging him to calm the populace's fears that his students were armed English insurrectionists who went about in disguise to check and measure the town's fortifications.[22]

But some of the paranoia was justified. On the feast of Candlemas, 2 February 1579, a former stationer's apprentice, Anthony Munday, and his friend Thomas Nowell arrived at the newly formed English College in Rome. Since William Allen's visit to the city four years earlier, the plans to open a seminary on the site of the old English pilgrim's hostel had come on apace. By the time of Munday's visit the college already held forty-two students, including the young Robert Southwell.[23]

Munday and Nowell were offered eight days' entertainment at the college, 'which by the Pope was granted to such Englishmen as come thither'. For Munday the invitation was followed by an awkward encounter. Earlier in his adventures he had been mistaken by a group of young Englishmen in Paris for the son of a prominent Catholic gentleman. This had afforded him a warm welcome and a number of letters of introduction to Rome so Munday had done nothing to disabuse his new friends of their notion. But now in Rome he was greeted by a priest who knew this Catholic gentleman well. Munday spent an uncomfortable evening parrying questions and 'was put to so hard a shift that I knew not well what to say'. When the supper-bell rang he fled with relief and thereafter did his best to avoid his interrogator.[24]

In the days that followed Munday had ample time to record in detail his impressions of seminary life, from morning study and prayers, through the daily tuition in divinity, logic and rhetoric, to the student chatter around the fireside at night. But Anthony Munday was a Protestant. In time he would become a professional informer.

Munday's diary of his stay in Rome is the first recorded memoir written by a spy. It must, however, be read with a certain scepticism: his target audience was a paying public eager to believe the worst about the seminary, and his literary credibility is dubious.* Moreover, his claims are suspicious. The students, he wrote, competed amongst themselves as to 'who shall speak worst of her Majesty', while their teachers were as insulting about her ministers: Francis Bacon appears as 'the Butcher's son, the great guts, oh he would fry well with a Faggot'; Ambrose Dudley becomes 'a good fat whoreson, to make Bacon of'. (Significantly, Sir Thomas Bromley, to whom Munday dedicated the work, does not feature in this list of gibes.†) And somewhat at odds with the ill-concealed relish of Munday's descriptions is his politic disclaimer that all these ministers were, of course, 'honourable personages, to whom the words do offer great abuse, and whom I unfeignedly reverence and honour'.[25]

But if much of Munday's account reads like the work of an entertaining profiteer looking to sell a few books at the Government's expense, certain facts in his story do ring true, particularly when viewed in conjunction with William Allen's syllabus at Douai and Reims. Each mealtime the students listened to readings from the Bible to arm them for the fight against heresy. They took part in daily disputations to sharpen their skills in debate. And encouraged by Allen, who explained that 'We must needs confess that all these things have come upon our country through our sins', they undertook public penance for the most minor infringement of the college rules. Munday, who in his brief stay 'was always apt to break one order or other', wrote with feeling about these penances. But it was his description of the students' self-flagellation that was most arresting. The penitent student entered the dining room dressed in a long canvas robe with a hole cut in its back, hooded to hide his identity and carrying a short-handled whip with 'forty or fifty cords

* Munday would later pass off his play about Sir John Oldcastle as being by William Shakespeare. In *Henry IV Part I*, the character of Falstaff was originally called Sir John Oldcastle. This was changed when Oldcastle's descendants complained about the slur on their ancestor's name. In Act I.ii.40 Hal addresses Falstaff as 'my old lad of the castle'.

† Cecil and Leicester, whose names also appear in the dedication to Munday's book, do feature in this list. Cecil received a veiled compliment on his 'wit'; of Leicester, Munday wrote that the comments made against him were 'not here to be rehearsed' – a tactful remark under the circumstances.

at it, about the length of half a yard: with a great many hard knots on every cord, and some of the whips have through every knot at the end crooked wires, which will tear the flesh unmercifully'. The student then walked up and down the room, whipping his back until the blood ran. Scourging was familiar among monastic orders as a means of discipline and it was still the recognized punishment for any priest found guilty of the disparate crimes of blasphemy, concubinage and simony (the selling of ecclesiastical privileges). Self-scourging was popular among the more ascetic orders as a means of mortification. But the picture Munday paints is reminiscent of the Flagellants, the fanatical sect that sprang up out of the plague-stricken thirteenth century and who whipped themselves until they bled in reparation for the sins of the world.* Allen's holy warriors, it seemed, were taking upon themselves the sins of the English nation.[26]

Munday's intrusion into life at the English College in Rome suggested Allen's missionaries-in-training could not long remain isolated from the outside world. Allen's own behaviour, however, had made a collision between priests and government spies inevitable. For in an age of high intrigue, William Allen was fast becoming an arch-intriguer.

On his journey to Rome in 1575 Allen coupled talks on the foundation of the new seminary with detailed discussions about a forthcoming Spanish-backed invasion of England; he only came away from the Holy City when it was felt his 'prolonged stay [there] might arouse suspicion in that woman [Elizabeth]'. He was also in contact with the imprisoned Mary, Queen of Scots, recommending a trustworthy courier as her go-between with the outside world. And his regular correspondence with New College exile Nicholas Sanders reveals the extent to which these two Oxford

* The Flagellants' movement spread throughout Europe, reaching England in the fourteenth century. There, they were regarded with interest, though very few could be persuaded to join their numbers.

graduates now valued their influence in the murky world of European affairs. Sanders wrote to Allen

> 'We shall have no steady comfort but from God, in the A [the Pope] not the X [Philip II]. Therefore I beseech you to take hold of A, for X is as fearful of war as a child is of fire, and all his endeavour is to avoid all such occasions. The A will give two thousand [troops], when you shall be content with them. If they do not serve to go to England, at the least they will serve to go to Ireland. The state of Christendom dependeth upon the stout assailing of England.'

Clearly, William Allen had begun to align himself with the more overtly political of the Catholic agitators, in addition to his own self-appointed task as director of missionaries. What was unclear was precisely how he intended to keep these two roles separate in the public mind. For separate they *must* be if his young priests were to be seen as agents of God rather than agents of a foreign power.[27]

His answer, if he and his Reims and Roman College graduates were to be believed over Anthony Munday, was this: all discussion of English affairs of state was banned among the seminary students. Elizabeth's name was never to be mentioned, neither in lectures nor in recreation. No student was to debate the extent of the Pope's authority over Christian rulers and no reference was to be made to the Pope's right to depose a monarch from their throne. William Allen was training his students 'so that they may serve the one side without offence to the other, which is the hardest thing in the world where the two contrary parties be man and God'. His solution was single-minded. No matter how sullied his own reputation was fast becoming, his missionaries-in-training would be political virgins. It was a theoretical distinction that might have made perfect sense in the classroom; what was less certain was whether it could ever make perfect sense in the outside world, particularly in England. Allen may have trained his students in the art of disputation, he may have schooled them in the scriptures, he may have fired them up with a hatred of heresy and inured them to physical pain. But he was sending them into a country whose Queen stood under sentence of deposition from the very Church they represented, and the only protection he had given them, their only defence against the charge of being agitators and secret agents, was that they had not been *allowed* to discuss politics during their training. More useful would have been detailed discussions about the realities of the political situation into which they were about to be dropped, about the theological doubts concerning Pope Pius' right to depose the Queen, about the impossibility of separating religion from politics in a country whose Church was a construct of Parliament. Had Allen's holy warriors been equipped to live for the cause or simply to die for it?* As they left the safe confines of Douai for the cold and treacherous waters of their homeland the answer would soon become apparent.[28]

On the night of 24 April 1576 the thirty-two-year-old Cuthbert Mayne, newly ordained into the Catholic Church, made the short Channel crossing to England, one of eighteen Douai graduates to make the journey that year. At daybreak he stepped ashore on the south coast, home again after an absence of three years. He was supplied with letters of introduction to the Catholic Sir Francis Tregian of Golden House in Cornwall, so, after taking leave of his fellow missionary John Payne, he set off for the West Country.[29]

The journey was long and nerve-racking as Mayne tried to avoid the ever present shire watches on the lookout for vagabonds and agitators. To be stopped meant to be questioned and to be questioned meant putting his cover story to unwelcome scrutiny.

* In 1587 a memorial was presented to the Pope recommending Allen for the cardinalship. The memorial read: 'He is unbiased, learned, of good manners, judicious, deeply versed in English affairs, and the negotiations for the submission of the country to the church, all of the instruments of which have been his pupils. So many amongst them have suffered martyrdom that it may be said that the purple of the cardinalate was dyed in the blood of the martyrs he has instituted.'

Keeping well to the south of Barnstaple, near which he had been born and where he was certain of being recognized, Mayne arrived at last at Golden House. Here, in his new disguise as the Tregian family's steward, he began working as William Allen had trained him, travelling the Tregian estates between Truro and Launceston, saying mass for the faithful and reconciling to the Church any who had faltered. Summer turned peacefully to autumn. In December that year news filtered slowly through the country of the Queen's clash with her new Archbishop of Canterbury and her displeasure at his Puritan leanings. Christmas and Easter were celebrated at Golden House with full Catholic ceremony. Spring turned to summer. On 8 June 1577 Cuthbert Mayne was sitting in the gardens of Golden House when a party of some one hundred men rode into view. At their head was the new High Sheriff of Cornwall, Richard Grenville, a ruthless naval adventurer with no love of Catholicism. Mayne rose quietly from his seat and left the garden 'where he might have gone from them', heading for his room.[30]

But Grenville was acting on inside information: 'the first place they went unto was M. Mayne's chamber, which being fast shut, they bounced and beat at the door. M. Mayne came and opened it'. To Grenville's question 'What art thou?' Mayne answered simply 'I am a man'. But when Grenville ripped open Mayne's doublet he found about his neck an Agnus Dei case. Agnus Deis were small wax discs made from the Easter candles, impressed with an image of the paschal lamb and blessed by the Pope. They had been outlawed by Parliament in 1571. The penalty for possessing one was death. Among Mayne's papers was found a copy of a papal bull, issued by Pope Pius' successor, Gregory XIII. These, too, had been outlawed by Parliament in 1571, in response to Pius' *Bull Regnans.* To bring any papal bull into the country was now a treasonable offence. So Cuthbert Mayne, former fellow of St John's College, Oxford and graduate of William Allen's seminary, was arrested and borne triumphantly away, first to Truro and then to the dank, underground castle gaol at Launceston.[31]

At the Michaelmas Assizes, Mayne was led out before Sir Roger

Marwood, Chief Baron of the Exchequer, and indicted on five counts, the most serious being the obtaining of a papal bull and the publishing of that bull in England. The sentence was death for high treason. It mattered little that the papal bull had expired, had no bearing on English affairs and had not in fact been distributed by Mayne since his arrival in England; Mayne claimed he had only brought it with him by mistake. It mattered less that the judges themselves were worried by the verdict and sent urgently to the Council for advice on how to proceed. The Council was by now extremely concerned by the reports it was receiving from its spies of an influx of Douai graduates into the country – some thirty priests had arrived home since the return of the first four pioneers in 1574 – and was in no mood for mercy. The sentence stood.[32]

Then on the morning of 29 November Mayne was offered his life. If he would swear on the Bible that Elizabeth was the supreme head of the Church of England he would be spared execution. Mayne refused. He went further: he reasserted his belief that England would soon be restored to the Catholic faith by the 'secret instructors' from Douai. And then, sealing his fate (and stepping outside the strictly apolitical role being claimed by Allen for his students), he declared that should 'any Catholic prince . . . invade any realm to reform the same to the authority of the See of Rome, that then the Catholics in that realm . . . should be ready to assist and help them'. The offer of a reprieve was rescinded.[33]

Cuthbert Mayne was 'drawn a quarter of a mile to the place of execution, and when he was to be laid on the sled, some of the Justices moved the Sheriff's deputy, that he would cause him to have his head laid over the car, that it might be dashed against the stones in drawing, and M. Mayne offered himself that it might be so, but the Sheriff's deputy would not suffer it'. This sheriff's deputy was a merciful man. He let Mayne hang until he was dead before disembowelling him, quartering him, and distributing his parts about the county for display. For his role in the affair, Sir Francis Tregian was sentenced to life imprisonment and his estates

were seized and given to Sir George Carey, a cousin of the Queen.*
John Stow, in his Chronicle of that year, recorded: 'Cuthbert Maine [*sic*] was drawn, hanged and quartered at Launceston, in Cornwall, for *preferring Roman power.*'[34]

Cuthbert Mayne had become the Douai seminary's first martyr. When his old master at Oxford learned of his death he exclaimed, 'Wretch that I am, how has that novice distanced me! May he be favourable to his old friend and tutor! I shall now boast of these titles more than ever.' Such was the power of dying for your faith, and not even the fact that Mayne had been executed as a traitor to his country could tarnish this. Yes, he had broken existing treason laws, but did anyone seriously believe that owning an out-of-date copy of a nondescript bull and a few wax discs posed a threat to national security?[35]

However, as the dust settled on Mayne's quartered remains and the political post-mortem began, it was soon clear that neither side had won a decisive victory in this opening skirmish. Catholics could claim that Cuthbert Mayne was a traitor only according to the most rigid set of definitions, in regard to his possession of a papal bull, or on the basis of hypothesis alone, in regard to his attitude towards Catholic invasions. But in regard to that same attitude, the English Government could claim that Allen's supposed political virgins were uncommonly quick to pronounce on matters apart from their faith. Blessed martyr of a persecuted Church, or secret agent of an enemy state? Cuthbert Mayne had become all things to all men. His foolishness in being caught with Agnus Deis and a papal bull, and his clumsy defence of the Pope's powers of deposition had left Catholics confirmed in their belief that they were being penalized for their religion, and the Government confirmed in its belief that Allen's seminarians were stirring for invasion. The battle lines had just been made clearer.

But for the young missionaries-in-training, Mayne's execution revealed to them that here was a war they might wage for the

* Tregian was held captive for twenty-five years (some accounts say twenty-eight) and only released after King Philip of Spain intervened. He died in Lisbon in 1608.

ultimate prize: the crown of martyrdom itself.* Just months after Mayne's death the Catacombs were unearthed beneath Rome, to great celebration among Catholics: here was proof that they and their Church were the direct descendants of those early Christian martyrs, sprung from their blood and their bones. And for a new generation the chance to save that Church was being offered to them again.

> 'Listen to our heavenly Father asking back his talents with usury; listen to the Church, the mother that bore us and nursed us, imploring our help; listen to the pitiful cries of our neighbours in danger of spiritual starvation; listen to the howling of the wolves that are spoiling the flock. The glory of your Father, the preservation of your mother, your own salvation, the safety of your brethren, are in jeopardy, and can you stand idle? . . . Do not, I pray you, regard such a tragedy as a joke; sleep not while the enemy watches; play not while he devours his prey; relax not in idleness and vanity while he is dabbling in your brother's blood . . . See then, my dearest and most instructed youths, that you lose none of this precious time, but carry a plentiful and rich crop away from this seminary, enough to supply the public wants, and to gain for ourselves the reward of dutiful sons.'

With such words ringing in their ears it was little wonder that, to their mentor William Allen, the student priests seemed 'like men striving with all their might to put out a conflagration. They cannot in any way be kept back from England'.[36]

During Elizabeth's first Parliament, Sir Thomas White, founder of St John's College, Oxford and a staunch Catholic, had exclaimed in fury and despair that 'it was unjust that a religion begun in such a miraculous way, and established by such grave men, should be abolished by a set of beardless boys'. Some twenty years on, the job

* In 1583, Niccolo Circignani, called Pomerancio, painted a series of thirty-four frescoes for the Roman College church, depicting the history of Christianity in England, and stressing the importance of martyrdom. Recognizable figures were shown being hanged, drawn and quartered, so that the students would be in no doubt as to the fate awaiting them. The originals have perished; those frescoes in the tribune of the new church are copies, painted in 1893.

of saving White's miraculous religion had fallen to another set of beardless boys. As William Cecil would write, with an old man's frustration at youth's idealism, 'The greatest number of papists is of very young men.' In a few years' time John Gerard and Nicholas Owen would be old enough to join their number. Meanwhile in Prague, a former fellow of White's college, and the author of that rallying call to the students at Douai's seminary, was about to step into the fray. His name was Edmund Campion.[37]

Chapter Four

'Campion is a champion, Him once to overcome,
The rest be well dressed
The sooner to mum.'
(Sixteenth century ballad)

DURING THE STATE VISIT to Oxford of 1566, before a packed house of royal dignitaries and university academics, Edmund Campion had impressed the young Queen Elizabeth with his skill at debating. Elizabeth, who admired a keen intellect every bit as much as the ability to hunt or dance, was delighted by Campion and the plaudits followed thick and fast. 'Ask what you like for the present', promised Oxford's Chancellor and Elizabeth's favourite, the Earl of Leicester; 'the Queen and I will provide for the future.'* At the age of twenty-six this son of a London bookseller had England at his feet.[1]

But Campion had taken a very different path from the one mapped out for him by the Queen and her courtiers. After his ordination into the Anglican Church in 1568 he had reportedly experienced great anguish of conscience. That same year it had been brought to the notice of the Grocers' Company of London, from whom he held an exhibition scholarship, that he was 'suspected to be of unsound judgement' in religion. The guild ordered

* Campion is reported to have asked for nothing but Leicester's friendship.

him to 'come and preach at Paul's Cross, in London' so they might 'clear the suspicions conceived of [him]' and, more importantly, so he might 'alter his mind in favouring the religion now authorised'. Otherwise, they added warningly, 'the Company's exhibition shall cease'. Campion declined their invitation and lost his scholarship. In 1569 he left Oxford for the more congenial – and more Catholic – shores of Ireland and in the summer of 1572 the man regarded by Sir William Cecil as 'one of the diamonds of England', with his own devoted group of followers known as Campionists, the man with an established reputation as a scholar and writer and an assured position in the hierarchy of the new English Church, threw it all away and sailed for Douai. 'It is a very great pity to see so notable a man leave his country,' wrote Cecil.[2]

At Douai, under the instruction of William Allen, Campion became a Catholic priest and in Rome, to which he travelled following his ordination, he joined the Society of Jesus.* The long and painful struggle with his conscience was over. In March 1580, eight years after his flight to the Continent, he was summoned back to England.[3]

Up to now the Jesuits had not involved themselves in the English mission. They were, though, ideally suited to the task. If William Allen's students were the ordinary foot soldiers in Rome's army of arguers then Ignatius Loyola's Jesuits were the special forces, physically toughened by strict, self-imposed hardships and vows of poverty, mentally strengthened by long periods of solitude and meditation, and well aware that education was the strongest weapon in the proselytizer's armoury. 'Give me a boy at the age of seven, and he will be mine for ever,' declared Loyola. Within a decade of their formation the Jesuits had established colleges throughout Catholic Europe and were ranging as far afield as

* Campion chose to walk from Douai to Rome as a poor pilgrim. On the way he was met by an Oxford contemporary who at first failed to recognize him and then assumed he had been robbed. When he learned it was voluntary mortification he dismissed the idea as un-English and fit only for a crazed fanatic, and he offered Campion a share of his purse. Campion refused.

Mexico and Japan, the front lines of Christian conflict. Their startling success aroused fear among Protestants and resentment among their fellow Catholics. But to Loyola's men this was holy war and in warfare the end justified the means.[4]

Having already lost many of his finest students, including Campion, to the elite new order, it was William Allen, always on the lookout for new ways to help England's beleaguered Catholics, who suggested the Jesuits widen their range of operations to include the English mission. Why sacrifice the lives of English priests in far-flung corners of the world when there was ample work for them to do in their own homeland? First, though, he had to persuade the unwilling Jesuit General, Everard Mercurian, that England was worth the venture.*[5]

Mercurian's reluctance to send his men to England was deep-rooted. He declared the Society already over-committed in other parts of the world. He 'found divers difficulties . . . about their manner of living there [in England] in secular men's houses in secular apparel . . . as how also their rules and orders for conservation of religious spirit might there be observed'. But most of all, he argued, as conditions in England now stood it would be impossible for his missionaries to maintain the kind of order, discipline and apoliticism in the line of fire on which the effectiveness of their work depended. How could he send his men into a political minefield like England and expect them to minister to Catholics while, at the same time, dodging the accusations of intrigue and treachery that would inevitably be hurled their way? And how could he ask them to do so in isolation, deprived of the support of their fellow Jesuits? Gradually, as the 1570s drew to a close, William Allen wore him down. He was helped in this by a fellow Oxford graduate and a Jesuit of some five years' standing, Robert Persons.[6]

Robert Persons was a 'fierce natured', 'impudent' West Countryman, born at Nether Stowey in Somerset in 1546. In 1564,

* Ignatius Loyola died in 1556.

at the age of eighteen, he went up to Oxford, where he discovered Catholicism, first as a student at St Mary's Hall, Allen's old college, and then as a fellow of Balliol. By 1573, his new allegiance to the old faith had brought him to the attention of the authorities and his abrasive manner had offended sufficient of his colleagues and he was summarily expelled, 'even with the public ringing of bells'. So Robert Persons took passage to the Continent. Once there he enrolled to study medicine at the University of Padua, but a chance meeting with a member of the Jesuits made a profound impression on the twenty-seven-year-old. After two years pursuing his medical studies, Robert Persons packed his bags and walked to Rome. On 25 June 1575 he joined the Society of Jesus, a day after his twenty-ninth birthday.[7]

Four years later Persons was writing privately to William Allen that among the English Jesuits there were 'divers to adventure their blood in that mission [to England], among whom I put myself as one'. Faced with such zeal Mercurian finally gave way. A first Jesuit mission to England was ordered; Robert Persons was named as its commander and Edmund Campion was selected to accompany him. 'The expense is reckoned,' wrote Campion, 'the enterprise is begun. It is of God, it cannot be withstood.'[8]

From Prague, where he was teaching Rhetoric at the university, Campion was ordered back to Rome to join Persons. Here, the pair were briefed for their mission. General Mercurian was at pains to stress the difficulties of living and working in disguise, of assuming and maintaining a false identity and of surviving alone without the support of the Society. He also pointed out the impossibility of retreat should the pressure grow too great. These hardships aside, their orders were clear. They were to work with those who were already favourable to the faith. They were to avoid all contact with the heretics. They were to 'behave that all may see that the only gain they covet is that of souls'. They were not to entangle themselves 'in affairs of State', nor to send back political reports to Rome. They were not to speak against the Queen, except perhaps among those 'whose fidelity has been long and steadfast and even then not without strong reasons'. And they were to carry with them nothing forbidden by English law: no papal bulls or Agnus Dei. This was a mission for 'the preservation and augmentation of the Faith of the Catholics in England' and it was not to be compromised by the amateurism that had tripped up Cuthbert Mayne.[9]

Campion and Persons departed Rome on 18 April 1580, waved off in triumph by the entire English colony there. With them rode a party of some twelve other English Catholics, including a lay brother of the Society, Ralph Emerson, who would act as their servant in England, and a group of young seminarians also on their way to join the mission. One witness, Robert Owen, a Welsh Catholic studying in Rome, wrote to his friend Dr Humphrey Ely at Reims, 'This day depart hence many of our countrymen thitherward, and withal good Father Campion.' Within days the letter had been intercepted by an English spy and its contents passed on to Sir Francis Walsingham in London. Edmund Campion was 'on the way to my warfare in England' and England was expecting him.[10]

The party travelled on foot, using false names. Heavy rain dogged their passage through Italy. From Turin they climbed steadily upwards, crossing the Alps at Mont Cenis before descending again into the rich pastureland of the Savoy. From here they continued on to Lyons and on 31 May they came at last to the French university city of Reims.[11]

But here some alarming news awaited them. Campion's was not the only Catholic expedition to the British Isles that month. At the same time the Jesuit and his fellows had left Rome for England, five Spanish ships containing arms and men had left for Ireland. They sailed at the request of Campion's Oxford contemporary Nicholas Sanders, now employed as a papal envoy. Their purpose was to assist the Irish rebel James Fitzmaurice unseat the 'tyrant' Elizabeth. And the man who had financed them was none other than Pope

Gregory XIII.* Robert Persons noted his party's reaction: 'we were heartily sorry . . . because we plainly foresaw that this would be laid against us and other priests, if we should be taken in England, as though we had been privy or partakers thereof, as in very truth we were not, nor ever heard or suspected the same until this day'.[12]

Their situation grew still worse with the second piece of news that now reached them. English agents had provided the Privy Council with a full description of every member of the group and the Channel ports were being watched for their arrival. It was testimony to their courage that only one of the party, Thomas Goldwell, Bishop of St Asaph, now wavered. Goldwell took to his bed and began writing to the Pope to ask whether he was the best man for the job of supervising Allen's missionaries. Indeed, he was not. He was seventy-nine years old, he had endured a gruelling journey from Rome and he was plainly terrified. His defection bore out William Allen's belief that this was young man's work. Allen himself remarked that 'it was better the old man should yield to fear now than later on, on the other side'.†[13]

* Gregory's attitude towards Elizabeth is controversial. In 1580 his Secretary of State gave the following answer to an enquiry by a group of English noblemen as to whether or not they would incur sin by assassinating the Queen: 'Since that guilty woman of England rules over two such noble kingdoms of Christendom and is the cause of so much injury to the Catholic faith, and loss of so many million souls, there is no doubt that whosoever sends her out of the world with the pious intention of doing service, not only does not sin but gains merit'. This judgement came with Gregory's approval. The logic behind it was clear: Elizabeth was a heretic; her actions imperilled the souls of her subjects; her killing was expedient (to cite 'thou shalt not kill' in objection ignores the fact that the Church was already busy burning heretics). But to extrapolate from this that the Vatican officially sanctioned the murder of its opponents is wide of the mark; indeed, Gregory's approval of the English noblemen's scheme had a hugger-mugger air to it, admission that Elizabeth's assassination was against the spirit, if not the letter, of contemporary moral reasoning. Of course, the net result of his dubious opinion was a propaganda coup for the Protestants.

† Goldwell's correspondence with the Pope took several months, by which time plague had broken out in Reims and he had grown desperate. One of his letters, dated 13 July 1580, began, 'Beatissimo Padre, – If I could have crossed over into England before my coming was known there, as I hoped to do, I think that my going thither would have been a comfort to the Catholics, and a satisfaction to your Holiness; wherein now I fear the contrary, for there are so many spies in this kingdom, and my long tarrying here had made my going to England so bruited there, that now I doubt it will be difficult for me to enter that kingdom without some danger.' In the end he dismissed himself without permission and returned to Rome to a chilly reception.

But while Goldwell panicked in Reims, his fellow travellers, joined by three students from William Allen's Reims seminary, pressed on with their journey, splitting up into groups of twos and threes and separating to the French ports, in preparation for finding their way across the Channel. Edmund Campion, Robert Persons and the Jesuit lay brother Ralph Emerson made their way to St Omer, a short distance outside Calais. For them rather more than for their fellows, the Pope's interference in Irish affairs had serious implications.

When Francis Drake sailed into Plymouth harbour on 26 September 1580 after successfully circumnavigating the globe, his ship laden to the gunwales with Spanish treasure, few doubted his success had just hammered another nail into the coffin of Anglo-Spanish relations. The grumblers were soon heard to complain that 'just because two or three of the principal courtiers send ships out to plunder in this way, their property must be thus imperilled and their country ruined'.* In reality Drake's actions and Elizabeth's evident delight in them – she attended a celebratory banquet in honour of his voyage at which she instructed the French ambassador to dub Drake a knight, and she happily pocketed her own share of the profits – were little more than an irritant to Philip II. By 1580 Spain's star was firmly in the ascendant. Decisive victories in the Netherlands by the Duke of Parma, Philip's new commander there, and Philip's surprise inheritance of the throne of Portugal had left Elizabeth commenting grimly, 'It will be hard to withstand the King of Spain now.'†[14]

And whereas in the past England had relied on France to help maintain the precarious balance of European power, this was now impossible, no matter how much Elizabeth and the French Duc d'Alençon flirted and spoke of marriage all that year. For France

* William Cecil, desperate to avoid provoking Spain further, had done his best to scupper Drake's adventure. He is even said to have placed one of his own agents among the crew to incite a mutiny. The agent was discovered and hanged from the yardarm.

† The old King of Portugal died in January 1580 without a direct heir and as the son of the dead King's eldest sister, Philip was quick to press home his claim to the title.

had religious divisions of its own to contend with. In February 1580 the smouldering embers of Catholic–Protestant conflict had reignited once again and the country was now embroiled in its seventh War of Religion. So while France imploded, Philip was free to fix England within his sights without fear of opposition. As a good imperialist the prospect of invasion was tempting (particularly as he now also commanded the powerful Portuguese navy), but as a good Catholic his duty was clear to him. In 1578 Philip had instructed his ambassador to 'endeavour to keep . . . [Elizabeth] . . . in a good humour and convinced of our friendship'. By 1580 he was openly backing the Irish rebel Fitzmaurice.[15]

With European stability deteriorating rapidly and the Spanish threat increasing daily, Pius's *Bull Regnans* was now more pertinent than ever. For if a good Catholic was, by definition, a bad Englishman, then the influx of the Douai missionaries alone – no matter the effect they were having on the populace as a whole – had certainly added to the number of good Catholics in England. And joining them now were the Jesuits, whose founder was no nice Oxford boy with an unfortunate weakness for the old religion, but an ascetically minded Spaniard. Worse still, the Jesuits pledged obedience directly to the Pope.

Before leaving Rome Campion and Persons had been granted an audience with Pope Gregory. From him they had received a fresh clarification of the current position of Pius's Bull in canon law to take with them to England. Gregory's *Explanatio* declared it lawful for English Catholics to obey Elizabeth in civil matters while she was still de facto Queen and unlawful for them to depose her – but only for the time being. For while Pius had been sufficiently foolish to publish his Bull without giving a thought as to the enforcing of it, Gregory regarded himself as a more astute tactician. As soon as the political and military conditions were right, he explained, Pius's Bull would be reactivated. He instructed Campion and Persons to deliver this ruling to England's Catholics and with that he gave them his blessing. All Mercurian's attempts to keep the religious aims of the Jesuits' mission separate and distinct from the political machinations of Rome had been compromised at a single meeting.[16]

So Campion, Persons and Emerson came to the Jesuit house at St Omer to consult with their superiors. Did General Mercurian wish their mission to continue or had Pope Gregory's interference in Ireland made it impossible for them to carry on safely? The discussions were tinged with doubt and anxiety but finally an agreement was reached: the mission would proceed as planned. They had all come far too far to stop now. Persons later wrote, 'as we could not remedy the matters, and as our consciences were clear, we resolved through evil report or good report to go on with the purely spiritual action we had in hand; and if God destined any of us to suffer under a wrong title, it was only what he had done, and would be no loss'. Behind the bravado, though, lay a very real appreciation of the increased perils now facing them.[17]

The decision made, Campion, Persons and Emerson were directed to the house of George Chamberlain, an English Catholic living in exile in France. There, they were equipped with new disguises for their onward journey and some time after midnight on 16 June 1580, dressed in a buff leather coat with gold lace trim and a feathered hat, 'under the habit and profession of a captain returned from the Low Countries', Robert Persons made the short sea voyage from Calais to Dover. The mission was begun.[18]

Close surveillance was being kept on the English seaports. When Persons arrived at Dover on the morning of 17 June he was brought before the port authorities and cross-examined. His cover story and performance held up under the scrutiny. Many Englishmen looking for adventure had gone abroad to fight for the Dutch rebels and Persons had taken to his role with ease – Campion described him to Mercurian as 'such a peacock, such a swaggerer, that a man needs must have very sharp eyes to catch a glimpse of any holiness and modesty shrouded beneath such a garb' – and so, after thorough interrogation, the Dover customs 'found no cause of doubt in him, but let him pass with all favour, procuring him both a horse and all other things necessary for his journey'.

One official proved sufficiently friendly for Persons to seize the initiative. He asked the man if he would forward a letter to his friend, a Mr Edmunds in St Omer, telling the 'jewel merchant' to come quickly to London where he would be met. And he asked the official to be sure to look out for his friend when he landed and see him safely on his journey. The letter was duly sent to the waiting Edmund Campion.[19]

From Dover, Persons rode north to Gravesend, arriving at nightfall. Here his luck continued. He boarded a waiting boat that took him upriver to London, depositing him at Southwark, on the south bank of the Thames, before dawn on the morning of 18 June. He had been on the move less than thirty-six hours.

But now his good fortune ran out. As Robert Persons came ashore in England's waking capital he found that 'the greatest danger of all seemed to be in London itself'. His immediate problem was that he could find nowhere to take him in, 'by reason of the new proclamations and rumours against suspicious people that were to come' from abroad. Every 'inn where he went seemed to be afraid to receive him, and so much the more for that they might guess by the fashion of his apparel that he was come from beyond the seas'. His mercenary's disguise had begun to work against him and for all the careful planning and for all that Allen's seminary priests had been returning to England for the past six years, there was still no system in place to help a new arrival make contact with anyone prepared to assist him. Persons spent a bleak few hours walking the streets of the city. Finally, he 'resolved to adventure into the prison of the Marshalsea and to ask for a gentleman prisoner there named Mr Thomas Pound', a former courtier turned devout Catholic.*[20]

Since the 1570s the number of Catholics arrested for attending

* The story of Pound's transformation from wealthy courtier to religious prisoner is remarkable (though quite possibly apocryphal). He is said to have performed a complicated *pas seul* before the Queen, who was so impressed she called on him to repeat the move. Pound did so, but this time he fell. To the ringing laughter of the Queen and her court he retired, with the words '*Sic transit gloria mundi*', and from then on he devoted himself to religion.

secret mass had increased steadily, so if you wished to meet an open and unrepentant papist there was always one place you were guaranteed to find one – prison. And though convicts were held at Her Majesty's pleasure, the ever parsimonious Elizabeth did not consider them her guests. Prisoners were expected to supply their own food and drink and their own beds and bedding, the latter to be donated to the gaoler at the end of their sentence. Wealthy prisoners prepared to pay for the privilege might entertain visitors, conduct business and even go out from time to time, so long as sufficient amounts of money changed hands.

Thomas Pound was delighted to receive his new guest. He introduced Persons to a young man named Edward Brookesby who also happened to be visiting the prison that day and soon Persons was following Brookesby to a house on Fetter or Chancery Lane. The Jesuit's luck had returned.[21]

Secrecy surrounded this house in the city. It was said to have belonged to Adam Squire, the Bishop of London's son-in-law and London's chief pursuivant, but by 1580 it had become the rented headquarters of a group of 'young gentlemen of great zeal', each one dedicated 'to advance and assist the setting forward of God's cause and religion . . . every man offering himself, his person, his ability, his friends, and whatsoever God had lent him besides, to the service of the cause'. This band of enthusiastic young Catholics (of which Edward Brooksby was one) was led by George Gilbert, a man already known to Robert Persons.[22]

Gilbert was a twenty-eight-year-old Suffolk man of enormous independent wealth, an accomplished athlete, horseman and swordsman.* He had been raised a strict Puritan but in Paris, where he had proved a great favourite at the French court, he

* An informer's report, dated 26 December 1580, describes Gilbert as 'bending somewhat in the knees, fair-complexioned, reasonably well-coloured, little hair on his face, and short if he have any, thick somewhat of speech, and about twenty-four years of age'. By this stage Gilbert was a wanted man.

had come under the spell of Catholicism. From Paris he travelled to Rome where, with religious instruction from Robert Persons, his confessor at St Peter's Basilica, Gilbert converted to the old faith. On his return to England in 1579 he began to gather about him a group of like-minded and equally wealthy young Englishmen, ready to devote their energies 'to the common support of Catholics'. Charles Arundel, Charles Basset (a descendant of Sir Thomas More), Edward Habington, Edward and Francis Throckmorton, Anthony Babington, Henry Vaux, William Tresham and John Stonor: all would give time and money to further the Catholic cause; several would give their lives. To what degree they had already begun working together as a secret society is the subject of dispute, but with Robert Persons' arrival in London their enthusiasm now found new focus.[23]

Once settled in George Gilbert's city headquarters, Persons began 'to acquire a number of friends and to arrange with inns, with a view to staying in the country for a few days'. Then, with Gilbert's aid and an escort to accompany him, Persons left London to 'employ himself in the best manner he could to the comfort of Catholics'.[24]

Meanwhile, in St Omer, Edmund Campion had received Persons' letter and was preparing for his own crossing to England. On the evening of 24 June the summer storms that had battered the Channel coastline for days finally let up and the waiting was over. Disguised as Persons' jewel merchant friend and with Ralph Emerson acting as his servant, Edmund Campion set sail from Calais.[25]

At daybreak the following morning the port of Dover stood at red alert. Word had reached the Council that Gabriel Allen, William Allen's brother, was returning to England to visit his family in Lancashire. Edmund Campion bore more than a passing resemblance to the wanted man. Campion and Emerson were dragged before the Mayor of Dover, cross-examined, then informed they were to be sent to London for further questioning. Then, for no obvious reason, the mayor changed his mind. Quickly, the two men left Dover, riding north to the Thames estuary before boarding a boat that took them upriver to the capital.[26]

Reaching London, they were still in some doubt as to what they should do next. Then a man detached himself from the waiting crowd at the quayside and stepped forward to greet them, saying, 'Mr Edmunds, give me your hand; I stay here for you to lead you to your friends.' The man's name was Thomas James. He was a member of George Gilbert's brotherhood of young Catholics and for several days now he had been keeping watch for the two men's arrival. By nightfall Campion and Emerson were safely installed at Gilbert's headquarters.[27]

On 6 April 1580, while Persons and General Mercurian discussed the details of their forthcoming mission and Campion hurried from Prague to Rome to join them, an earthquake hit London and the southern counties of England. 'The great clock bell in the palace at Westminster strake of itself against the hammer with the shaking of the earth.' Stones tumbled from St Paul's Cathedral. In Newgate an apprentice was killed by falling church masonry. Meanwhile, at Sandwich in Kent the sea 'foamed . . . so that the ships tottered' and at Dover 'a piece of the cliff fell into the sea'.[28]

In the weeks and months that followed, strange visions appeared in the skies above Cornwall, Somerset and Wiltshire – ghostly castles and fleets of ships, three companies of men all dressed in black, a pack of hounds whose cry was so convincing it drew men from their houses in readiness for the chase. In Northumberland hailstones rained down in the shape of frogs, swords, crosses and, worse, the 'skulls of dead men'. And in Yorkshire and Huntingdonshire strange births were reported, monstrous creatures part human, part beast, to signify 'our monstrous life', wrote Holinshed, who chronicled the year with a baleful gloom. The arrival of the Jesuits, like the arrival of the Spanish eight years later, was preceded by many ominous portents (not surprisingly, perhaps, when the prevailing view among William Allen's circle was that 'two Jesuits should do more than the whole army of Spain').[29]

And with the coming of these portents, the fear that had haunted the nation throughout the preceding decade grew stronger still. A future war with Catholic Europe now seemed a foregone conclusion. It was really only a matter of when and, specifically, with whom that war would be fought. Would it be with the Pope, who was already sending invasion forces to Ireland? With the Spanish, who seemed invincible? Or, closer to home still, with Scotland? In 1578 the pro-English and Protestant Regent to the Scottish throne, the Earl of Morton, had been forced to resign. Now the country was ruled jointly by the Earl of Arran and Esmé Stuart, the boy-king James VI's favourite cousin, both of whom were pro-Catholic, pro-Mary and, worst of all, pro-French. Wherever you looked as an Englishman in 1580, to all points of the compass and to the very skies above your head, there were signs to trouble the bravest of souls. And set beside these general fears of imminent conflict was the more specific fear that while eyes and minds had been otherwise distracted England's Catholics had been growing stronger.[30]

At the close of the 1570s the Spanish ambassador reported home to Philip II that 'The number of Catholics, thank God, is daily increasing here [in England], owing to the college and seminary for Englishmen which your Majesty ordered to be supported in Douai.' If this was designed to flatter, soon there were other reports flying backwards and forwards supporting the ambassador's claim. Henry Shaw, one of Allen's four proto-missionaries, wrote back to his mentor, 'The number of Catholics increases so abundantly on all sides, that he who almost alone holds the rudder of state [Sir William Cecil?] had privately admitted to one of his friends that for one staunch Catholic at the beginning of the reign there were now, he knew for certain, ten.' From Warwickshire the Earl of Leicester wrote to Sir William Cecil in alarm, to 'assure your Lordship, since Queen Mary's time, the Papists were never in that jollity they be at this present time in the country'. Meanwhile, the new Bishop of

London, John Aylmer, warned Sir Francis Walsingham 'that the Papists do marvellously increase, both in number and in obstinate withdrawing of themselves from the Church and service of God'.* A nationwide census of convicted recusants, drawn up in 1577, offered the shocking proof that there was not a diocese in England that did not contain a number of Catholics steadfastly refusing to attend the Anglican Church. And for every Catholic who openly defied the law, it was believed there were many more still who publicly conformed to the 1559 Settlement while attending Roman mass in secret.[31]

It is difficult to know what to make of these reports. Offered in isolation, without other annual figures against which to compare them, there is little way of telling whether the Government's 1577 findings show there to have been a significant growth in Catholic numbers, a change in Catholic behaviour (thanks to the influence of Allen's missionaries), or simply (and most likely) an invigoration of the investigative process by which Catholics were being identified. Add to the mix a measure of Protestant paranoia and Catholic pride and it becomes still harder to get at the facts. But as the new decade dawned the widely held perception was that the number of practising – and thereby dissenting and potentially treacherous – Catholics had increased substantially. Here was a threat more specific and far closer to home than the potential invasion forces of Rome or Spain. And now, too, that threat could be personified. It had a name: a traitor and a turncoat's name, the name of a former royal favourite and a courtier's protégé, of the one-time ablest man in Oxford. As Privy Councillor Sir Walter Mildmay later testified in the Star Chamber, of all the 'rabble of runagate friars' there was 'one above the rest notorious for impudency and audacity, named Campion'.[32]

* * *

* Walsingham was Elizabeth's new Principal Secretary of State since Sir William Cecil's appointment as Lord Treasurer in 1573. Cecil was also created Baron Burghley, in recognition of his service to Elizabeth.

News of Edmund Campion's arrival spread quickly through the Catholic community, the buzz that had surrounded his name up at Oxford undiminished by his many years abroad. On 29 June at the Feast of St Peter and St Paul, just three days after his appearance in London, a huge audience assembled at the Smithfield house of Lord Norris, hired for the occasion by Lord Paget, to hear him speak.* As a precaution, gentlemen 'of worship and honour', members of George Gilbert's association, were set to guard the doors of the house against intruders, but no guard could halt the whispers now rippling through the busy London streets. It was not long before those same whispers had reached the ears of the Government's informers.[33]

Soon spies were at work across the capital, detailed to 'sigh after Catholic sermons and to show great devotion and desire of the same, especially if any of the Jesuits might be heard'. When Robert Persons returned from his preliminary tour of the country in early July he found Campion 'retired for his more safety' to Southwark and the situation a grave one, the searches 'so eager and frequent . . . and the spies so many and diligent'. Clearly for Campion to remain longer in London was courting danger.[34]

But first the two priests had another problem to address, for it was not just the English Government which harboured suspicions about the Jesuits' intentions. Some of England's Catholics, too, though desirous to hear Campion and Persons preach, were less than happy to welcome the pair home for a prolonged stay. At a secret conference held in Southwark, near St Mary Overies (now Southwark Cathedral), Persons and Campion met with a panel of leading Catholic laymen and priests. Persons opened the meeting. He declared under oath that neither he nor Campion had been forewarned of the Pope's Irish invasion – they had learned of the

* Henry Norris was a favourite of Elizabeth's – his father had been executed on a manufactured charge of adultery with Elizabeth's mother, Anne Boleyn (thus sealing Anne's downfall), and the family had forfeited their lands. At her succession Elizabeth restored those lands and later ennobled Norris. Thomas Paget was a staunch Catholic who endured frequent terms of imprisonment for his faith.

expedition only at Reims. Next, he read out the instructions for their mission, emphasizing that their orders strictly prohibited them from dabbling in 'matters of state'. But his protestations failed to convince one of the attending priests, who now argued that the Catholics to whom he had spoken feared the Jesuits' mission could only ever be viewed as political by the English Government. For the good of the faith, therefore, the pair should leave the country at once. Persons refused. The Jesuits had been expressly called to the mission. If they turned back now it would represent a decisive propaganda victory for Elizabeth and her Council. His argument won the day, but in the decades to come this conflict would become a full-scale and enervating war of attrition between the rival Catholic factions.[35]

The conference broke up not a moment too soon. Government agents were closing in on the venue. Charles Sledd, a former student at Rome who had found it profitable on his return to England to turn Protestant informer, recognized a face familiar from his college days – that of law student and known Catholic, Henry Orton, Persons' guide on his earlier tour of the country, now travelling to Southwark to take part in the secret meeting. Sledd fell into step behind him, but before Orton could reach his destination Sledd had him apprehended. When, just a short while later, Sledd spotted the elderly Marian priest Robert Johnson making the same journey to Southwark, the informer's suspicions were aroused. Once again Campion and Persons had a lucky escape. Sledd's constable, growing impatient of the hunt, broke cover too soon and arrested Johnson some distance short of the meeting house. The time had come for the two Jesuits to leave London for the comparative safety of the open road.[36]

Each equipped with a pair of horses, a servant, travelling clothes suitable for a gentleman and sixty pounds of spending money – all provided by George Gilbert, who accompanied them on the first leg of their journey – Campion and Persons headed north out of the city to Hoxton, where they spent the night, possibly at the house of Sir William Catesby, a landowner with Catholic sympathies. The following morning they were surprised by Thomas Pound, who had successfully bribed his way out of the Marshalsea prison and had ridden through the night to intercept them. The prisoners had been talking among themselves, said Pound. If Campion or Persons were captured it would be an easy matter for the Government to paint them as traitors and political agitators. They must each, therefore, set down a declaration of their aims and the precise purpose of their mission, which Pound would safeguard for them. It seemed a sensible idea and the two Jesuits duly wrote out their statements, handing them to the waiting Pound before heading on their way. Persons sealed his paper; Campion left his open: a small character distinction that would have huge repercussions.[37]

Back in the Marshalsea, Pound read Campion's document. He showed it to his fellow prisoners. Soon, copies of the text were circulating through the gaol, smuggled from cell to cell. Visitors to the prison carried transcripts away with them. The pages fanned out across London and to the countryside beyond, landing indiscriminately in the hands of friend and foe. Campion's testimony, intended as a defence of his case *only* in the event of his arrest, was now blowing through England like a campaign manifesto.[38]

Campion addressed the Privy Council directly and in measured tones at first. His return home to his 'dear Country' was 'for the glory of God and the benefit of souls'. He was 'strictly forbidden by our Father that sent me, to deal in any respect with matter of State or Policy of this realm'. He begged for a chance to defend the Catholic faith before the Privy Council and an assembly of judges and theologians, so certain was he that no one could fail to be persuaded of the rightness of his argument if they would only give him an 'indifferent and quiet audience'. But then, in a flourish of rhetoric familiar from his Oxford days, Campion laid down a challenge that horrified his Protestant readers: 'be it known to you that we have made a league – all the Jesuits in the world . . . – cheerfully to carry the cross you shall lay upon us, and never to despair your recovery, while we have a man left to enjoy your Tyburn'. Even now, gathered beyond the seas, were 'many innocent hands', all of whom were 'determined never to give you over, but either to win you to heaven, or to die upon your pikes'. To Catholics it was a blast of hope. To Protestants, and to Elizabeth's Government in particular, it was a war cry. If Campion had been a wanted man before, now he had become the official spokesman of the Catholic mission and a voice to be silenced at all costs.[39]

The immediate aim of the Jesuits' mission, highlighted by the awkward few hours spent by Persons on his arrival searching for a Catholic contact to aid him, was to bring order to the local efforts of Allen's seminary missionaries. Successful the seminarians had undoubtedly been, but their successes were isolated. What was required was a coherent, national scheme, providing a country-wide network of priests capable of administering to all England's Catholics. Secondary to this aim was the need to ensure a regular supply of students to the Reims and Rome seminaries. This latter objective could be dealt with straightforwardly. It was arranged for a Father William Hartley and a Father Arthur Pitts to take up discreet residence at Oxford and Cambridge respectively, to search out and approach likely recruits for the mission. Hartley's job was the easier – Oxford students were still proving highly 'responsive to the ancient faith', even by 1580.* Pitts, though, had his work cut out for him. Cambridge's proximity to the trading routes of the East Coast, along which Protestant theories had first flooded into England, meant that the country's second university was never as defiantly pro-Catholic as the more isolated Oxford.

* On 16 March 1583 William Allen observed: 'Great complaints are made to the Queen's councillors about the university of Oxford, because of the number who from time to time leave their colleges and are supposed to pass over to us.'

Nevertheless, a number of students were still eager to listen to Pitts' arguments and 'within a few months he gained a harvest of seven young men of very great promise and talent, and now they are on the point of being sent to the Seminary at Rheims', reported Persons afterwards. More would soon follow.*[40]

But to achieve their primary objective Campion and Persons needed to travel – to divide up the country between them and cover it, county by county. Their first tour of duty lasted three months, with Persons taking in Gloucester, Hereford, Worcester and on through to Derbyshire, and Campion visiting Berkshire, Oxfordshire and Northamptonshire. A second round of journeying found Persons moving in and about the London area, and Campion going north for six months, up to Lancashire and Yorkshire.[41]

With the travelling, though, came all the pressures of isolation and nervous exhaustion General Everard Mercurian had warned them of back in Rome. 'I cannot long escape the hands of the heretics,' wrote Campion; 'the enemy have so many eyes, so many tongues, so many scouts and crafts.' He was forced to switch disguises continuously to keep ahead of the pursuivants, but still this offered him little sense of security: 'My soul is in mine own hands ever.' And as fast as the pursuivants chased him so the rumour mills turned: 'I read letters sometimes myself that in the first front tell news that Campion is taken, which, noised in every place where I come, so filleth my ears with the sound thereof, that fear itself hath taken away all fear.' Persons wrote simply: 'We never have a single day free from danger.'[42]

As the manhunt intensified so too did the means used to flush the two Jesuits from their hiding places. Campion informed Mercurian 'at the very writing hereof, the persecution rages most cruelly. The house where I am is sad; no other talk but of death, flight, prison, or spoil of their friends'. Persons wrote: 'the violence

* Both Hartley and Pitts were eventually caught by the authorities. They were banished from England in 1585. Hartley returned soon afterwards and was recaptured. He was executed at Shoreditch in London on 5 October 1588, one of the many Catholics executed in the aftermath of the Armada.

. . . is most intense and it is of a kind that has not been heard of since the conversion of England. Everywhere there are being dragged to prison, noblemen and those of humble birth, men, women and even children'. He described sitting at table when 'there comes a hurried knock at the door, like that of a pursuivant; all start up and listen, – like deer when they hear the huntsmen; we leave our food and commend ourselves to God . . . If it is nothing, we laugh at our fright'. Too often, though, it proved not to be nothing. Ralph Sherwin, a young seminarian and former Oxford student who had accompanied the two Jesuits on their journey from Rome, was arrested on 13 November, preaching at the house of Mr Roscarock just twenty-four hours after he had been with Persons. Edward Rishton, another former Oxford undergraduate and one of the first English students at Allen's Douai seminary, was captured during a raid at the Red Rose Tavern in Holborn. Persons was expected at the inn, but he had lost his way en route and only arrived when the search was over.[43]

On 16 January 1581 Parliament met to consider the Jesuit peril. Unsurprisingly, Sir Walter Mildmay's opening speech was full of invective against the newly arrived priests in particular and the Catholic population in general. The Jesuits crept 'into the houses and familiarities of men of behaviour and reputation . . . to corrupt the realm with false doctrine', and 'to stir sedition'. Meanwhile, 'the obstinate and stiff-necked Papist is far from being reformed as he hath gotten stomach to go backwards'.[44]

Invective alone could never be enough, though, and for the next few months both Houses debated how best to counter the perceived threat. As so often before, Elizabeth acted as a restraining influence on her ministers and the legislation finally passed that session, entitled an 'Act to retain the Queen's Majesty's subject in their due obedience', was far milder than had at first seemed likely. It declared it treason to withdraw Elizabeth's subjects 'from their natural obedience' to her, or to convert them '*for that intent* to the Romish religion'. All those who willingly allowed themselves to be converted would also be adjudged traitors.[45]

It was the wording 'for that intent' that was significant here. It represented an attempt to wrest the problem of English Catholicism away from the religious, back towards the political, ground on which the Government knew itself to have surer footing. For the act failed plainly to define *conversion* to Catholicism as treason. Rather, it suggested, it was the withdrawal of allegiance to the Queen that such a conversion, by necessity, implied that was the real crime. Even as Campion and Persons declared their aim to be purely spiritual, Parliament was further enshrining opposition to the official English Church as a political act. Here was an argument that would run and run, but more immediately the new Treason Act and the anti-Jesuit vitriol that accompanied it merely served to reinvigorate the pursuivants trailing Campion and Persons. The hunt was closing in.

On Tuesday, 11 July Campion bade farewell to Persons and set off from their safe house at Stonor in Oxfordshire on yet another round of travelling. He was scheduled to go east into Norfolk, a county as yet unvisited by the Jesuits, calling first at Houghton Hall in Lancashire to collect some papers he had left there. It was a roundabout route but by now the pair had begun to build up a network of safe houses between London and Lancashire where he could stagger his journey. First, though, he had a favour to ask of Persons. For some time now he had been begged by the owners of Lyford Grange near Wantage, the Yate family, to come and stay with them. As the Yates were known to be defiantly Catholic – Mr Yate was then a prisoner in London for recusancy, while his mother supported a community of two priests and eight nuns at the house – he had always felt it unwise to call there before. Now, though, he was passing close by Wantage. Would Persons give him permission to stay at Lyford? He would not preach. Nor would he call attention to himself. And he would leave immediately the following morning. On these terms Persons agreed to his request.[46]

The visit went according to plan and early the next day Campion was on the road again, heading towards Oxford. But back at Lyford the house was alive with whispers, the familiar little currents that sucked and eddied around Campion wherever he went. Visiting Catholic neighbours were dismayed to learn they had missed the famous Campion. They were more dismayed still to learn he had not even preached. He must be made to return to them at once and a rider was dispatched to deliver this message. He intercepted Campion at an inn outside Oxford where the Jesuit was talking with a group of students and masters who had journeyed out from the university to meet him. With the rider's announcement, the voices of the students' rose up in unison – Campion must go back to Lyford and speak. The lay brother Ralph Emerson could ride on to Lancashire and collect Campion's papers. Indeed it would be safer for Campion that way, for hadn't Persons been worried about him revisiting Houghton? Campion and Emerson could arrange a rendezvous point in Norfolk. The pair were no match for this barrage and Campion was borne triumphantly back to Lyford Grange.[47]

The next couple of days passed peacefully. Campion was introduced to a steady flow of Oxford students and local Catholics all eager to meet him. Word filtered quickly through the district that Campion was staying with the Yates. On the morning of Sunday, 16 July a Mr George Eliot arrived at Lyford. Eliot had served in a number of Catholic households across southern England and he was an old friend of the Yates' cook, Thomas Cooper. He was also, it was said, a convicted rapist and murderer who had bargained his way out of gaol by turning informant. If this was true, it was known to very few, and soon Cooper whispered to Eliot that a secret mass was about to begin. Eliot's companion, David Jenkins, who was not a Catholic (though he was, said Eliot, sympathetic to the faith), was left drinking beer in the buttery and Eliot, himself, was ushered through to a 'fair, large chamber' beyond, where Campion was preaching. Immediately the service ended Eliot and Jenkins, along with several others who had attended the celebration, rode away, leaving Campion to dine with members of the household and those few stalwarts who had stayed on to talk with him. At one o'clock the house was surrounded by a company of soldiers. At their head was the neighbouring magistrate, Mr Fettiplace. Beside him rode George Eliot and David Jenkins.[48]

The pursuivants were held at the gate while Campion was hidden away. Then the doors were opened and the soldiers began their search for him. In the hours that followed they discovered 'many secret corners' but no sign of Campion. Mr Fettiplace grew apologetic at the inconvenience he was causing his neighbours; George Eliot grew more resolute – now he and Jenkins took charge of the search. That night a guard was set about the house and next day the hunt resumed.[49]

It was chance that finally led to Campion's discovery. As the despondent pursuivants made to leave after a fruitless morning's search, by now 'clear void of any hope', Jenkins 'espied a chink in the wall of boards' over the stairwell, 'which he quickly found to be hollow'. Seizing a crowbar he broke through to a small chamber beyond, where Campion and two other priests lay concealed.*[50]

It was chance, too, that had led Eliot to Lyford in the first place. He was, he later admitted, on the trail of the seminary priest John Payne.† When he saw a servant keeping watch on the roof of Lyford he had simply decided to investigate further. With two such simple instances of chance the Jesuit mission lay in ruins.[51]

Campion was led up to London under armed escort. With him were Fathers Ford and Collington, the two Lyford priests discovered with him in the hiding place, nine laymen, accused of aiding and abetting him and attending his forbidden mass, and the luckless Father William Filby, who had arrived at Lyford in secret only to find the place overrun with pursuivants and the magistrates

* During restoration works at Lyford in 1959 an Agnus Dei blessed by Pope Gregory XIII and papers dated 1579 were found in a wooden box nailed to a joist under the attic floorboards.

† John Payne was Cuthbert Mayne's travelling companion to England in 1576. He was eventually captured and executed at Chelmsford, Essex on 4 April 1582.

in possession. At Henley, where the party spent the first night, Robert Persons, now in hiding at nearby Stonor, was able to send a servant to see how the captives were being treated. The servant reported back that Campion appeared in good spirits and was on friendly terms with his guards. It was George Eliot who was treated with disdain by soldiers and magistrates alike. Members of the watching crowds were even bold enough to shout out 'Judas' at the informer as he passed by.[52]

As the party neared London, though, the procession took on a different aspect. At the Council's request the prisoners were pinioned in their saddles, their arms strapped tightly behind them and their legs bound together by a rope slung beneath the belly of their mount. Campion, himself, rode at the front of the cavalcade, a sign about his head reading 'Edmund Campion the Seditious Jesuit'. In this fashion they passed through the streets of London to the Tower.[53]

The trial of Edmund Campion took place on 20 November 1581, four months after his arrest and imprisonment. The charges laid against him, and those arraigned with him, were that on specific dates in Rome and in Reims the previous year, Edmund Campion, Robert Persons, William Allen, the layman Henry Orton and the entire haul of priests then in custody, including Ralph Sherwin, Edward Rishton, Robert Johnson, the Lyford priests, Thomas Ford and John Collington, and William Filby, had conspired to murder Queen Elizabeth. Further to this they had been privy to foreign invasion plans – they, themselves, were the advance party for that invasion, sent to stir up rebellion. It is unclear why the original indictment, which invoked Parliament's new Treason Act, was dropped in favour of these accusations. Perhaps Elizabeth's reluctance to make martyrs had something to do with it – political enemies of England deserved execution; so many priests dying solely for their faith smacked strongly of religious persecution for its own sake. More likely, though, it was a calculated attempt by the Government to turn what might have become an intellectual argument about the lawfulness of the Anglican Church – in which Campion might have triumphed – into an emotive debate about national security. Either way, the Council dropped what would have been a legal, if unpopular, arraignment on the grounds of converting the Queen's subjects to Catholicism, in favour of these trumped-up charges of mass conspiracy to murder.* It was entirely in keeping with the paranoia of the age.[54]

Also in keeping was the procession of shady characters brought out to testify against the accused. The informant Charles Sledd swore that while in Rome and in Reims he had learned of the invasion plans from William Allen and one of the prisoners, Luke Kirby. George Eliot claimed that Campion, in his Lyford sermon, had spoken of 'a great day' that was soon to come, and that another of the prisoners had sworn him to secrecy about the plot. And the arch-fabricator Anthony Munday was brought in to announce to a packed courtroom that the English seminary students were schooled in treason, that Henry Orton had told him at Lyons that Elizabeth was not the rightful Queen of England – Orton vehemently denied ever having set eyes on Munday before – and that Edward Rishton was a skilled maker of fireworks who was planning to burn Elizabeth in her royal barge with 'a confection of wild fire', an event to be followed by a general massacre of all those not in possession of the password 'Jesus Maria'.† The verdict was a foregone conclusion.[55]

Campion had always believed he was coming home to England to die. The night before his departure from Prague a colleague had inscribed on the door above his cell *P. Edmundus Campianus, Martyr.* Earlier, another priest had painted a garland of roses and

* No doubt many of the Council believed in Campion, Persons and Allen's guilt but both Ford and Collington had been in England a number of years before the Jesuits' arrival, while poor Filby had only bad timing to thank for his presence at Lyford.

† Quick to jump onto the bandwagon, Munday published an account of Campion's capture within a couple of days of the Jesuit's imprisonment. Eliot later complained that the book was 'as contrary to truth as an egg is contrary to the likeness of an oyster'.

lilies on the wall above his bed – the symbol of martyrdom. On the morning of 1 December 1581 Campion was led out from the Tower, through the driving rain and the mud-choked London streets, to the scaffold at Tyburn. There he was hanged, drawn and quartered before the assembled crowds. With him were Father Alexander Briant, a close friend of Robert Persons, and Father Ralph Sherwin, the young seminarian who had set off from Rome with Campion and Persons in such high spirits the year before.[56]

In May the following year seven more priests were executed, including Thomas Ford, Luke Kirby, Robert Johnson and William Filby. Edward Rishton and the layman Henry Orton, though both found guilty of treason and sentenced to death, were not executed. They were kept prisoner in the Tower until January 1585 when they were forcibly deported to France. Father John Collington was able to find a witness to confirm he had been resident in England since July 1576 and therefore could not have been in Reims and Rome on the dates specified. Like Orton and Rishton he was exiled to France in January 1585, having spent the intervening years in the comparative comfort of the Marshalsea prison.

After Campion's execution the lay brother Ralph Emerson escaped from England and made his way safely to Rouen. He joined in exile George Gilbert, the Jesuits' friend, guide and self-appointed financier, whose activities had placed him in grave danger of arrest and who had been persuaded to leave England shortly before Campion's capture. As for Robert Persons, with Campion's arrest the Government now turned its attention wholly on him. Clearly, he could not elude the pursuivants for long and in August he made his way to France, disguised as one of a number of Catholic refugees fleeing persecution. He would never see England again.[57]

The savagery of Campion's death had taken people's breath away. It was not just that he had been tortured while in the Tower – so severe were the bouts of racking he endured that when his keeper asked him how he felt, he allegedly answered 'Not ill, because not at all'; witnesses to his trial reported he was unable to raise his hand to take the oath and witnesses to his execution reported 'that all his nails had been dragged out'. It was not just that so many had been executed with him – since Cuthbert Mayne's execution in 1577, only two other priests and one scholar had suffered the same fate. It was more the realization that the Government had turned against one of its own, and such a one as the scholar Campion, that shocked onlookers.[58]

Some felt Elizabeth had sacrificed Campion as a sop to those Puritans concerned by her proposal to wed the Catholic Duc d'Alençon. Others, that Campion had been silenced by a Government unable to defend its new faith against the theological reasoning of the Catholic Church. Ballad-mongers were soon singing:

If instead of good argument,
We deal by the rack,
The Papists may think
That learning we lack.

Many were even more direct in their criticisms:

Our preachers have preached in pastime and pleasure,
And now they be hated far passing all measure;
Their wives and their wealth have made them so mute,
They cannot nor dare not with Campion dispute.

What was clear to all, though, was that with Campion's death, the Jesuit mission to England had been stopped in its tracks. The question was, could it ever regain its momentum?[59]

Seven years later, in October 1588, Father John Gerard was setting out to answer this question. Campion had written of a 'league' of 'all the Jesuits in the world' – a league dedicated to restoring England to the Catholic Church, no matter how brutal the cost. For Gerard the time had come to make good that promise.

Chapter Five

'And better it were that they should suffer, than that her highness or commonwealth should shake or be in danger.'
(Device for the Alteration of Religion, 1558)

TO THE NORTH OF the city of London, beyond the walls and the great gates, Moorgate and Aldersgate (Aldgate), lay the lordship of Finsbury and Finsbury Fields. Once used as a site for archery tournaments and wrestling matches, by 1588 the fields had fallen victim to urban sprawl. Contemporary commentator John Stow complained 'there is now made a continual building throughout of garden houses and small cottages'.[1]

A description of one of these cottages remains. The ground floor contained a kitchen and a dining room. The first floor was given over to a chapel that doubled at night as a sleeping loft. The cellar beneath held sufficient storage space for logs, coal and beer barrels. And behind the carefully piled provisions was a hiding place with room for six or seven men. In the autumn of 1588 this small, three-roomed cottage served as the London headquarters of the Jesuit mission to England.[2]

It was here that John Gerard made his way towards the end of November, as the first snows of an unseasonably bitter winter blanketed the country. Gerard recorded his journey south in perfunctory style – 'there was no incident on the way' – and for the length of time it took him to appear in London he gave no explanation. But there was one man to whom an explanation was owing: the man who had sent for him.[3]

Father Henry Garnet was thirty-three, cheerful, scholarly, the son of a Nottingham grammar school master. Unlike Gerard, Garnet's family had conformed to Elizabeth's nationalized Church after 1559. In Nottingham and those parts of Derbyshire bordering the city there had been little opposition to the change in religion. Then in 1567 Garnet won a scholarship to Winchester School and there he came under a more Catholic influence. Winchester was among the last of the schools to accept the new faith. In 1561 the then headmaster had been arrested for his refusal to conform to Protestantism. In protest the boys had boycotted the school chapel, locking themselves in their dormitories and accusing the replacement headmaster of destroying 'the souls of the innocents'. The military commander of Portsmouth Harbour was called to break up the strike and a dozen boys were expelled soon afterwards. It was into this world of Catholic defiance and classical scholarship that the twelve-year-old Henry Garnet was soon immersed.[4]

Avoiding the usual passage from Winchester to New College, Oxford (by now he was no longer prepared to pretend to be a Protestant) Garnet headed for London to become 'corrector for the press' at Richard Tottel's printworks in Temple Bar. He was in London in June 1573 for the execution of Thomas Wodehouse at Smithfield. Wodehouse had the unhappy distinction of being the first Catholic priest to be executed in London under Elizabeth; more than that, he was rumoured to have joined the Society of Jesus while a prisoner.* In the summer of 1575 Henry Garnet left London for Rome to join the Jesuits himself.[5]

Garnet's return to England in July 1586 was followed with uncomfortable swiftness by his promotion to Superior of the English Jesuits just a couple of weeks later, after the arrest of the

* Although Wodehouse was ordained during the reign of Queen Mary and was therefore entitled to amnesty under Elizabethan laws, he repeatedly denied Elizabeth's claim to the throne – a treasonable offence. As a further irritant to the Privy Council he wrote copious letters in defence of the Catholic faith from his prison cell in the Fleet, which he tied to stones and threw to passers-by in the street below.

man under whom he had come to serve. The death of Campion still hung albatross-like around the neck of the Jesuit mission; moreover, it seemed those destined to pay the price for that death were not Campion's killers, but the men struggling to follow in his footsteps. In the six years between Campion and Persons' first landing and Garnet's arrival the Jesuits had failed to establish a permanent bridgehead in England. Aborted attempt had followed aborted attempt. And now with the threat of the Spanish Armada drawing nearer, Garnet's own efforts to revive the mission seemed destined to flounder in the wave of anti-Catholic feeling sweeping over the country.

Garnet's first action as Jesuit Superior had been to write to Rome for more men to be hurried over to help him. In the spring of Armada year he was informed that reinforcements – in the shape of John Gerard and Edward Oldcorne – would soon be on their way. So as the events of 1588 unfolded about him Garnet prepared for their coming, choosing the cottage in Finsbury as a base for the mission, for 'since it was believed that no one was actually residing there, it was never molested by the officers whose duty it is to make the rounds of every house to enquire whether the inmates are in the habit of attending the [Protestant] church'. He was in London to witness the spate of executions that took place that autumn, as post-Armada relief gave way to bloodlust and revenge. In all, seventeen priests, nine Catholic laymen and one woman were executed over a three-month period. The one woman to be killed, Margaret Ward, was charged with supplying the priest William Watson with a rope to escape from the Bridewell prison – sympathetic onlookers claimed that as she climbed to her death they could see she was 'crippled and half-paralysed' by torture. Elizabeth, herself, was said to have been appalled by Ward's death and 'it was for that reason that recently she pardoned two other women who had borne themselves before the tribunal with singular courage', wrote Garnet in October. Two months later Garnet was in the crowds to watch Elizabeth's triumphal procession to St Paul's Cathedral to give thanks for England's victory.

And throughout this time still he waited for news of Gerard and Oldcorne's safe arrival.[6]

There are few details of Gerard's meeting with Garnet in Finsbury that winter – Gerard himself is significantly quiet on the subject – but the following March Garnet wrote to Claudio Aquaviva, Jesuit General since the death of Everard Mercurian in 1581. His letter is characteristically cryptic, but in it he mentions 'that without consulting me at all [Gerard] did things which [he] had no authority to do and which manifestly [he] should never have done'. During his brief stay in Norfolk it seemed not only had Gerard freely dispensed spiritual guidance and religious pardons, but he had also passed information – probably details of the privileges granted by the Pope to the English Jesuits – to 'certain priests who, during their time in Rome, were not considered well disposed to us'. With memories of priests-turned-informers like Charles Sledd and Gilbert Gifford still fresh in mind, such an action was risky at best. At worst it threatened to destroy the mission. And although Garnet concluded his letter 'It is not a serious matter but it might have been', it must have seemed to him now that his new recruit was uncommonly confident in his own abilities and apparently lacking in discipline. It was not an auspicious start.[7]

Before Christmas the fourth and final Jesuit priest on the mission arrived back in Finsbury from a tour of the country.* Robert Southwell was twenty-seven years old, the son of an old East Anglian family. His grandfather had been one of the commissioners for the dissolution of the monasteries, his mother had been a childhood companion of the then Princess Elizabeth and his father was a prominent courtier. Among his more important relations were Lord Burghley, Sir Francis Bacon and the Attorney General, Sir Edward Coke. Southwell's family serves as a

* Edward Oldcorne arrived in Finsbury some time before Gerard. It appears he too behaved indiscreetly during his journey to London, though the precise nature of that indiscretion remains unclear. In his letter to Aquaviva Garnet begged the Jesuit General to ensure 'that those who are sent hereafter . . . should properly understand their accounts and seek out some veteran as quickly as possible'.

particularly colourful reminder of just how few assumptions can be made about the religious complexion of England at this period.[8]

Southwell left home for the English College at Douai in the summer of 1576, aged fourteen. In November that year he moved to the Jesuit College of Cleremont near Paris and in the spring of 1578 he applied to join the Society of Jesus. Political upheavals in the Spanish Netherlands and an initial rejection by the Society on the grounds of his youth meant it was not until October, and then only in Rome, that Southwell was admitted to the Jesuit noviceship, around the time of his seventeenth birthday. There, he quickly won praise for his skills as a writer and respect for his vivid intelligence. When, in 1586, Henry Garnet was chosen to leave for England to join the mission it was the poet Southwell who was picked to accompany him on the journey.[9]

Since that time, like Persons and Campion before them, Garnet and Southwell had travelled England in secret, labouring to rebuild the network of Catholic families willing to maintain a priest; labouring, too, to enlarge that network in the face of increased government persecution.* They had become a formidable team. With the arrival of John Gerard and Edward Oldcorne that team had doubled. What remained unclear, though, was how many of their countrymen would still be prepared to welcome them in, now that Catholicism had been linked so strongly with un-Englishness in the public consciousness. For if to be Catholic was to be an unnatural Englishman, then to draw attention to that unnaturalness in the weeks and months following the Spanish Armada was tantamount to signing your own death warrant. This being the

* The best description of Garnet and Southwell's work so far – and of how they believed the mission should develop – is contained in a letter by Garnet to Claudio Aquaviva, written in June 1588. 'This is the plan we have agreed on for the greater glory of God, when there shall be a greater number of [Jesuits] here. Two should be stationed in London – or one in London and one in the environment. The others should have assigned to each one a province or county in which each can work for all he is worth to promote religion. There will not be lacking other priests, men of outstanding holiness and learning who will come to their assistance – and to this we most of all can testify by experience. The field will be theirs to take over from our labours and the harvest from it will be beyond measure, owing to Him who guides the work of our hands unceasingly.'

case the new and expanded Jesuit mission seemed destined to have only limited appeal.[10]

And for the English Government and the Catholic Church both, religion had become a numbers game. To play for were the souls of all those who still stood wavering in the middle of the great religious divide. For the Government 1588 had proved a windfall year, drawing to the Church of England all those obedient to the pull of patriotic fervour, like the Earl of Shrewsbury (viewed by most as 'half a catholic'), who now redoubled his efforts to rout out seminary priests on his estates.* The challenge for the Jesuits was to reverse this process.[11]

'Christmas was drawing near', wrote Gerard, 'and we had to scatter. The danger of capture was greater at festal times and, besides, the faithful needed our services. I was sent back, therefore, to the county where I had first stepped ashore.' As December 1588 drew to a close and England wearily prepared to celebrate the Nativity, Gerard retraced his steps to Norfolk.[12]

Norfolk seems to have turned its back on English affairs throughout much of history. While other shires engaged in the civil disturbances that marred the reigns of weaker monarchs, in the county-league rivalry of the Wars of the Roses, Norfolk scanned the horizon for the distant sails of potential invaders, for the safe return of its merchant ships and indulged in a style of lawlessness all its own. The county records bristle with stories of Norfolk nobles and squires holding their neighbours to ransom, of houses defended 'in a manner of a forcelet' and of bands of armed

* For many years Shrewsbury was gaoler to Mary, Queen of Scots. Always vulnerable to accusations of leniency on account of his Catholic sympathies – his mother Mary Dacre was from a devoutly Catholic family – he was finally relieved of this duty in 1585. Many of the accusations made against him came from his second wife, the notorious Bess of Hardwick, who also accused him of having an affair with the Scottish Queen. The pair separated soon afterwards.

retainers roaming the district in search of someone to terrorize.[13]

Hanseatic merchants brought the writings of Martin Luther to Norfolk early in the sixteenth century. Luther's ideas were greeted with enthusiasm by the academics of nearby Cambridge; soon a band of them, including a number of Norfolk men, were meeting regularly in the town's White Horse Tavern (behind King's Parade), which quickly became known as Little Germany. It was a Norfolk man, Thomas Hilton, who helped smuggle home William Tyndale's forbidden English Bibles; he was burnt for heresy in 1530. And it was a Norfolk woman whose charms led to schism with Rome in the first place: Anne Boleyn was the daughter of the Norfolk squire Sir Thomas Boleyn and a niece of Thomas Howard, Duke of Norfolk. Later, when Anne's daughter Elizabeth laid claim to being England's most English of monarchs, the countrymen of Norfolk could nod their heads in agreement; after all, she was one of their own.[14]

Yet when the Duke of Northumberland seized the English crown on behalf of his daughter-in-law Lady Jane Grey, it was to Norfolk that Mary Tudor fled to rally her supporters. Mary came to Kenninghall, west of Thetford, from where she wrote to the House of Lords on 9 July 1553 asserting her claim to the English throne. She was joined there by the loyal Norfolk gentry, all of them Catholic to a man. It was into the care of the Norfolk Catholic Sir Henry Bedingfeld that Mary entrusted her recalcitrant younger sister; Bedingfeld escorted Elizabeth from the Tower of London to house arrest at Woodstock in May 1554.* And it was of Norfolk that a Spanish agent wrote in 1586, as Philip II scouted for suitable landing sites for his Armada fleet, 'the majority of the people are attached to the Catholic religion'. If England was a country still divided by religion then the county of Norfolk was England in microcosm. And few families illustrated this divided

* There seems to have been little love lost between the twenty-one-year-old Elizabeth and her warder. Bedingfeld found the princess so confusing he was at a loss to know 'if her meaning go with her words, whereof God only is judge'. Elizabeth was reputed to have said to Bedingfeld that 'if we have any prisoner whom we would have hardly and strictly kept, we will send him to you'.

country and county better than the Yelvertons, to whom John Gerard now returned.[15]

Edward Yelverton, Gerard's Catholic contact from Norwich Cathedral, was the eldest son from the second marriage of William Yelverton of Rougham, Norfolk. On his father's death Edward inherited the family's estate at Grimston, extending well over two thousand acres. There he lived with his family, his younger brother Charles, a committed Protestant, and his newly widowed half-sister Jane Lumner, viewed by Gerard as 'a rabid Calvinist'. (Gerard also mentions a half-brother, Sir Christopher Yelverton, who was 'one of the leaders of the Calvinist party in England'.) It was a dangerous place for Gerard to begin his mission, despite Edward Yelverton's keenness for Grimston to become a Jesuit base. Such religious differences as divided the Yelvertons could stretch family loyalty to breaking point, as a contemporary poem revealed:

> . . . your husbands do procure your care [imprisonment],
> And parents do renounce you to be theirs;
> . . . your wives do bring your life in snare,
> And brethren false affright you full of fears;
> And . . . your children seek to have your end,
> In hope your goods with thriftless mates to spend.[16]

In years to come many Catholics would learn that their nearest was all too often their dearest enemy, particularly when an inheritance was at stake. Gerard, himself, had already experienced the Yelverton family's mistrust. 'On my first arrival [at Grimston]', he wrote, 'the Protestant brother was indeed suspicious – for I was a stranger, I had come here in company with his Catholic brother, and he could think of no reason why his brother treated me so kindly.' The priest's easy familiarity with the occupations of an Elizabethan country gentleman had soon allayed Charles Yelverton's fears: 'When I got the opportunity I spoke about hunting and falconry, a thing no one could do in correct language unless he was familiar with the sports.' Freshly equipped as 'a gentleman of moderate means' – in clothes provided by Henry Garnet, who 'was anxious that I should not be a burden to my host at the start' – Gerard now took up his role of sporting squire once more.[17]

The methodology of the Catholic mission to England had changed little since William Allen's seminary priests first began arriving home in 1574. The protomartyr Cuthbert Mayne had clothed himself as the Tregian family's steward; Robert Persons as a returning soldier of fortune – the key to a missionary's success or failure lay in his ability to inhabit his new identity fully. This was best evinced by Father Richard Blount, who returned to England in the spring of 1591 disguised as a homecoming prisoner-of-war. Blount was interviewed by Admiral Lord Howard of Effingham and provided him with enough information – all fabricated – about the deployment of the Spanish fleet to earn himself a naval pension in recompense, or so Blount's friends reported afterwards.[18]

In his play *The Taming of the Shrew*, written *c.*1592, Shakespeare introduced 'a young scholar that hath been long studying at Rheims . . . cunning in Greek, Latin and other languages'. The role was cover for the amorous suitor Lucentio, enabling him to woo the 'fair Bianca'. But among the audience watching the new comedy, there would have been those who recognized in Shakespeare's words an allusion to an altogether different form of deception. The young Reims scholars *they* knew, seminary students fresh from their lessons in Greek, Latin, Hebrew and English, were even now being deployed across the country disguised as tutors, stewards and visiting poor relations.* John Gerard was now Mr Robert Thompson, 'attired [according to a later spy's report] costly and defensibly in buff leather, garnished with gold or silver lace, satin doublets, and velvet hose of all colours with cloaks correspondent, and rapiers and daggers gilt or silvered'. 'It was thus', wrote Gerard, 'that I used to go about before I was a Jesuit and I was therefore more at ease in these clothes than I would have been if

* A contemporary spy's report refers to a suspected seminary priest who 'under colour of teaching on the virginals goeth from Papist to Papist'.

I had assumed a role that was strange and unfamiliar to me . . . [Now] I could stay longer and more securely in any house or noble home where my host might bring me as his friend or acquaintance.' More importantly, now he could 'meet many Protestant gentlemen' and bring 'them slowly back to a love of the [Catholic] faith'.[19]

There exists a memorandum dated 1583, written by George Gilbert, leader of the band of young Catholics who had assisted Campion and Persons during the first Jesuit mission to England. It is called *A way to deal with persons of all sorts as to convert them and bring them back to a better way of life – based on the system and methods used by Fr. Robert Persons and Fr. Edmund Campion* and, as its title suggests, it is a proselytizer's handbook. Gilbert was ideally placed to advise the new missionaries. It had been his money and his connection with most of the major Catholic families in England that had enabled Campion and Persons to travel the country in relative safety, setting up their network. And until his escape to France in 1581 (at Persons' entreaty), Gilbert had been adept at cheating capture. Arrested and brought before the Bishop of London in midsummer 1580, he was quickly released when Norris, the bishop's pursuivant, attested to his honesty; Norris was said to be in Gilbert's pay. This memorandum was Gilbert's last contribution to the English Catholic cause – he died in Rome on 6 October 1583 aged just thirty-one, having been admitted into the Society of Jesus on his deathbed. But it was Gilbert's instructions that John Gerard now followed as he began his Norfolk apostolate.[20]

'As soon as any father or learned priest has entered an heretical country', wrote Gilbert,

> 'he should seek out some gentleman to be his companion. This man should be zealous, loyal, discreet and determined to help him in this service of God, and should be able to undertake honourably the expenses of both of them. He should have a first-rate reputation as a good comrade and as being knowledgeable about the country, the roads and paths, the habits and disposition of the gentry and people of the place, and should be a man who has many relations and friends and much local information.'

In Edward Yelverton, Gerard had found just such a companion. This primary contact made, the newly arrived priest could then 'mix freely everywhere, both in public and in private, dressed as a gentleman and with various kinds of get-up and disguises so as to be better able to have intercourse with people without arousing suspicion'. And so it proved for Gerard: 'I stayed openly six or eight months in the house of that gentleman who was my first host. During that time he introduced me to the house and circle of nearly every gentleman in Norfolk, and before the end of the eight months I had received many people into the Church.'[21]

Among his first converts were three members of Yelverton's own family, including the Protestant Charles Yelverton and the Calvinist Jane Lumner. Before Gerard's arrival at Grimston Jane had reportedly expressed certain anxieties about the state of her soul. A consultation with the ill-famed Dr Perne of Peterhouse had left her more confused than ever. Hearing Gerard say 'time and again that the Catholic faith was the only true and good one, she began to have doubts, and in this state of mind, she brought [him] one day an heretical book which more than anything had confirmed her in her heresy'. Here was a chance for the priest to vindicate William Allen's training methods. Allen had prepared his students for a war of words, schooling them in reasoning and rhetoric. Now Gerard proceeded to demonstrate 'all the dishonest quotations from Scripture and the Fathers, the countless quibbles and mis-statements of fact'. Few English theologians were equipped to engage in dialectical combat of this sort single-handedly. Certainly no layperson was. Jane Lumner would prove an easy and a lasting convert. From then on her name was always among the annual lists of 'obstinate' Catholics returned to the Bishop of Norwich for the purpose of fining; the last entry dates from 1615, probably the year of her death.[22]

Catholicism bound the Norfolk gentry together; marriage bound them tighter still. Yelvertons wed Bedingfelds, Bedingfelds wed Southwells – theirs was a network of such interconnectedness as the Jesuits could only dream of, criss-crossing the county, extending deep into Suffolk and Essex too. Many had benefited from the wholesale disposal of church lands at the Dissolution, seizing the chance to extend their estates and entrench their position as the traditional power brokers of old England. Robert Southwell grew up at Horsham St Faith near Norwich, within sight of the Benedictine priory acquired by his grandfather Sir Richard. It was one of four such properties bought by the family as some of the best real estate in England changed hands.*[23]

So at Elizabeth's accession, with their new lands and old influence, the Yelvertons, the Bedingfelds, the Southwells seemed set to enjoy the peace and prosperity the new Queen was intent on pursuing. By the time of Gerard's arrival at Grimston it was clear that for them, such peace and prosperity would never again be possible. The leniency of the 1559 settlement may not have given way to martyr-making in the style of Mary Tudor but it had, by 1588, crystallized into something altogether harder and more finely focused. There were no public tribunals or baying mobs, no calls to recant and save your soul, no crisis of survival in fact – rather, the slow, sapping efficiency of English law.

From the very first, the peace of the Catholic Norfolk gentry had been threatened. In 1561 it had come to the Council's attention that Sir Edward Waldegrave's estate at Borley in Essex was serving as a mass centre for a number of visiting Norfolk gentry. Waldegrave was arrested and dispatched to the Tower where he later died; the others were imprisoned in Colchester gaol. To the Government its policy was clear: such high profile arrests served as an example to the rest of the county, obviating the need for further action.

* The curiosity of English Catholics benefiting from the Dissolution of the Monasteries was noted by outsiders: Simon Renard, ambassador to the Holy Roman Emperor, would report home at the time of Queen Mary's succession that the papists held far more of the plundered church property than the heretics.

Seventeen years on it seemed Norfolk's landowners were due another lesson.[24]

In the summer of 1578 Elizabeth embarked upon her annual royal progress, this time through Suffolk and Norfolk. 'The truth is', wrote Thomas Churchyard, who observed the proceedings, 'they had but small warning . . . of the coming of the Queen's Majesty into both those shires.' Once word of Elizabeth's impending arrival leaked out, though, the preparations took on a frantic air: 'all the velvets and silks were taken up that might be laid hand on, and bought for any money, and soon converted to . . . garments and . . . robes'. On Sunday, 10 August Edward Rookwood, now suitably attired, welcomed Elizabeth and her entourage to Euston Hall near Thetford. On the morning of Monday, 11 August, as Elizabeth took her leave, Rookwood was arrested. Royal hanger-on Richard Topcliffe described the scene to the Earl of Shrewsbury. The Lord Chamberlain 'demanded of [Rookwood] how he durst presume to attempt [Elizabeth's] real presence, he, unfit to accompany any Christian person; forthwith said he was fitter for a pair of stocks; commanded him out of the Court, and yet to attend her Council's pleasure; and at Norwich he was committed'. Rookwood was charged with refusing to attend his parish church, 'contrary to all good laws and orders and against the duty of good subjects'. Joining him in the dock were nine other Norfolk gentlemen, all guests at a dinner in honour of the Queen hosted by Lady Style of Braconash near Norwich; among them were Sir Henry Bedingfeld, Humphrey Bedingfeld and Robert Downes, John Gerard's contact from the Norwich inn. Rookwood and Downes, already excommunicate for non-attendance, were imprisoned until such time as they should conform. The others were placed under house arrest in Norwich, fined 200 pounds and instructed to conform or face a lengthy gaol sentence. So much for their hopes of peace.[25]

Soon their prosperity would be forfeit too. For although there was no way of measuring whether the actual number of Catholics was increasing under the influence of William Allen's missionaries, there was a sure way of determining the effect the priests were having on existing Catholics. From 1574 onwards the number of people fined for refusing to attend their parish church grew steadily. Before long these dissenters had earned a name for themselves: recusants, from the Latin *recusare*, to refuse. Before long, too, a simple truth had dawned on many in the Government: recusants meant revenue.

In 1577 the Bishop of London wrote to Sir Francis Walsingham with a proposal. In view of this spread of recusancy, he and his colleagues were devising a scheme to 'procure the Queen a thousand pounds by year to her Coffers'. Since the scheme involved bypassing existing law and imposing 'round fines' on all those refusing to receive Communion (not in itself a criminal offence), the bishop suggested Walsingham keep from Elizabeth the precise details of his proposal, 'or else', he added feelingly, 'you can guess what will follow'. This small difficulty aside, the bishop reckoned his scheme would 'weaken the enemy and touch him much nearer than any pain heretofore inflicted'. The proposal was never implemented. Evidently, though, it set Walsingham thinking. When the Privy Council carried out its nationwide census of recusants that year, included among the list of returns were precise details of how much every recusant was worth.[26]

It took until 1581 for the Government to decide fully to profit from papistry. The arrival of the Jesuits the year before had removed even Elizabeth's objections to sterner anti-Catholic measures. If her new Church of England was not to become a mockery, then no more English Catholics could be permitted to break the law and absent themselves from it. Up to now she had prevented Parliament from raising recusancy fines – no longer. Included in that year's Treason Act was a clause designed to hit recusants hard. Overnight, the fines for non-attendance jumped from twelve pence up to twenty pounds per month. And for the purposes of calculation, the year was deemed to contain thirteen lunar months, rather than the customary twelve. William Cecil explained the new policy with unusual political candour: 'The causes that moved the renewing of this law was for that it was seen that the pain being no greater than xii pence, no officer did seek to charge any offender therewith, so that numbers of evil disposed persons increased herein to offend with impunity.'[27]

Now the full moneymaking potential of Catholic recusancy could be realized – and legally too. Soon a Hampshire clerk of the peace was writing to the Council complaining that 'The number of recusants which at every session are to be indicted is so great that [I am] driven to spend . . . a great deal of time before and after every session . . . in drawing and engrossing the indictment[s].' In 1587 new legislation was introduced to make the collection of recusancy fines more efficient. To ease the pressure on the courts recusants could now be convicted in their absence, on the evidence of informers, disgruntled relatives or jealous neighbours. Henry Garnet wrote to the Jesuit General in fury that 'any utterly base creature can cast it in our teeth that we are unfitted to have our share of life in common with them'. Another clause allowed the Exchequer to seize two-thirds of the recusant's estate in default of payment. They 'build their houses with the ruins of ours', wrote Robert Southwell angrily, and 'by displanting of our offspring adopt themselves to be heirs of our lands'.[28]

It was not just fines from which the Government was determined to benefit. In 1585 recusants were assessed for contributions 'towards the providing of horses and furniture for her Majesty's present services in the Low Countries'. As the Council explained, given the religious nature of the up-coming conflict 'her Majesty seeth so much the less cause to spare them in this'. In Norfolk, County Sheriff Sir Henry Woodhouse began the unpleasant process of extracting forced donations from his recusant neighbours. Evidently he was sympathetic to their plight. In December that year he wrote to Walsingham apologizing for his delay and naming only the unfortunate Robert Downes as capable of furnishing a light horse for the cavalry.* Other Catholics in other counties

* The Woodhouses were a prominent Norfolk family with several Catholic members. Francis Woodhouse was forced to sell his Breccles estate in 1599 to pay his recusancy fines.

were not so lucky. Londoner Mary Scott wrote to the Council begging to be excused payment. Her husband – on account of whose recusancy the family was being assessed – had died just four days earlier, she explained, and now she was 'plunged in many cares'. Mrs Margaret Blackwell of Sussex wrote with an even more heartfelt plea. She had never been absent from church, she protested, and she enclosed a certificate signed by Arthur Williams, parson, and all the churchwardens of the parish of St Andrew's to prove her statement. And in March 1585 John Gerard's father would write to the government, apologizing for his failure to pay the military levies (on account of his debts) and, instead, 'offering his person to serve Her Highness in any place in the world'.[29]

These new anti-Catholic measures were entirely in tune with a financial policy grown increasingly desperate as the looming war with Spain began to drain the Exchequer dry.* Moreover, they upheld Elizabeth's determination that the extirpation of Catholicism from England should never be seen as religious persecution. Yes, she explained, there were 'a Number of Men of Wealth in our Realm, professing contrary Religion', but none was 'impeached for the same . . . but only by Payment of a peculiar Sum, as a Penalty for the Time that they do refuse to come to Church'. English Catholics were still only being punished for breaking the law; that the law forbade the profession of their faith was really neither here nor there. However, behind closed doors at Westminster something else was happening that was altogether more invidious. It revealed itself in a bid to make Catholics turn in their weapons, in a motion that they be expected to pay double rates as foreigners did, in MP Dr Peter Turner's demand that they be forced to wear an identifying badge so 'that by some token a

He died in poverty in 1605. His wife Eleanor and son John were still 'obstinate recusants' in 1615. Sir Philip Woodhouse married Edward Yelverton's sister Grisell – both were converts of John Gerard.

* It was not always Catholics who suffered in this way – on occasion Elizabeth was compelled to extort forced loans from wealthy Anglicans too. In 1598 rumours that another such loan was imminent saw citizens 'shrink and pull in their horns'; some Londoners even fled to the countryside to avoid payment.

Papist may be known'. Catholics were different; Catholics were dangerous; Catholics were *other*. Not content that Catholicism had become un-English by imputation, Parliament was attempting to make it un-English by force of law.[30]

For Norfolk's gentry, for the Yelvertons, Bedingfelds and Southwells, the effects of such a policy were devastating. In 1574 Edward Rishton had written: 'the greater part of the country gentlemen was unmistakably Catholic; so also were the farmers throughout the kingdom . . . Not a single county except those near London and the Court . . . willingly accepted the heresy'. If the city and the Court, both frantic and fast-moving, spoke for the new religion, then the Norfolk gentry and their ilk, farmers, countrymen and landowners, rooted in the slower rhythms of the soil, spoke for the old – and they saw no reason to change. For them Catholicism was the traditional religion of Englishmen and women since Christianity was first introduced to the isle almost a thousand years before; it was the stripling Germano-Swiss construct Protestantism that was the foreign interloper. For the new merchant-class Members of Parliament to tell them they were un-English impugned their rank, for their fellow Englishmen to tell them they were traitors impugned their patriotism, but for them to change their beliefs impugned their very identity. Moreover it imperilled their mortal souls.[31]

Small wonder that on the eve of the Spanish Armada, they looked on in baffled disbelief as many were rounded up, along with every foreigner in the country, and summarily interned as a threat to national security. As the Council explained to those charged with arresting them, the Queen was 'hardly ventured to repose that trust in them which is to be looked for in her other good subjects'. No doubt their response was similar to that of Northamptonshire Catholic Sir Thomas Tresham, who begged for the chance 'to witness to the world and leave record to all posterity of our religious loyalty and true English valour'. His plea fell upon deaf ears.[32]

* * *

If families like the Yelvertons, Bedingfelds and Southwells were finding it hard to accept the new religion, then they had now become the focus not only of the Government's attention, but also of the Jesuits'. When Henry Garnet and Robert Southwell sailed for England in 1586, among their instructions from Rome was an order to deal only with the gentry.* It would be wrong to view this as snobbery on the part of the mission. Rather it offered a realistic appraisal of Elizabethan society. As Gerard wrote, 'in the districts I was living in now Catholics were very few. They were mostly from the better classes; none or hardly any, from the ordinary people, for they are unable to live in peace, surrounded as they are by most fierce Protestants'. Tudor England may have seen the unstoppable rise of the middle classes, but there was no concomitant improvement in the situation of, or attitudes to, the working classes. Bluntly, the ordinary people were considered of little importance in the subtle game being fought for religious supremacy. Shortly before leaving for England Edmund Campion would summarize the position thus (in terms immediately recognizable to his contemporaries, if distasteful today): 'Of their martyrs they brag no more now. For it is now come to pass that for a few apostates and cobblers of theirs burned, we have Bishops, Lords, Knights, the old nobility.'[33]

But it was the practical advantage gained by concentrating on the gentry that was of most concern to the Jesuits. The great country houses of Tudor England were still kingdoms within a kingdom, despite the decline of feudalism. As late as the 1580s the Earl of Derby's household comprised some one hundred and forty members, excluding his family. According to thirteenth century ecclesiastical law a landowner was responsible for the spiritual probity of his vassals. In Elizabethan England this law might no longer apply, but the sentiment it encapsulated remained intact

* This order echoed Everard Mercurian's advice to Persons and Campion: 'As regards dealing with strangers, this should, at first, be with the upper classes rather than with the common people, both on account of the greater fruit to be gathered and because the former will be able to protect them against violence of all sorts.'

and it was a sentiment the Jesuits were quick to exploit. Convert the master and the servant would follow; reconcile a landowner to the Catholic Church and provide him with a seminary priest to minister to his household and the entire mini-kingdom could be won back to the faith. John Gerard summarized the policy thus: 'The way, I think, to go about making converts . . . is to bring the gentry over first, and then their servants, for Catholic gentle folk must have Catholic servants.' Practical, too, from the point of view of safety, was the fact that in the coming and going of a large house, one more face among so many could pass unnoticed. A new tutor, a steward who spent little time working as a steward, a visiting sportsman like John Gerard would not attract unwelcome attention. But there was another and even more specific safety aspect that, in the end, would make the country houses of England valuable to the Jesuit mission, that would turn them into the nation's new Catholic churches. They had legions of rooms. They had capacious attics. They had many staircases and miles of corridors, they had underground sewers and solid, thick brickwork. In short, they had space. More importantly, they had sufficient space not just for the constructions of legitimate architecture, but for the concealments of an entirely different kind of building work. Upon their walls might hang the finest art; *behind* their walls lay craftsmanship no less masterly. 'When God will protect,' wrote John Gerard, 'He can hide a Felix between two walls, and make spiders His workmen to cover the entry with their webs.' Except neither God nor the Jesuits needed spiders as workmen: they had Nicholas Owen.[34]

The village of Oxborough lies about fifteen miles from Grimston. On a day probably in 1589, some time during the six to eight months he remained with the Yelvertons, John Gerard took the road south, riding through countryside studded with villages, small stone parish churches, dense coppices and damp marshland. It was a journey he must have made several times before, accompanied by his guide Edward Yelverton, but this time Gerard had specific work in hand and new company beside him. Nearing Oxborough, the spire of St John the Evangelist loomed out of the trees on the left. Ahead lay walls, outbuildings, a glimpse of formal gardens and the tall twin redbrick turrets of the medieval gatehouse of Oxburgh Hall, home to the Bedingfeld family.

Oxburgh had once belonged to those same rapacious Norfolk nobles who had roamed the county terrifying their neighbours, before passing into the hands of the more respectable Bedingfelds early in the fifteenth century. Edmund Bedingfeld had built the present house in the 1480s complete with moat and battlements, but his fortifications were more for decoration than defence, sign of the changing times.

Cross over Oxburgh's bridge today and pass through the wide brick archway and you come out into a large quadrangle, doorways opening off it. A porter's lodge at the foot of the gatehouse's western turret leads you to a circular brick staircase, climbing all the way up to the crenellated rooftop. From these stairs you can access the King's Room, on the first floor over the archway, named after Elizabeth's grandfather Henry VII, who stayed at Oxburgh in 1487. A pale northerly light streams in through its vast central window, picking out the huge fireplace opposite, and the fine silken bed hangings. In the far corner beyond the window stands a door that opens onto a small octagonal chamber set in the eastern turret. This, in turn, leads into a garderobe (lavatory) with its draughty opening down to the moat below.

On that day in *c.*1589, with the then owner Thomas Bedingfeld's blessing, this garderobe became the centre of concerted if highly clandestine activity. Nicholas Owen, Oxford joiner and hide-builder of genius, the Jesuits' secret weapon in their holy war against the English Church, had just been put to work.

* * *

While John Gerard had been travelling Europe, tasting prison and training as a Jesuit priest, Nicholas Owen had remained in Oxford. In 1577 his father had enrolled him on an eight-year apprenticeship, under the eye of Oxford joiner William Conway. Here he had learned carving and turning, the creating of seamless joints, the constructing of wide-beamed wooden staircases that curled around landings, untwisting from floor to ceiling, of sinuous, linenfold panelling that enveloped a room, of coffers, chests, bedsteads and joint-stools. In 1587 the commentator W. Harrison observed, 'The furniture of our houses . . . is grown in manner even to passing delicacy and herein I do not speak of the nobility and gentry only, but likewise of the lowest sort in most places of our south country.' Nicholas Owen was trained in the deceptive art of conjuring delicacy out of dense English oak.[35]

Nothing is known of Owen in the four years between his completing this apprenticeship and starting work at Oxburgh Hall, though according to Gerard, he, too, was a newcomer to the mission. But if the details of Owen's childhood and early adulthood remain an enticing blank, we do know that in 1588 or 1589 an unknown Oxford joiner in his mid to late twenties presented himself to Jesuit Superior Henry Garnet, asking to be his servant. And however this meeting came about, Garnet leapt at the offer.[36]

In 1585, the year before Garnet's arrival in England, a clause had been introduced into English law that had horrified Catholics. The clause was part of the new Treason Act against Jesuit priests; it stated that any layman who 'shall wittingly and willingly receive, relieve, comfort, aid, or maintain any such Jesuit, [or] seminary priest . . . shall . . . for such offence be adjudged a felon'. It was small comfort that the House of Lords rejected an earlier draft of the bill, attempting to make the crime an act of treason rather than felony: the penalty in either case was death. English Catholics now found themselves in potentially the same position as England's first martyr St Alban, killed *c.*304 for offering refuge to a Christian priest.[37]

Left Sir Francis Walsingham, who succeeded William Cecil as Elizabeth's Principal Secretary of State and was head of the Elizabethan spy network.

Below Sixteenth-century London. To the right of the image stands the Tower of London. To its immediate left is London Bridge, on which the heads of traitors were exhibited. The bridge leads over the river to Southwark, with its theatres and the Clink prison. To the far left of the image, round the bend in the river, are Whitehall Palace and Westminster.

Above Henry Garnet. As a student in Rome he was known as the 'poor sheep', on account of his shyness.

Above right The Jesuit General, Father Claudio Aquaviva. Aquaviva was the son of a Neapolitan Duke. As a young man he volunteered to learn English and cross to England with Edmund Campion. His instructions to his English Jesuits were explicit: 'They are not to mix themselves in affairs of State, nor should they recount news about political matters in their letters to Rome or to England; and in England they are to refrain from talk against the Queen, and not to allow it in others.'

Right Robert Southwell. His father was a prominent courtier, and his mother a childhood friend of Queen Elizabeth; he was related to Sir William Cecil, Sir Francis Bacon, and the Attorney General, Sir Edward Coke.

AN HVM-BLE SVPPLICA-TION TO HER MAIESTIE.

MOST MIGHTIE and moſt merci-
full, moſt feared, and beſt beloued
Princes, they are at the bottome of a help-
leſſe miſerie, whome both a condemned e-
ſtate maketh common obiects of abuſe, and
an vnpittied oppreſsion barreth from diſ-
couering their griefes to thoſe, that onelie
are able to afford them remedy. Euery one
trampleth vpon theyr ruine, whom a Princes diſgrace hath once ouerthrowne. Soue-
raigne fauours being the beſt foundations
of Subiects fortunes, and theyr diſlikes the
ſteepeſt downe-falles to all vnhappineſſe.
Yet a Prince ſupplying the roome, and re-
ſembling the perſon of Almightie GOD,
ſhould be ſo indifferent an arbitrator in all
cauſes, that neyther any greatneſſe ſhould

A 2 beare

The opening page of Robert Southwell's *An Humble Supplication to Her Majesty. . .*, written in the winter of 1591. This expurgated version, printed in secret in 1600 (though falsely dated 1595), was used by the Appellants in their battle to have the Jesuits expelled from England.

Left A Catholic priest being tortured. At the back sits a panel of examiners asking questions, while, in the foreground, three more priests stand and wait their turn on the Rack. The illustration comes from William Allen's 1583 book *A Brief History of the Glorious Martyrdom of Twelve Reverend Priests*, and was intended to convey to the reader the full extent of the English Government's cruelty. This, as much as the secrecy surrounding the practice of torture, has led to inaccuracies in the depiction. Priests were not examined four at a time, neither were ordinary gaolers (the men straining on the ropes) allowed to be present when prisoners were revealing potential State secrets.

Right An image of Catholic priests being hanged, drawn, and quartered, taken from the same book. There are sufficient eyewitness accounts of executions to confirm that this depiction is accurate. At the back of the picture a priest is being cut down from the scaffold still alive. In the foreground a second priest is being disembowelled and his entrails thrown into a cauldron, while, behind the cauldron, an executioner is hacking a third priest's body into quarters. The head of this last victim has been stuck on a pole immediately to the right of the scaffold, completing the gruesome process.

Left Iron shackles used for the form of torture known as the Manacles, in which a victim was suspended by the wrists, often for hours at a time. The invention of this torture has been credited to Richard Topcliffe. Among those put to it were Robert Southwell, and John Gerard, who vividly described the 'gripping pain' of it. Those who hung too long lost all use of their hands.

Below The hiding-place at Oxburgh Hall, Norfolk. This priest-hole is almost certainly the work of Nicholas Owen, the Oxford joiner who became the Jesuits' hide-builder in *c.*1588–89. When the trapdoor is closed it is indistinguishable from the tiled floor surrounding it. The hide beyond is big enough for a man to stand in, and is remarkably comfortable.

Above The underground hide at Baddesley Clinton, Warwickshire. On 19 October 1591 ten men hid for over four hours in this old converted sewer, while the house above them was searched. At its highest point the channel measures only four feet, making it impossible to stand upright.

Right The swinging beam hide at Harvington Hall, Worcestershire, one of the most extraordinary hiding-places still in existence (*see Appendix*). Again, it is almost certain that it is the work of Nicholas Owen, built some time after 1600. It was discovered by accident in 1897 by boys playing in the, then, derelict hall.

The Tower of London, from a survey made in 1597, the year of John Gerard's imprisonment there. Gerard was held in the Salt Tower, marked M, and John Arden in the Cradle Tower, marked Q. It was from the Cradle Tower that the two men were able to make their escape, clambering down a rope slung over the moat, scaling the wall at the edge of the moat, and then crossing the wharf, where a rowing boat was waiting for them.

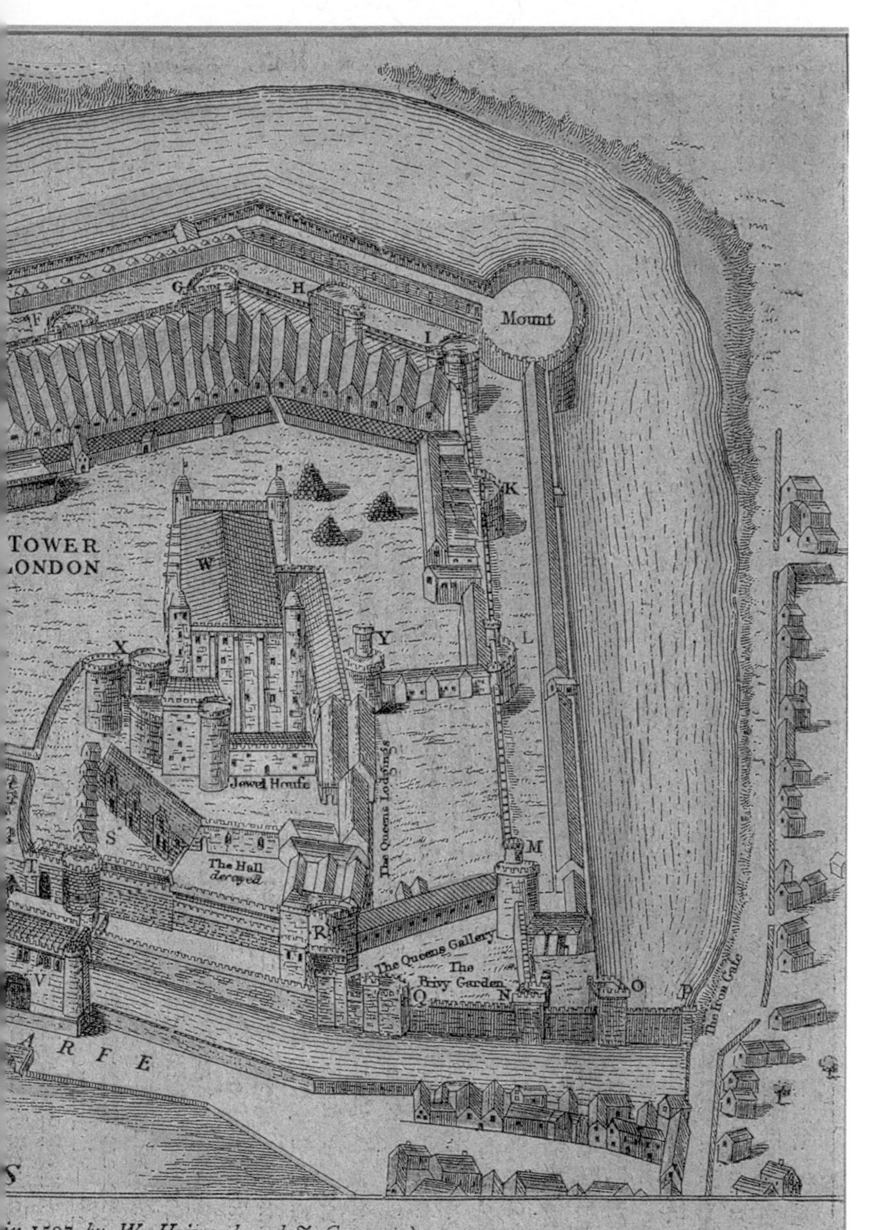

in 1597 by W. Haiward and J. Gascoyne.)

er. F. Flint Tower. G. Bowyer Tower. H. Brick Tower. I. Martin Tower. K. Constable
e. P. Tower above Iron Gate. Q. Cradle Tower. R. Lantern Tower. S. Hall Tower. T. Bloody
drobe Tower. A B. House at Water Gate, called the Ram's Head. A H. End of Tower Street.

An inscription in the wall of the Salt Tower, carved by Henry Walpole during his imprisonment there in 1594. John Gerard, who was held for a time in the same cell, wrote of it: 'It was a great comfort to me to find myself in a place sanctified by this great and holy martyr'.

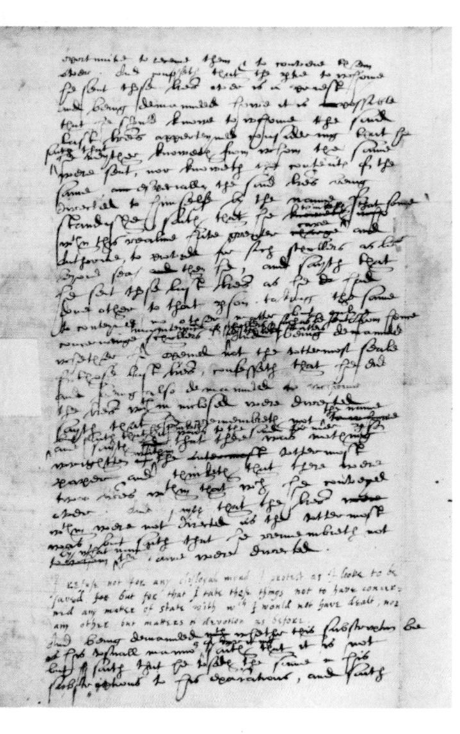

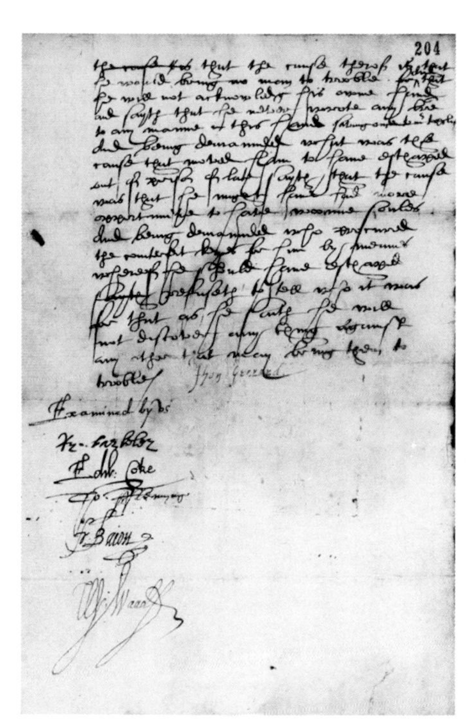

The report of John Gerard's first interrogation in the Tower of London, dated 14 April 1597. On the right hand page can be seen Gerard's signature; on the left hand page towards the bottom, can be seen the only reply he was permitted to set down for himself (in disguised handwriting). The other signatories are Richard Berkeley (the Lieutenant of the Tower), Edward Coke, Thomas Fleming (the Solicitor General), Francis Bacon, and William Waad. Immediately after this examination Gerard was led away to be tortured.

Left Attached to the original document of Gerard's first interrogation is this playing card, the seven of spades, with a list of names on the back written by Sir Edward Coke. Gerard's name is at the bottom of the list. Above it are the names Polwhele, Walpole, Pat. Cullen, Annias, Williams, and Squire, all of whom had been linked by the Government to a fantastical plot to kill the Queen (the last two by poisoning the pommel of her saddle).

Above Pope Clement VIII, a staunch opponent of religious tolerance.

Right Pope Paul V. Paul never spoke out publicly against the Gunpowder Plot, preferring to indicate his dislike of it in letters to his Nuncio. His seeming reluctance to make audible pronouncements prohibiting Catholic resistance – despite letters indicating that he condemned violence – left Henry Garnet painfully isolated in the months preceding the plot. Neither did Paul come to Garnet's defence after the plot.

Above Sir Robert Cecil, Principal Secretary of State both to Elizabeth and to James. He was made 1st Earl of Salisbury in 1605. Cecil was short in stature and suffered from slight curvature of the spine, a significant handicap in an age in which one's outer physique was thought to reflect one's inner psyche. Elizabeth called him her 'elf', James his 'little beagle'. Cecil's motto, featured in this portrait, was 'Sero, sed serio', 'late but in earnest'.

King James I of England and VI of Scotland. This portrait was completed in *c.*1606, just months after the Gunpowder Plot to kill him.

King Philip III of Spain, who, like James, was eager to end the long-running war between their two countries. Although Philip tried to make religious tolerance for England's Catholics a requirement for the peace, widespread diplomatic pressure forced him to drop the demand from the agenda. This left English Catholics more isolated than ever, an important factor in Robert Catesby's decision to blow up Parliament.

The signing of the Spanish Peace Treaty at Somerset House in 1604. Robert Cecil is on the immediate right of the picture; the Constable of Castile is on the immediate left; next to him is Juan de Tassis, the man charged with trying to buy tolerance for England's Catholics.

Below A seventeenth-century engraving of Westminster from the River Thames, showing Parliament House. It was this that Robert Catesby planned to destroy.

EYGENTLICHE ABBILDUNG WIE ETTLICH ENGLISCHE EDELLEUT EINEN RAHT
schlußen den Konig sampt dem gantzen Parlament mit Pulfer zuvertilgen.
Bates
Robert Winter
Christopher Wright
John Wright
Thomas Percy
Guido Fawkes
Robert Catesby
Thomas Winter
Abbildung wie und welcher gestalt etliche der furnemsten Verrahter in Engellant von leben zum tod hingerichtet worden

Anno 1606.
Mense. Feb.
comme les conspirateurs contre le Roy D'angleterre sont executez a mort. Premierement
et pendus a une potence,
et les testes mis sur des perces de fer, et les entreailes brulé

Left Sir Edward Coke, the Attorney General who led the prosecution at the trial of his cousin, Robert Southwell, and at that of Henry Garnet. Coke also conducted John Gerard's examination under torture. It has been suggested that this portrait was painted after Coke's death, hence the inclusion of the skull beneath his left hand.

An entirely fictitious depiction of the torture of Edward Oldcorne and Nicholas Owen by the Flemish engraver Gaspar Bouttats. The image, from the mid- to late-seventeenth century, is evidence of Europe's keen interest in English Catholic affairs, particularly the events surrounding the Gunpowder Plot. The subject of English barbarism made good copy on the Continent.

Opposite above The Gunpowder Plot Conspirators. This engraving shows eight out of the thirteen plotters. Of those depicted, Catesby was a long-time supporter of the Jesuits, and Robert Wintour kept a Jesuit chaplain at his house. Of those missing – Digby, Tresham, Rookwood, and Keyes – Digby and Tresham were both well known to the Jesuits, while Rookwood's wife had accompanied the Jesuits on their pilgrimage to St Winifred's Well in the weeks preceding the plot. Each connection made it easier for the Government to implicate the Jesuits in the plot.

Opposite below A seventeenth-century Dutch engraving showing the execution of the Gunpowder Plotters.

Left A silver case containing Edward Oldcorne's right eyeball. When Oldcorne was decapitated at his execution, having already been hanged and disembowelled, the axe-man struck with such force that Oldcorne's eyeball flew from its socket. A witness picked it up and preserved it as a relic. The silver case, made specially to hold the eyeball, bears the Latin inscription *Oculus dexter P. Ed. Olcorni Soc. Jesu* on the back.

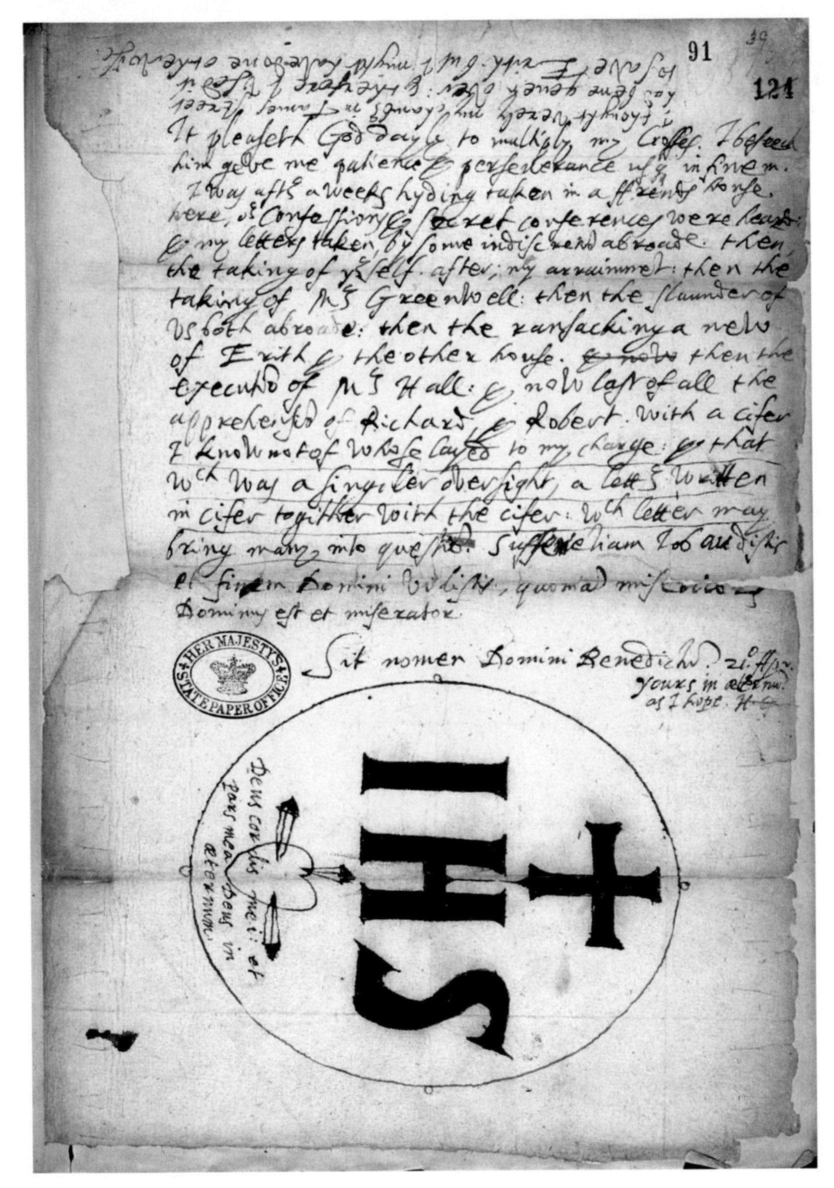

91

121

It pleaseth God dayly to multiply my Crosses. I beseech
him geve me patience & perseverance usq in finem.
I was after a weeks hyding taken in a ffrends house,
where or Confessions & secret conferences were heard:
& my letters taken, by some indiscretion abroade. then
the taking of yrself. after, my arraignment: then the
taking of Mr Greenwell: then the slaunder of
us both abroade: then the ransacking a new
of Erith & the other house. ~~& now~~ then the
execution of Mr Hall: & now last of all the
apprehension of Richard & Robert. with a cifer
I know not of whose layed to my charge: ~~&~~ that
wch was a singler oversight, a letter written
in cifer together with the cifer: wch letter may
bring many into question. Sufferentiam Iob audistis
et finem Domini vidistis, quoniam misericors
Dominus est et miserator.

Sit nomen Domini Benedictum. 21 Apr.
yours in aeternum
as I hope. H.G.

IHS

Deus cordis mei: et pars mea Deus in aeternum.

Henry Garnet's last letter to Anne Vaux, dated 21 April 1606, twelve days before his execution. In it he writes: 'It pleaseth God daily to multiply my crosses. I beseech Him give me patience and perseverance.' The letter is evidence of the lies being fed to him to get him to confess his approval of the plot. He tells of Tesimond's arrest, the capture of his servants, and of the 'slander' being spread about him abroad. The letter ends: 'Yours in eternum, as I hope, H.G.'

In April that year, soon after the clause became law, an emergency meeting was called at Hoxton to the west of the city of London, home of Mr Wylford. Attending the meeting were leading members of the Catholic laity, including Sir Thomas Tresham, Sir William Catesby and Lord Vaux, along with representatives of the seminary priests. Also present was Garnet's immediate predecessor as Jesuit Superior, Father William Weston.* The talk was angry and urgent. In view of this change in legislation, against which all of them had protested, was it fair to expect Catholic householders to give aid to the missionaries arriving from the Continent? No matter how desperate the priests' need for shelter, should families risk their lives and their lands by taking them in? No doubt many views were aired, but finally it was agreed 'that the priests shall shift for themselves abroad, as in inns or such like places, and not visit any Papists . . . except they be sent for'. The young priests were not to be left entirely unprovided for, though: 'it was ordered the Lord Vaux should pay to the relief of priests that would tarry, one hundred marks'. Tresham, Catesby and Wylford would pay 'one hundred marks the piece'. Certain 'other gentlemen' were assessed 'at lower sums'. All this money was to be placed into a central fund and administered by Vaux's son Henry, a former cohort of George Gilbert, on behalf of the priests. The meeting broke up soon afterwards and the members departed to break the unhappy news to the rest of the missionaries. All except one man. He sat down to write to Sir Francis Walsingham. Within hours the Council was in possession of a detailed account of all that had taken place at Hoxton. Government agent Nicholas Berden – who earlier had written 'I profess myself a spy, but I am

* Weston was born in Maidstone, Kent in 1550. He was an Oxford contemporary of Edmund Campion. On his return to England in September 1584 he took the alias Mr Edmunds in tribute to his friend. Weston was arrested in August 1586. Rather than risk the international opprobrium of killing another Jesuit, the Council imprisoned him. Friends took up his case, but no one was brave enough to intercede for him at court. One was said to comment: 'If he were a common thief, or a murderer or buccaneer, or something of the kind, I would not hesitate one moment to obtain a pardon, or at least to ask for it. But where it is a matter of a Jesuit, I cannot; I am afraid to ask.' He was finally released in 1603 and died in Spain in 1615.

not one for gain but to serve my country' – had earned his retainer that day.*[38]

The solution reached at Hoxton was only intended as a stopgap measure while the Catholic laity petitioned Elizabeth to repeal the offending clause. Should that fail, wrote Berden, they were determined 'to adventure the danger of the statute'. The following year William Weston, along with the newly arrived Henry Garnet and Robert Southwell, reconvened at Hurleyford on the banks of the River Thames near Marlow, to decide how best to break the new law safely. Hurleyford belonged to the Catholic Richard Bold. It was a secluded house with the added advantage of standing on the Berkshire/Buckinghamshire border. Should it be raided in the general search of Catholic properties ordered that month, Weston would try to escape by crossing the county line; pursuivants had no authority outside their own shire.[39]

Present at the meeting were familiar faces from Hoxton and each day, after a sung mass conducted by 'Mr Byrd, the very famous English musician and organist', they sat down to discuss business.† The only absentee from before was Nicholas Berden, so sadly the precise details of this business have not been recorded; Weston only notes 'I told them what I knew about conditions in England'. The figures would have made poor listening. Since 1574 some three hundred priests, seminarians and Jesuits, had returned home, but of them, only about one hundred and thirty were still at work; thirty-three had been executed, fifty were in prison, several

* Berden was one of the Council's best spies. In 1586 he wrote to Walsingham: 'I humbly thank your honour for that it pleased you to spare Christopher Dryland's [a seminary priest] life at the last sessions . . . assuring you that it hath much increased my credit amongst the Papists . . . I protest I abhor the man in regard of his profession, [but he] . . . is singularly well persuaded of me, supposing me to be a most apt man to serve the Papists' turn.' Dryland was imprisoned until 1603, when he was banished. He went to Rome and became a Jesuit. When Berden's treachery was finally exposed Walsingham arranged for him to take up the more lucrative if less adrenaline-charged post of purveyor of poultry to the Queen.

† This was composer William Byrd. Byrd was educated under Thomas Tallis at St Paul's Cathedral music school. By 1570 he had become a chorister of the Chapel Royal and by 1575 he was the Queen's organist. Despite his Catholicism he seems to have enjoyed Elizabeth's favour and was rewarded with a monopoly to print and publish sheet music.

had died from natural causes and over sixty had been banished or fled voluntarily. Since it took between two and seven years to train a priest and the seminaries operated on a tight budget, this was not a good return on investment; a new policy was in order. Weston noted, 'Then we discussed our future methods of work and the prospects that lay before us.' It seemed these future methods were to depend heavily on men like Nicholas Owen.[40]

Cuthbert Mayne had been arrested in his bedroom at Golden House. Edmund Campion had been discovered with two others in a rudimentary shelter visible from the outside; during the same raid his host's brother had been forced to hide in the dovecote. Robert Persons once took refuge in a haystack.* It was clear that though basic hiding places existed in some Catholic houses – to conceal forbidden priests and forbidden books – there had as yet been no real attention paid to hide-building as a necessary and improvable craft. Weston, himself, could attest to this. Once he had hidden in a barn; on another occasion he had fled underground, to a purpose-built hide. 'Catholic houses have several places like this,' he wrote, 'otherwise there would be no security.' This one, though, 'was constructed with no particular cunning or ingenuity'. It 'was dark, dank and cold, and so narrow that I was forced to stand the entire time'. Weston was lucky. Had his hide been tucked into the side of a chimney-stack, or positioned in the hole under a garderobe turret, as so many of these early ones were, he would soon have been found (as they, themselves, have been). Such hides were built to a formula and discovered formulaically. With the Hurleyford conference this was to change. If Catholic families were safely to be provided with a priest, then they must also be provided with a safe priest-hole. Five days after the meeting broke up Weston was arrested on the streets of London near

* A spy's report of 1591 reveals that the priests made use of every sort of hiding place available to them. 'As you go forth of Mr Wynshcomb's house towards Newbury, in the first close without the gate, upon the left hand in the hedgerow, there is a great oak that is hollow, and by knocking upon it you shall find it to sound.' It continued: 'Oliver Almon is a priest and did lie at Mr Wynshcombe in Berkshire, near Newbury . . . If he be not in the house, there is a great tree wherein he is hidden.'

Bishopsgate and Henry Garnet was left in charge of implementing this policy. No wonder he was happy to welcome Nicholas Owen to the Jesuit mission.

As John Gerard opened up the county of Norfolk for the Jesuits, carrying out the work for which Campion had been destined at the time of his arrest, Nicholas Owen rode with him. They made a contrasting pair: the aristocratic Gerard, to whom Garnet had given the nickname Long John of the Little Beard, and the diminutive Owen, who soon became known as Little John. With Edward Yelverton vouching for Gerard to neighbours sympathetic to the faith and with Gerard vouching for Owen as the Jesuits' hide-builder, a chain was quickly formed leading from Norfolk, via London (which served as a general sorting office), all the way to the colleges on the Continent. Henry Garnet would later describe this chain, and the others like it, criss-crossing the country, to Jesuit General Claudio Aquaviva: 'When the priests first arrive from the seminaries, we give them every help we can. The greater part of them, as opportunity offers, we place in fixed residences. This is done in a very large number of families through our offices.'[42]

The individual links in this chain were comparatively simple to forge. 'It will be necessary . . . for the priests to be stationed in various parts of the country and for each of them to stay at the house of some gentleman or other,' George Gilbert had written. John Gerard's job was to reconcile such a gentleman and then persuade him to take in a priest. Then came Nicholas Owen's contribution.[43]

Under cover of carrying out legitimate building or repair work Owen would craft a hiding-place, working in secret and as near to silence as he could manage, for any attention drawn to the location of the hide rendered it useless for its purpose. Even loyal servants were kept from learning the whereabouts of a hide, for fear torture might turn them. Owen's genius was to exploit the main structure of a house, burrowing deep into the masonry of its interior, lodging his hide within the very framework of the building, within what, to the practised eye of the pursuivants, could only be solid wall or ceiling. His hides are three-dimensional puzzles of Max Escher-like complexity. And, for maximum safety, every one of them was different. John Gerard summed up Owen's career: 'he was so skilful both to devise and frame the [hides] in the best manner, and his help therein desired in so many places, that I verily think no man can be said to have done more good of all those that laboured in the English vineyard'.[44]

It was impossible for Owen to build every hide for every priest stationed across England by the Jesuits. It seems he concentrated on those hides used by the Jesuits, themselves, and on those hides destined for 'the chiefest Catholic houses', while elsewhere acting as an adviser and discoursing 'of the fashion of [hides] for the making of others'. In this way he came to know 'the residences of most priests in England, and of all those of the Society'; he also knew 'the means and manner how all such places were to be found, though made by others'. It was a heavy burden of knowledge. But it was a burden Owen seemed more than capable of carrying. 'One reason that made him so much desired by Catholics of account, who might have had other workmen enough to make conveyances in their houses, was a known and tried care he had of secrecy.' 'He was', wrote Gerard, 'so careful that you should never hear him speak of any houses or places where he had made such hides.' Thomas Bedingfeld had reason to be grateful for this secrecy as Owen now set to work at Oxburgh Hall.[45]

In the garderobe off the King's Room, set against the interior wall, was a small recess of door height. It suggested the room might once have had another purpose; perhaps the recess held an altar or *prie-Dieu* and the room had served as a private chapel. At the foot of this recess Owen chiselled away a small section of the tiled floor. Today, if you place your foot close to the recess's wall and tread heavily on this section, it swings open, revealing the entrance to the hide beneath. The 'lid' is some nine inches thick, made of two solid oak blocks bolted together and with a layer of tiles to camouflage it. In place, it is seamless and unnoticeable. It is rap-proof, ring-proof, and its balance when it pivots is perfect. Slip through it and you enter the hide itself, an irregularly shaped brick vault that opens above and beyond you, like a capital L stretched out on its side, measuring about three feet at its widest point, two feet at its narrowest and at its tallest over seven feet high. It lies between the walls of the King's Room, to its west, and a collection of smaller rooms and stairs to its east. One of these rooms served as the Bedingfelds' secret chapel and at the end of the L's vertical, close to this chapel, was the second entrance to the hide, now filled in. This entrance made use of the natural features of the house: one of the treads on a small set of stairs in an adjoining passageway hinged open to reveal a gap wide enough for a man to squeeze through to the hide behind. In the base of the L is a ledge with a wooden seat and evidence of a communication hole with the King's Room beyond it. The whole is remarkably comfortable, relatively soundproof and utterly undetectable.[46]

It is improbable that Owen hollowed out the entire vault from scratch – it is unnecessarily large and unnecessarily irregular for its purpose. More likely there was some sort of empty space there already, created when the house itself was built; the family name for it – The Dungeon – points to a possible former usage. Owen's genius was to conceal whatever original entrance there may have been and build two new ones, invisible to the most avid of searchers. In February 1590 his work was put to the test.

That month a seemingly anonymous letter was sent to Henry, Lord Crumwell, a Norfolk justice. The letter gave information against Henry Bedingfeld, Thomas' son and owner of Oxburgh since his death that year, of 'some treasonable designs in conjunction with the Papists and Recusants'. The informant may have been right about Henry's connection with papists and recusants, but the accusation of treason smacks more of self-interest than accuracy. Henry was only eight years old. Nonetheless Crumwell ordered a diligent search of the house. Nothing suspicious was found and Crumwell duly reported this fact to Sir Francis Walsingham, enclosing a copy of the letter delating Henry as a traitor to add to the ever increasing body of evidence held against the

Bedingfelds. Nicholas Owen's hide had withstood its first challenge. No priest had been found. The link in the chain had held.

As the years went on, more and more links would be added to this chain as more and more priests arrived to take their place on the mission. Campion had spoken of a 'league' of men, of 'many innocent hands', gathered beyond the seas, determined to reconcile England to the Catholic Church. With the Jesuits' new policy in place and with men like Garnet, Gerard and Owen to implement it, these words no longer had such a hollow ring. But if Campion had promised priests in profusion, he had also made another promise: 'never to despair your recovery, while we have a man left to enjoy your Tyburn'; 'to win you to heaven, or to die upon your pikes'. As the network grew, as it lengthened and spread across the country, so more and more priests would keep their grisly side to this bargain. The links in the chain might be simple to forge, but they were also simple to break, and one man had made it his personal mission to do precisely that.[47]

Chapter Six

'. . . they do come into the [Realm] by secret Creeks, and Landing Places, disguised, both in their Names and Persons'.

Queen Elizabeth I, November 1591

DURING THE CLOSING DAYS of the 1586–7 Parliament, the MP for Old Sarum, Richard Topcliffe, revealed to the House that in the building 'joining to the Cloth of Estate' – next door but one in fact – a quantity of 'weapons and all massing trumpery, with books papistical' had just been uncovered. Such a viper's nest of Catholicism on Parliament's very threshold was intolerable. Immediately the Commons appointed a party 'to search certain houses in Westminster suspected of receiving and harbouring Jesuits [and] seminaries'. There is no record of how the session ended.[1]

Neither, sadly, is there any record of how the search party proceeded, but it is unlikely Topcliffe would have been happy with anything less than a lynch mob. No man took the outrage of the continued existence of English Catholicism more personally, it seemed, than Richard Topcliffe. His critics wondered whether he might not be a former Catholic himself.*

Richard Topcliffe was born in 1532, the son of Robert Topcliffe of Somerby, Lincolnshire and Margaret, daughter of Lord Borough.

* Topcliffe's father and father-in-law had both come to the attention of the Government during the 1536 Pilgrimage of Grace, the ill-fated Catholic uprising in protest against Henry VIII's religious reforms.

His mother died when he was an infant, his father when he was twelve, and he was raised by an uncle, Sir Anthony Neville; he was educated at Gray's Inn in London. In 1570, the year after the Northern Rebellion, he sued for the lands of one of the rebels, the Catholic Richard Norton of Norton Conyers, Yorkshire. Three years later he was in Lord Burghley's pay for services unspecified. In 1578 he was reporting back with relish the unhappy events of Elizabeth's Norfolk progress and by 1584 he was revealing the full fervour of his anti-Catholicism in a letter to the Privy Council. 'My instruments have learned', he warned, of bands of seminary priests roaming the capital. 'They walk audaciously, disguised in the streets of London. Their wonted fears and timorousness is turned into mirth and solace.' By the 1590s, not content with seeing Catholics wherever he went, he had become their judge, gaoler and executioner. He had also turned his Westminster house into a private torture chamber, the better to serve this end. '*Homo sordidissimus*,' spat the usually measured Henry Garnet, when called upon to describe him.[2]

The career of Richard Topcliffe is perplexingly, paradoxically linked to one woman, Queen Elizabeth I herself. Moderate, pragmatic, uncharacteristically squeamish for a Tudor, Elizabeth gave licence to a monster. It was not just Catholics who found him so. In the fashionable in-speak of Court the word *Topcliffizare* soon came to mean to go recusant hunting. Meanwhile, Sir Anthony Standen, praising the manners of the new royal favourite the Earl of Essex, noted that 'Contrary to our Topcliffian customs, he hath won more with words than others could do with racks.' Throughout much of the 1590s Richard Topcliffe could be found at the head of an army of pursuivants raiding Catholic houses, in the torture chamber extracting evidence, in the courtroom cross-examining the accused, or on the scaffold overseeing executions. For all these activities he possessed only one title: that of Queen's Pursuivant. It seemed, though, Topcliffe had something better than a title: he had the willing ear of Elizabeth.[3]

The pair first came into contact in January 1570. On the tenth of that month the Earl of Leicester wrote to Elizabeth expressing his longing to hear from her. With bad weather making travelling difficult, Leicester employed as his messenger 'a Mercury' prepared 'to take the more pains' to deliver his letter. Mercury's name, it transpired, was Richard Topcliffe.* And in a postscript Leicester described how Topcliffe had recently supplied '30 horse and men, all well appointed, at his own charge' for the campaign against the Northern rebels. With Leicester vouching for him and with such a solid demonstration of loyalty to recommend him, Topcliffe evidently stuck in Elizabeth's memory. Twenty years on she had few compunctions about turning to him for assistance and even fewer about doing so in private. 'This most unclean of men', wrote Garnet, 'has attained . . . such favour with Her Highness that he always has easy access to her, and need not fear the power or influence of any Councillor or Minister.'[4]

From the outset Elizabeth's attitude to English Catholics had tended towards leniency; as much as her ministers had champed at the bit and demanded harsher measures against Catholics, so Elizabeth had held tight to the reins and refused to budge. When, in 1563, her pressing need for Parliamentary subsidies forced her to accept legislation extending the death penalty to anyone twice refusing to take the Oath of Supremacy, she promptly sabotaged the act by commanding her bishops never to tender the oath a second time.† 'Compel them you would not; kill them you would

* Topcliffe's connection with the Earl of Leicester came about through his wife's family, the Willoughbys, who were allied to Leicester's family. These useful contacts aside, Topcliffe's marriage was not a happy one, clouded by allegations that he failed to pay his wife adequate maintenance; and the product of this marriage, Topcliffe's son Charles, spent most of his adult life in trouble with the authorities. In 1602 Sir Robert Cecil wrote to Topcliffe, chiding him for not having had Charles 'cleansed'.

† This was not an isolated incident. In late 1579/early 1580 Sir Francis Walsingham wrote unofficially to a number of county justices, instructing them not to 'persecute' recusants, 'for that if you shall proceed therein, you shall not prevail to do that good you desire, but shall rather . . . fail through some commandment from hence, prohibiting you to surcease in proceeding in that behalf, which would breed no less discredit unto you than encouragement to the papists'. This perception that Elizabeth did not wish the laws to be enacted against Catholics at this period (probably because of her ongoing marriage negotiations with the French Duc d'Alençon) was supported by independent reports from the French and Spanish ambassadors.

not . . . trust them you should not,' admonished a frustrated William Cecil. Beside her clemency, her dislike of bloodshed and her continued favour of many prominent Catholics, Elizabeth's support of Richard Topcliffe sits like an uninvited guest at a wedding feast, unwelcome and impossible to ignore. Equally, it is almost impossible to explain.[5]

In view of Elizabeth's leniency, was Topcliffe's ulterior purpose – a purpose assuredly kept from him – to serve as a sop to the Protestant hard-liners about her and prevent harsher anti-Catholic penalties being introduced in Parliament? Was his very lack of official status Elizabeth's way of avoiding more official religious legislation? If this were the case, then were she to be accused by her ministers of being soft on papistry, Elizabeth need only point to Topcliffe to demonstrate the opposite was true. For someone who detested being bounced into hasty decisions as much as Elizabeth, Topcliffe's presence provided a sure way of keeping in check the headlong drive for religious reform favoured by many in her Government and by most of her Parliaments.

There is another possibility that presents itself. Sir William Cecil's maxim, one to which first he then Francis Walsingham would adhere as they set about developing their far-reaching spy network, was that the only people a ruler could never trust were those they had irreparably harmed.* By this reasoning Elizabeth could never trust England's Catholics, no matter that, as a religious conservative, she had more in common with them than with the Protestant extremists and radical Puritans about her. Indeed,

* The Elizabethan spy network mainly consisted of an amorphous body of paid informants, reporting straight to the Government. There was no formal structure to it, but what little there was can be credited first to Cecil, then to Walsingham, who has been called the 'father of the British Secret State'. It was not surprising that a paranoid age produced England's first attempt at a secret service. Francis Bacon would note: 'there is nothing makes a man suspect much, more than to know little'. The practice of paying informants for intelligence produced many dubious results, as a story told by the eighteenth century playwright Sheridan illustrates: an informant was paid a retainer of 1 guinea a week for general information and 2 guineas for serious information; of course all his information was serious. Little is known about these informants, but their letters reveal that the majority of them were impecunious debtors seeking relief from prison. On that fact alone the reliability of their intelligence can be judged.

the number of assassination plots over the course of her reign of which rogue Catholic terrorists were found to be at the bottom gave the ring of truth to Cecil's claim. The Ridolfi Plot of 1571, the 1583 Throckmorton Plot, the 1585 Parry Plot and the Babington Plot of 1586 were all Catholic-supported attempts to remove Elizabeth from the English throne and replace her with Mary, Queen of Scots. When Pope Pius sanctioned Elizabeth's overthrow in 1570 – giving rise to Ridolfi, Throckmorton *et al.* – one of the Queen's first actions had been to issue a proclamation restating her intention not to examine her subjects' 'consciences in causes of religion'. Richard Topcliffe, unlicensed by anyone but Elizabeth, was a useful counter-balance to this leniency. He was, perhaps, her private weapon in the battle against religious terrorism, while publicly she continued practising conciliation. There were few if any instances during the reign in which Elizabeth waged open warfare with a happy heart. Conflict was costly, bloody and, above all, uncertain in its outcome – three factors guaranteed not to endear it to the cautious Queen. Rather than making war on every Catholic in the land, the presence of Topcliffe on her staff allowed Elizabeth to make lightning strikes against likely insurgents, at minimal expense, with minimal loss of life and with minimal danger to herself. Such a luxury could not be underestimated.[6]

So with Elizabeth's blessing and a supporting cast of similarly unsavoury characters – Anthony Munday was an early employee of his – Richard Topcliffe appears to have levelled his sights on bringing down the Catholic mission to England. Certainly his name appears with increasing regularity among the State papers of the period as he engaged in this pursuit. In January 1590 he received his first warrant from the Privy Council; it instructed him to subject the seminary priest and scholar Christopher Bales to 'such torture upon the wall as is usual for the better understanding of the truth of matters'. Bales had sailed to England on the same ship as John Gerard and Edward Oldcorne two years earlier; he was also a former pupil of Robert Southwell's at the English College in

Rome. Perhaps this link with three out of the six Jesuits then at liberty in England exacerbated Bales' fate.* He was manacled and strung up by his hands from an iron staple fixed to the prison wall, hanging there for twenty-four hours with the tips of his toes just touching the floor. In spite of this, he seems to have admitted little more than that he had been ordained overseas and had returned to England to minister as a Catholic priest. On these two counts he was indicted for treason – in which case, he was reported to have asked at his trial, was St Augustine not a traitor too? No, replied the presiding judge, Justice Anderson, for Augustine had fortuitously arrived before the introduction of these new laws. Bales was executed on 4 March in Fleet Street.[7]

Two months later Topcliffe, himself, was on the Fleet Street scaffold to extort an eleventh hour confession from the seminary priest Anthony Middleton. Pointing about him, in the direction of nearby Gray's Inn Lane and Shoe Lane, Topcliffe charged Middleton with taking refuge in Catholic houses in the area. Middleton replied, 'You know, Mr Topcliffe, I never approached any man nor confessed any place.' He died without naming those who sheltered him.[8]

But if Bales and Middleton's unwillingness to talk had saved the lives of many, then the case of Henry Walpole revealed just how dangerous Topcliffe might prove to the mission if once he truly targeted a victim. Henry Walpole was born in 1558, the eldest son of Christopher Walpole of Anmer Hall, Norfolk, on today's Sandringham estate. He was educated at Peterhouse, Cambridge and in 1578, aged twenty, he moved to London to study at Gray's Inn. Under different circumstances he would in time have qualified as a lawyer and returned home to take up his considerable inheritance, but on 1 December 1581 Walpole stood in the mud and rain of Tyburn fields to watch the execution of Edmund Campion.

* These six Jesuits were Garnet, Southwell, Oldcorne and Gerard, and two newcomers, Richard Holtby and John Curry, who arrived in England in the spring of 1589. Curry, a Cornishman, was stationed in the southwest of England, while Holtby, who earlier had served as Edmund Campion's guide during his travels through the north of England, was sent to Yorkshire.

As the hangman hurled Campion's entrails into the cauldron of boiling water, it was said that a gob of blood spattered Walpole's white doublet: his fate was sealed. The following year Henry Walpole fled England for William Allen's seminary at Reims and by February 1584 he had transferred to Rome to join the Society of Jesus. On 17 December 1588 he was ordained a Jesuit priest.[9]

For the next five years Walpole travelled the Low Countries, for a time acting as chaplain to the Spanish forces there. Back in Norfolk his family was in contact with the recently arrived John Gerard. Soon Henry's youngest brother Michael Walpole had become Gerard's confidential servant, riding with him, wrote Gerard, 'whenever I went to a house where my assumed status made it necessary for me to have one'. In 1589, under Gerard's guidance, Michael left England to join the Jesuits himself. The next year Henry's cousin Edward Walpole followed suit. Twelve months later Gerard had 'persuaded another Walpole, Christopher by name, to come from Cambridge and see me. Then I received him into the Church, and, giving him the money for the journey, sent him to Rome'. 'Gerard doeth much good,' wrote Henry Walpole jubilantly. Meanwhile, a fourth brother, Thomas, had left England to serve as a mercenary overseas.*[10]

In 1593 Henry Walpole was sent as a minister to the recently established English seminary at Valladolid, to the north of Madrid. The violent, feverish heat of the Spanish summer, the red dust that filled the air of the narrow, shuttered streets were a far cry from the marshy flatlands of Norfolk. Many Englishmen, in the four years since Robert Persons first suggested the foundation of the college, had succumbed to disease and death in these unaccustomed conditions. Nicholas Owen's youngest brother Walter had

* There were six Walpole brothers in all: Henry, Richard (who became a Jesuit soon after Henry), Geoffrey, Thomas, Christopher and Michael. Geoffrey was the only one who never ventured abroad; indeed, he appears never to have moved far from home, not even to attend university. In 1608, aged forty-six, he married a cousin, Dorothy Beckham of Dersingham, and from then until his death in 1622 he remained at Dersingham. It was his younger brother Thomas who finally inherited Anmer Hall on their father's death.

arrived in Valladolid fresh from Reims and had been newly ordained a deacon in October 1590. A year later he was dead, aged just twenty-three. But it seemed the extreme climate of Valladolid was no deterrent to those young men arriving 'faster out of England than their rooms can be made ready'. Like its Reims and Roman counterparts, the Spanish seminary had no shortage of students willing 'to stand and die in the Catholic cause'. What it needed now was a protomartyr of its own.[11]

Ever since his ordination Henry Walpole had been bidding to return to England and take up a place on the Jesuit mission. Until now his superiors had rejected his application, though their reasons for this are unknown, but when Robert Persons visited Valladolid in June 1593 he brought with him good news: 'suddenly he told me he was resolved I should go into England if I did not refuse', wrote Walpole. Walpole did not refuse. Stopping only in Madrid to beg money for the seminary from King Philip II – a priest destined for England was felt to be the best ambassador for such a delicate job – Walpole arrived at the French coast in October and began hunting for a boat to take him over the Channel. His luck was out. London was in the grip of plague and no ships were sailing from Calais to England 'by reason of the sickness'. Walpole spent a dismal November in St Omer, frustrated at his lack of progress. While there he encountered his brother Thomas and another mercenary, Edward Lingen, both seeking passage to England, and the three men decided to join forces. It was Lingen who found them a fleet of French warships leaving for Scotland and prepared to carry them into English waters. Making up the passenger list was a Scotsman, reportedly a prisoner of the French.[12]

Strong winds carried them swiftly up the English coast, past Norfolk, which was Walpole's intended landing site, to Flamborough Head in Yorkshire. The unknown Scottish prisoner was put ashore first, to raise money for his ransom. No doubt the information which he provided the English authorities – that a Jesuit priest was attempting a secret landing – paid him suitably well, for by the time Henry Walpole and his party were rowed into land at Bridlington, the watch had been alerted. That night the three men kept together, blundering through woodland in the dark until, by first light, they had reached the village of Kilham about nine miles inland. There, they took refuge in an inn while they planned their next move. Before long news had spread of the appearance of three travellers in the district and by sunset of their first day, 7 December 1593, they had been arrested.[13]

They were taken to the gaol at York Castle, where, towards the end of January 1594, Richard Topcliffe arrived to question them. It seemed 'young Walpole', as Topcliffe referred to Thomas, was disposed to talk and he was subsequently released, but Henry Walpole and Edward Lingen revealed little other than their identities. 'Much more lieth hid in these two lewd persons, the Jesuit and Lingen', wrote Topcliffe to the Council; they 'must be dealt with in some sharp sort above, and more will burst out.' Over the next few weeks Walpole would be subjected to repeated examinations in an effort to make him speak; 'I marvel that my very common condition makes the Crown so interested in me,' he noted. In fact, the Crown – in the person of Topcliffe – was more interested in anything that might lead the pursuivants to Walpole's superior and head of the Jesuit mission in England, Henry Garnet, and Walpole would later be charged at his trial with failing to provide this information. And bearing the Jesuit's inheritance in mind Topcliffe confidently promised 'more in this service than ever I did in any before to her Majesty's . . . purse'. Walpole, meanwhile, responded to Topcliffe's threats by remarking, 'I am much astonished that so vile a creature as I am should be so near, as they tell me, to the Crown of Martyrdom.'[14]

In February Topcliffe escorted Henry Walpole to London, to the Tower. There he was held for eight weeks in solitary confinement. Then, on 27 April, he was brought before Attorney General Edward Coke, the ubiquitous Richard Topcliffe and an officer of the Tower named Sergeant Drewe for his first bout of questioning. At his next two examinations, spread out over the month of May,

Coke was absent and Walpole was left to the tender mercies of Topcliffe and Drewe. At the beginning of June Coke returned to the Tower, but evidently Walpole had not been broken yet. On the tenth of the month Coke wrote to Lord Keeper Puckering admitting he had little against the Jesuit other than his priesthood. But by 13 June Topcliffe was in possession of a signed testimony in Walpole's hand, which, in stark contrast to his earlier examinations, contained a flood of information, including the two aliases – Walley and Roberts – by which Henry Garnet was known and details of the families with whom he sometimes resided. Walpole also promised to recant his Catholicism. The following day this testimony had stretched to include the names of twelve scholars and priests currently at Valladolid, twelve students in Seville, 'as also of five sent to England'. What had happened to Walpole in those few days can only be imagined.[15]

Walpole's trial at York's Lent Assizes was a shabby affair. Evidently he had now retracted his desire to recant for he was indicted with abjuring the realm without a licence, with receiving holy orders overseas and with returning to England to minister as a priest. For a while it seemed he would be denied the chance to speak in his own defence until one of the presiding judges, Justice Beaumont (father of the playwright Francis Beaumont and the son of a devoutly Catholic mother), overruled this decision. The court transcripts reveal a flash of Walpole's legal training in the exchange that followed. '[Beaumont:] Our Laws appoint that a Priest who returns from beyond the Seas, and does not present himself before a Justice, within Three Days . . . shall be deemed a Traitor. [Walpole:] Then I am out of the Case . . . [for I] was apprehended before I had been one whole Day on English Ground.' Such a technicality could not save him, though, and Henry Walpole was duly found guilty of treason. He was executed on 7 April 1595. Valladolid had its first martyr, Topcliffe had his man and Walpole, himself, had the hero's fate it seemed he had been seeking ever since, as a twenty-three-year-old law student, he had stood at Tyburn watching Campion die. There was in this a gruesome symbiosis. The question was who or what did such a symbiosis serve?[16]

Fourteen years before Walpole's death, on 14 October 1581, Claudio Aquaviva, the Jesuit General, had written to Robert Persons acknowledging the news of Edmund Campion's arrest and imprisonment. His letter reached Persons at Rouen, where the priest had taken refuge after his hurried escape from England. In it, Aquaviva stated for the record that should the English Jesuit mission ever be revived, then all those sent on it in the future had an obligation not to be caught. Campion, he implied, had not been careful enough. Later, when Persons proposed to write a biography of his former colleague, Aquaviva was quick to stress that it should, indeed, be a life of Campion, rather than a glorification of his martyrdom. Amid the confusion, obfuscation and sheer lack of information that surrounds the actions of many characters in this story, there is something refreshingly candid about Aquaviva's response. For the Jesuit General the missionaries' job was simple: they were to survive and serve others, not die and serve themselves. Yes, he wrote to William Allen, martyrdom was a noble end, but the prospect of bringing aid to English Catholics was nobler still. And for that one had to stay alive. According to this assessment Henry Walpole, unlucky though his capture was, had failed the mission.[17]

Set against Aquaviva's pronouncements, though, was the ineluctable fact that the martyrdoms of first Campion and now Walpole were having an extraordinary effect on new recruits. Shortly after Campion's death the Oxford Regius Professor of Divinity had written to the Earl of Leicester observing, 'It used to be said, "Dead men bite not"; and yet Campion dead bites with his friends' teeth.' And, he warned, 'in the place of the single Campion, champions upon champions have swarmed to keep us engaged'.

William Cecil, writing to Queen Elizabeth, identified the same phenomenon. 'Putting to death doth no ways lessen them', he explained, because persecution was 'the Badge of the Church'. Persons, too, in the flurry of correspondence provoked by Campion's execution, noted 'Walsingham declared lately that it would have been better for the Queen to have spent 40,000 gold pieces than to kill publicly those priests'. For his own part, he added triumphantly, 'it cannot be told . . . how much good their death has brought about'.* All three men referred, the first two explicitly, to the hydra-effect of martyrdom: 'upon cutting off one [head]', wrote Cecil, 'seven grow up' to take its place. Among these new heads was that of student John Owen – elder or, possibly, twin brother of Nicholas Owen – who left Trinity College, Oxford for Reims in the summer of 1583, accompanied by the fifteen-year-old Walter Owen.† By October 1584 John had been ordained a priest and was back in England to take up his place on the mission.[18]

Meanwhile, in Walpole's case a painting of the new martyr was quickly commissioned to hang in the Great Hall of his old college at Valladolid. It showed him 'with his left hand upon the rack, whereon he had been nine times tormented, with a rope about his neck, & his breast opened with the knife wherewith he was embowelled, and in his right hand he held his heart, which he offered up to Christ'. It was, wrote one commentator, 'a very lively picture . . . [that] moved all to devotion that beheld it'. Certainly, it moved many to emulate his fate and not just from within the college. Reading of Walpole's execution in his newly published biography, brought out soon after his death, the Spanish noblewoman Doña Luisa de Carvajal determined to go to England, serve

* An unlikely convert in the wake of Campion's execution was the Jesuit's former gaoler. According to Persons, 'The man who had been Fr Campion's private warden in the Tower of London is now a very fervent Catholic, though previously he was obstinate in his heresy.'

† John Owen was born in 1561, the earliest probable date of Nicholas Owen's birth. His university career began at Corpus Christi College; then in 1579 he transferred to Trinity. On 1 December 1581 Walter Owen matriculated at Trinity as a college servant, the standard way for poorer students to fund their education.

alongside the missionaries and seek martyrdom herself, by any lawful means possible. In 1605, all attempts to dissuade her having proved unsuccessful, she was granted leave to sail, with Walpole's brother Michael in attendance as her confessor. After a brief stay at Battle on the Sussex coast she moved to London and took up residence with the Spanish ambassador, but if this was meant to contain her it failed. In May 1606 she fell into vociferous argument with some local shopkeepers over a matter of religion and was arrested. Diplomats negotiated her release and the Spanish ambassador begged her to return to Spain, but to no avail. For a time she moved to north London, from where she wrote to her brother outlining the possibilities of converting the heretics of Highgate village. He, meanwhile, had set sail for England to fetch her back home – again to no avail. By 1613 English patience had worn thin. In October of that year the Sheriff of London raided Doña Luisa's Spitalfields house and she was carried off to Newgate prison. The Spanish ambassador began the long process of renegotiating her release, the Spanish ambassador's wife demanded to be arrested, herself, unless her friend was freed, and the then King of England, James I, sent to Madrid, imploring Spain to recall her.[19]

In the end Doña Luisa would have her wish and die on English soil. Her health had always been poor and on 2 January 1614, her forty-sixth birthday, it failed – before she could answer the Spanish King's summons home. Michael Walpole accompanied her body back to Spain and she was buried at the Chapel of the Augustinian Convent of the Incarnation in Madrid. Hers was a singular story, yet it illustrated how the cult of martyrdom, so strong within the Christian Church since its infancy, had lost none of its power at the Reformation. With the Church newly divided, with the very interpretation of Christianity in crisis, never had the call for martyrs been greater than now it seemed. It was the early Christians' willingness to die for their faith that had helped turn an obscure little sect into the dominant world power it had become. That same willingness to die for the Catholic mission now appeared, at one level at least, to be its greatest strength. For what government-forged weapon was strong enough to defeat a man already seeking death?[20]

At the same time, though, it was also becoming its greatest weakness. After Campion's martyrdom, among the recriminations and celebrations of both sides, Claudio Aquaviva's voice had sounded a note of cool, clear pragmatism. Campion had had a job to do and he had failed. With the emergence of Richard Topcliffe as the Queen's personal priest-hunter, leading an increasingly organized force against the mission, Aquaviva's words had gained a fresh imperative – an imperative brought sharply into focus by the treatment of Henry Walpole in the Tower. For those charged with running the mission, the challenge now lay not just in keeping their priests alive long enough to bring comfort to England's Catholics, but in keeping them out of the torture chamber so they could not imperil the entire operation. The logistical obstacles in the way of this challenge, not least the fact that many missionaries seemed drawn to death like moths to a flame, were legion.

As early as 1577 William Allen had outlined the hardships faced by his students in returning home to England. 'I could reckon unto you the miseries they suffer in night journeys in the worst weather that can be picked, peril of thieves, of waters, of watches, of false brethren.' For Allen and Robert Persons, to whom the job of selecting potential missionaries had devolved, their first task was to pick men capable of withstanding these pressures.[21]

This was not easy.* For every John Gerard sent to England, clear-headed and apparently thriving on the danger, was another for whom the sheer terror of their situation seemed the cause of instant paralysis. On 28 December 1581, just weeks after Campion's

* The correspondence between Persons and Aquaviva concerning the Jesuit Thomas Marshall gives some indication of the Society's requirements. Initially Persons rejected Marshall's request to join the mission because of a 'sluggish nature of which he gave evidence'. Marshall pleaded to go. 'I think', wrote Persons to Aquaviva, 'that this good father's importunity will compel me to let him go to some district in England, in the hope that God will co-operate with his holy simplicity to a greater extent than, humanly speaking, can be expected.' Aquaviva refused his assent. Later Persons wrote, 'I am still of the same opinion that [Marshall] is not very suitable as regards talents though extremely suitable in the matter of spiritual zeal.'

execution, the young seminary priest William Bishop arrived at Rye harbour on the Sussex coast. Robert Persons takes up the story:

> Bishop was examined at the port and as he answered in rather a hesitating way he was detained, whereas the other two priests on the same occasion, owing to the fact that they spoke briskly and promptly, were allowed to go. They asked Bishop what calling in life he followed. He answered that he was a merchant. Again he was asked what sort of merchandise he dealt in. He was silent. And on being pressed a little more severely, he confessed that he was a priest. From there he was conducted next day to the Royal Council and made a most uncompromising confession of his faith, and was thrown into prison. To many people, however, such simplicity in the face of these very cunning wolves does commend itself. But what shall we say? God is wonderful in his providence and it is not understood by us. Bishop was warned about this at the time when he was about to go on board, but he seemed so absorbed in meditating on heavenly things as to be quite oblivious of human affairs.*[22]

It was into the ever widening gulf between heavenly things and human affairs that many priests would stumble and come to grief.

This gulf had been clearly identified at Campion's trial. There, the Queen's Counsel had quizzed the Jesuit on his use of a disguise and alias and it is worth quoting the court transcripts to illustrate the dilemma now faced by the mission's organizers.

> [Counsel:] your deeds and actions prove your words but forged; for what meaning had that changing of your name? Whereto belonged your disguising in apparel? Can these alterations be wrought without suspicion? Your name being Campion, why were you called Hastings? . . . No, there was a further matter

* Bishop was imprisoned in the Marshalsea on 11 February 1582. The following year he was condemned to death on a generic charge of plotting against Elizabeth while in Reims and Rome. The sentence was eventually commuted to exile and he was banished in January 1585. In May 1591 he returned to England.

> intended; your lurking and lying hid in secret places concludeth with the rest a mischievous meaning. Had you come hither for love of your country, you would never have wrought a hugger-mugger; had your intent been to have done well, you would never have hated the light; and therefore this budging deciphereth your treason.*[23]

Here, then, was the missionaries' quandary. If a priest entered the country in disguise and lived in secret he laid himself wide open to accusations of treachery – for what need had an honest man for secrecy? If, on the other hand, he entered the country openly he would, like William Bishop, be dragged off to prison – and still be accused of treason, by the simple virtue of his being a Catholic priest. Given the impossible nature of this situation, it made increasing sense to send to England priests who were at least capable of behaving like the spies the Government believed them to be, howsoever this risked damaging the spiritual reputation of the mission long-term. Men like Bishop, enthusiastic, idealistic, utterly breakable, were a danger to every missionary and to every Catholic who took them in.

Always assuming the newly arrived priests could make it through customs and past the many checkpoints designed to prevent their entry into the country, then their troubles had really started. The actual landing had been adrenaline-fuelled and immediate, with little time for reflection on their position, but life underground seldom offered these comforts.† Soon most were acquainted with feelings of creeping despair and numbing loneliness. The transcript of the interrogation of seminary priest John Brushford gave some indication of just how dispiriting life

* Campion's response to this accusation was remarkably controlled under the circumstances. He declared, 'I am not indicted upon the Statute of Apparel.'

† In March 1598 returning Jesuit Oswald Tesimond was able to bully a fellow passenger – a Puritan – into stalling the enquiries of a galleon sent out to intercept their ship as it sailed up the Thames. The Puritan answered the galleon's crew 'so excellently and with such details, that they on board the galleon, instead of examining us, were examined by us'. Earlier, Tesimond had been forced to swim ashore when Dutch warships attacked the Spanish vessel carrying him to France. His capacity to improvise his way through an emergency would stand him in good stead during his years on the mission.

on the run could be. Brushford arrived in England in the spring of 1585, just before the introduction of the new law making sheltering a priest a felony, 'and, by reason thereof, I found everybody so fearful, as none would receive me into their houses'. There followed a catalogue of rented accommodation across London as Brushford attempted to shift for himself – in Tottenham, Clerkenwell, Bishopsgate and 'also a chamber in Gray's Inn Lane, at one Blake's house, unto the which I resorted, when I knew not whither to go else'. For a time Brushford left London for lodgings in Monmouth, 'until the gentleman [his landlord] began to suspect what I was'. At Winchester, where he also travelled, he was unable to find shelter at all. A Mrs Coram, recognizing him as a priest, turned him away, saying, 'Her husband was not at home; her house was full of strangers; and she had sheep to shear: wherefore she prayed me to depart.' Brushford headed back to London 'where I remained until I had opportunity to depart the land, which I earnestly desired'. This opportunity came when he learned of a ship at Southampton preparing to sail to France and he fled the country gratefully. His had been a depressing ordeal from start to finish.*[24]

But if life on the run could be thoroughly enervating, then a life in hiding was no less stressful, as a contemporary report revealed. The priests 'lived for the most part in the upper stories or attics of the house; as remote as possible from the observations of both domestics and visitors'. Remoteness was not sufficient a safety measure on its own, though. 'Great caution had to be observed as to the windows, whether to admit or exclude light; by day they were careful in opening them, lest the passers-by might observe that someone lived in the room; at night they were more careful still in shutting them, lest the light might betray the inhabitant.'

* Sadly, Brushford never had the chance 'to forsake the world, and to serve God quietly in religion' as he had hoped. En route from Reims to Spain in 1590, as a newly ordained Jesuit, he was captured off the Scilly Isles by an English warship and taken to Launceston prison in Cornwall. From there he was transferred to a London gaol, where he died early in 1593.

Similarly, priests were to take care to step 'lightly . . . when pacing the room, or [to] proceed cautiously along the beams', while at 'certain hours all movement in the room was prohibited, that no noise might be heard'. If they left the house, the priests were advised to do so between 'the second or third hour of the night, and return . . . when the domestics had retired to rest', for fear they might be betrayed. So, except for the hour of mass, which every Catholic would attend who could be spared, a priest might spend 'almost entire days, weeks and months alone'. This 'constant solitude' was 'oppressive', wrote the unnamed priest who put together the report.* With this in mind, Henry Garnet, writing to Claudio Aquaviva about the kind of men he needed in England, was quick to request priests 'reliant on divine providence, equipped with virtue' and 'outstanding for their piety'. Without piety, believed Garnet, no man had a hope of surviving the pressures of the mission.†[25]

Henry Garnet was uniquely placed to recognize this. From his own experiences of life underground he knew well what chimeras a climate of fear and suspicion could breed in the mind. As a new arrival in the summer of 1586 he had been interviewed by the daughter of the house at which he had stopped for directions. Her manner began to cause him concern and soon he was convinced she believed him to be a priest. 'So cleverly and cunningly were [her] questions put that I might have been in a court of law,' he wrote. It took him some while to realize the child was simply curious. But, as head of the Jesuit mission – and thus de facto head of the entire network of priests across England – to Garnet had fallen the duty of caring for the missionaries' welfare. It was to him that both seminarians and Society men turned for assistance, advice and accommodation and on him that they unburdened

* A copy of this report exists in the handwriting of Henry More, Jesuit historian and great-grandson of Sir Thomas More. It is dated 1616. However, More did not join the mission until the 1620s and the original author is unknown.

† Garnet specified that missionaries 'should also have good health and average intelligence' and should not be perturbed 'by the clamour and shouting of the enemy, since all the noise they make is to be accounted nothing more than the barking of dogs'.

their anxieties. If William Allen and Robert Persons were charged with selecting suitable priests for England, then Garnet was charged with keeping them that way: obedient to the mission's aims, responsible to the families who sheltered them and, above all, self-possessed. 'A wart on the face is observed instantly,' Garnet noted cryptically, adding that 'by endurance of hardship lay-Catholics become sharp-sighted both in observing and in assessing the actions of priests'. If the missionaries were to expect those same lay Catholics to risk their lives and livelihoods by breaking the law, then their own behaviour had to be without fault, both procedurally and psychologically. It was this realization that had caused Garnet to react so strongly to John Gerard's unauthorized activities immediately after his landing; now it saw him adopting the role of father confessor to every man on the mission.*[26]

The case of Thomas Lister revealed just how difficult this task could be. Lister was a Lancashire man and widely considered a brilliant scholar. In 1589, aged thirty, he returned to England from Rome and was quickly posted to Worcester, to Hindlip Hall three miles from the city, to serve as Edward Oldcorne's assistant there. Oldcorne was opening up the west of England for the mission in precisely the same way as Gerard was the east. In 1590 Garnet received Lister into the Society of Jesus, but soon the new Jesuit was complaining of health problems, in particular of neuralgia. In 1594 he was recalled from Hindlip to join Garnet in London. It seemed Lister had developed claustrophobia and now found it impossible to enter any of the hiding places with which Nicholas

* A letter to Garnet from the seminary priest John Pibush gives some indication of the esteem in which he was held by those he helped. 'Dear Father, with all due affection I commend myself to you as to the first and best friend I have met since my return to my country . . . Well I remember your care for me, your advice that so greatly helped me, and I humbly thank you for your friendly conduct towards me.' The letter also reveals Pibush's thoughts about life on the mission: 'those who purpose to come to this country and to work profitably therein, must bring along with them vigorous souls and mortified bodies. They must forgo all pleasures and renounce every game but that of football, which is made up of pushes and kicks, and requires constant effort, unless one would be trampled under foot; and in this game they have to risk their lives in order to save souls'. Pibush was arrested in Moreton-in-Marsh, Gloucestershire, in July 1593. He was executed at Southwark in February 1601.

Owen had equipped Hindlip. He had become a liability to the family sheltering him and it is testimony to Garnet's courage that he kept Lister with him for two years, at risk to his own life, while attempting to find a solution to the problem. Lister, meanwhile, wrote bitterly to Claudio Aquaviva begging him for help and complaining that Garnet was unsympathetic. Garnet noted 'we hope for better things from him in the future'.[27]

In autumn 1596 Garnet found transport to take Lister out of England and he wrote to Aquaviva informing him, 'I have sent to Flanders Thomas who for a long time earnestly asked to be sent into voluntary exile.' But in February the following year Garnet was once again writing to Aquaviva about Lister, this time with uncharacteristic anger.

> 'The Thomas about whom I wrote I sent away to Flanders. But just outside the very walls of Antwerp he had such a compulsive desire to return home, that I have him with me here again . . . I am deeply distressed that his departure came to nothing. He has had no regard for the interests of the mission: both in my judgement and in his own it was essential that he should go . . . I am tortured in mind over him, uncertain and hesitant how I should deal with him, for the source of his disease is not so much weakness of character as a disturbed mind and lack of responsibility'.

It took many more months for Garnet to persuade Lister to leave England and seek help on the Continent and then, it seems, only on condition the priest could return again when his health was improved. By 1602 Lister had rejoined the mission and Garnet was noting that 'Thomas . . . is almost completely cured of his complaint'. But with the spectre of discovery and arrest stalking every priest at every turn, it was unlikely Lister, and all those others for whom the pressure of their situation brought with it nightmares and neurosis, could ever fully be free from disturbances of mind. Little wonder, given the nature of the death awaiting them.[28]

> You must go to the place from whence you came, there to remain until ye shall be drawn through the open City of London

> upon a hurdle to the place of execution, and there be hanged and let down alive, and your privy parts cut off, and your entrails taken out and burnt in your sight; then your head to be cut off and your body divided into four parts, to be disposed of at her Majesty's pleasure. And God have mercy on your soul.[29]

Such was the sentence for all those found guilty of high treason.

Hanging, drawing and quartering was first introduced in 1241, specifically, it was said, for the pirate William Maurice. By the end of that century David, last prince regnant of Wales, had been executed in similar fashion by the conquering King Edward I of England and the punishment was officially recognized as the lawful penalty for treason against the monarch and the State. It would remain so until 1870, when it was struck from the statute books.* The term *drawing* referred both to the means of transport to the scaffold – the prisoner was laid on a hurdle tied to a horse's tail and drawn through the streets – and to the process of disembowelment. The body of the still-living victim – having been castrated to signify the accused was unfit 'to leave any generation after him' – was now sliced open and the stomach, entrails and heart slowly drawn out. The prisoner's severed head, 'which had imagined the mischief', was held aloft to the watching crowds with the traditional cry 'Behold, the head of a traitor! So die all traitors!' From thence it was taken to Newgate prison, to a room called Jack Ketch's kitchen, and parboiled in a mixture of salt water and cumin seed. Then it was hoisted above the battlements of London Bridge on one of a number of long, jagged poles adorning the southern gate, as a deterrent to other traitors. The salt water and cumin seed were, themselves, a deterrent to scavenging seagulls.[30]

But, in the case of the Catholic missionaries, the sentence of hanging, drawing and quartering having once been passed, the guilty priest was offered a reprieve: if he recanted his faith he would live. Even Edmund Campion, convicted of plotting to kill

* The sentence remained in place under Scottish law until 1950.

the Queen, was afforded this chance to save himself. And here was the chief flaw in the English Government's campaign against the mission, a flaw of which both sides were aware and yet which increasing numbers were happy to ignore in the face of the more convenient myth being peddled by the Crown, that a good Catholic was a bad Englishman. For if the Government seriously believed the missionaries to be traitors, then clearly it was illogical to release them on the grounds that they now conformed to the State religion. And if such conformity was sufficient to save a priest's life, then the Government's insistence that no one in England suffered for his faith, but only for endangering national security, rang resoundingly hollow. And, perhaps, in this last ditch attempt by the Government to avoid making martyrs there was tacit acknowledgement of this. But for those priests now faced with the choice between living in the sin of heresy or dying in the agony of live evisceration, ahead of them lay the bitterest of struggles.[31]

Indeed, every step towards the scaffold was designed to make the struggle bitterer still. If the clinical brutality of the death sentence's wording was not in itself enough to awake the survival instinct, then what followed surely was. The summons usually came in the still grey hours of early morning. At first light, Tower prisoners were led from their cells to Coldharbour, an open space near Tower Green. There, they were bound two to a hurdle and the cavalcade would set off at a slow walk; any faster and the victims' heads would be dashed to pieces on the cobbles.[32]

At the place of execution a cart would be made ready beneath the gallows. The first man was lifted onto the back of the cart and a noose was placed about his neck. Then the horses were whipped up and the cart would lurch away. Those waiting their turn to be hanged watched what happened next, as the following description, from a contemporary account of the executions of Robert Johnson and John Shorte, revealed. Johnson 'being brought from the hurdle, was commanded to look upon Mr Shorte who was hanging, and then immediately cut down. And so being [helped] into the cart, was commanded again to look back towards Mr Shorte who was then in quartering'. All the while the priests were exhorted to recant. Over the years many of them would.[33]

In September 1588 John Owen, Nicholas's brother, would renounce his faith at the Chichester assizes, agreeing to take the Oath of Supremacy and any other oaths of the judges' devising. Court official Thomas Bowyer promptly set about creating a new formula and soon Owen was swearing to make known to the authorities 'all such parties and practices as shall any way tend to the endangering' of Elizabeth's life; in other words, all details of the mission. Owen was then placed in the custody of the Bishop of Chichester.* The following day, 1 October 1588, another of the priests with whom Owen had been convicted, Francis Edwards, would also recant upon the scaffold. What agony of conscience this cost them both is unknown, but the case of Catholic layman John Thomas suggested few men took the decision easily. Thomas was sentenced to death at the Winchester assizes of March 1592. Immediately, he recanted his faith, but, as Henry Garnet described to Aquaviva, 'After his return to prison, [Thomas] sent a message to inform the judges . . . that he regretted his cowardice and determined henceforth to do nothing unbecoming a Catholic.' Learning of some thieves, convicted at the same assizes, who were due to be executed, Thomas presented himself to the sheriff at the gallows, explaining he was condemned and requesting to be killed in their company. '"Since you are so keen to be hanged", said the sheriff, "assure yourself that I should be delighted to oblige you if your name were on my list. But since it is not here, go away."' Thomas was finally executed at the autumn assizes.[34]

It is hard to say just how many men recanted during the course of the mission, or what happened to them all afterwards, but every time news trickled down of a priest renouncing his faith a cold

* It is unclear what happened to Owen next. In a letter of 1596 Henry Garnet mentioned that Nicholas Owen had a brother, a priest, then in prison. Afterwards, the name John Owen appears in a number of documents relating to the mission, indicating that – provided this was the same John Owen – he was still active in the Catholic cause. The last entry is dated 1618 when a John Owen was convicted of treason. At the intervention of the Spanish ambassador the sentence was commuted to one of banishment.

shiver of apprehension ran through the Catholic community. In 1586 the seminarian Anthony Tyrrell, a close associate of the then Jesuit Superior William Weston, was captured and imprisoned. Terrified of the torture to come, Tyrrell promptly sold his services to the Government and from his cell in Southwark's Clink prison, crammed full of Catholics, he worked hard to discover the whereabouts of Garnet and Southwell for his interrogators. In 1592 James Younger – having already inadvertently compromised the location of Garnet's base at Finsbury Fields by blundering there in daylight hours, seeking advice – was arrested and sent to the Counter prison. There he, too, broke down in terror. In a series of letters to Lord Keeper Puckering, Younger repeatedly offered to turn Queen's evidence, naming vast numbers of priests then in England and assuring Puckering he could quickly learn where each one was stationed if he were set free. Just as repeatedly, he also craved pardon for the errors of a young man led astray through bad counsel. By June the following year he had been released from prison, but to his dismay he now found himself shunned by fellow Catholics. Bitterly he wrote to Puckering asking if he might leave London for the north of England where his treachery might not be suspected. From there his trail goes cold, until, in March 1594, the priest Robert Barwise revealed under interrogation that Younger was now back at Douai, teaching at the English College there.* Like Anthony Tyrrell before him – who, when called upon to read his recantation in public, retracted his statement – James Younger appeared to have regretted his activities as a Government spy and returned to the Catholic Church. In neither case was this regret lasting. In time Anthony Tyrrell would re-recant and become a minister in the Church of England, while James Younger would join the growing anti-Jesuit faction at Douai, prepared to betray priests of the Society to the English Government in

* In February 1594 the Catholic Benjamin Beard, a prisoner in the Fleet, wrote to Puckering offering his services in locating Younger. He took care, though, to ask for anonymity in his task – so as not to bring disgrace on his mother and his kindred, all of them being papists and recusants.

return for greater freedoms of its own. Their stories provided ample proof that the Catholic mission was only as strong as its weakest member.[35]

But if the selection of priests possessed of the right balance of piety and practicality to serve the mission's aims was an imperfect science, then there were other measures more exact that could be adopted to secure operations. The information is fragmentary, as typifies any secret organization. Indeed, as in the case of Nicholas Owen's hides, it was often only the unsuccessful methods that were discovered; those that were successful are seldom mentioned in the letters, State papers and spies' reports that document the period. Nonetheless, from the few facts that are available it is possible to piece together an outline sketch of the underground network in place across England.

It had been identified early on that returning missionaries were at their most vulnerable during the voyage home and in the period immediately after landing. In transit there were too many factors outside their control that could go wrong, namely the weather and the unpredictable intentions of their crew and fellow passengers. On arrival, until they could be placed in a secure house, they were prey to every suspicious watch, inquisitive villager or opportunistic informer in the neighbourhood; it was also during their first few hours ashore that priests were most likely to make mistakes, through fear, through ignorance of local conditions, through sheer bad luck. So, in the four years he spent in Rouen immediately after his flight from England, Robert Persons had supervised the development of a number of safe routes home for missionaries. Rouen was 'a most convenient town', he noted, 'on account of its nearness to the sea, so that there some can make trips to the coast to arrange for boats to convey people across'. Assisting him was Ralph Emerson, Campion's servant from their time together in England, and in August 1584 Persons wrote, 'Ralph . . . has done wonders by contriving two new ways of crossing over.' It is possible these were the two routes used by William Weston in September 1584 and by Garnet and Southwell two years later. Weston sailed from Dieppe to a point just south of Lowestoft on the Suffolk coast in a privately chartered boat. He was accompanied, on Persons' instructions, by Henry Hubert, 'an English gentleman who was staying [in Rouen] and [had] properties on the English coast'; Hubert's job was to escort Weston ashore and guide him to his own house, before returning to France. Garnet and Southwell sailed from Calais to the coast of Kent, about a mile east of Folkestone, again in a privately chartered vessel and with an escort, a Flemish lay brother, to see them onto dry land. It was these two rules – secure a private boat and a guide – that Walpole, in his haste to reach England, had flouted with such devastating consequences. Indeed, Walpole's storm-racked journey and his few hours on land became a textbook case of precisely what not to do as a returning missionary. In the wake of his capture every effort would be made to ensure that new arrivals did not fall into the same trap.*[36]

While Walpole still languished in York Castle, Henry Garnet had written to Aquaviva asking to be informed of priests' impending departures: 'help can be sent over without incurring any hazard,' he explained, 'if I am given warning beforehand, so that I can determine what place they are to be put ashore'. In response to Persons' efforts to develop safe routes home, part of Garnet's job had now become the development of secure landing places. This was easier in the more remote parts of the country. The Jesuit Richard Holtby, busy expanding operations in the north of England since his return home in 1589, was able to establish two

* A spy's report of 1611 reveals how this system of safe routes home had grown over the years. 'Three sundry courses they use in conveying seminaries to England; the one is by way of Rotterdam, by means of one, Mr Skult, a merchant and a great Papist, being employed by the King of Spain etc., who, by confession of Fr. Gardiner to me, hath conveyed into England above sixty priests, seminaries and Jesuits and he doth it by means of his shipping, for he attires them as mariners with thrum caps, and fishermen. The other is by way of St Valeris in France upon the River Somme, where divers Frenchmen are that continually go to Newcastle for coals. The other is by way of Hamburg.'

new landing sites for priests on the River Tyne at Hebburn and St Anthony's, just upstream from South Shields. They remained undiscovered for twenty years. On the more heavily defended southern shores of England it was harder to conceal clandestine activity and here Garnet was dependent on the assistance of Catholic families. Henry Hubert, William Weston's guide in 1584, owned land bordering the Suffolk coast at Kitley, Kessingland and Pakefield; elsewhere, the powerful Arundell family owned Chideock on the Dorset coast, six miles to the east of Lyme Regis. Both these estates were quickly made available to the mission and soon Garnet was in contact with many more Catholic families in possession of coastal property.* With their cooperation, it seems he was able to establish a number of safe entry points for his returning missionaries and with Walpole's arrest serving as an added impetus to the mission to maintain these sites carefully, it now became increasingly rare for a priest to be captured on landing.[37]

Having landed, though, the next challenge for any priest was to make contact with the mission's organizers. Again the evidence is fragmentary, but it appears missionaries were supplied with only the first name in a chain of connections leading them underground and into hiding. In most cases the name was that of a prominent Catholic family in or near London. William Weston was given the name of the Bellamy family of Uxendon near Harrow, well known to Robert Persons from his time in England. In addition, he was given 'some small articles, tokens of friendship' that Mrs Bellamy would recognize; these precautions were necessary to safeguard the family from entrapment by spies claiming to be priests. Neither

* This evidence comes from a secret meeting held in September 1590 at which Garnet was called upon to advise Catholics on the issue of the ownership of shipwrecked goods. According to common law anything washed ashore from a shipwreck belonged to the Crown. 'Now a large number of Catholics with whom I have occasional dealings seize this property without scruple,' Garnet informed Aquaviva. Despite many families' evident belief that this wreckage was rightfully theirs in lieu of the recusancy fines they were paying, Garnet – a stickler for the law, despite his fugitive status – ruled in favour of the Crown.

Henry Garnet nor Robert Southwell provides any clues as to the contact name they were given, but Garnet later told Aquaviva that on their first day in London 'by chance we met the man we were looking for . . . [and] . . . we were safely hidden away'. Even John Gerard, who in his first hours ashore had found himself a safe house and a future base for his apostolate in Norfolk, eventually made his way to London to contact 'some [unnamed] Catholics'. The logic was simple. London, with its crowds and its bustle, was the easiest place for a returning priest to pass unnoticed; prominent Catholic families were among those most likely to have taken precautions against Government informers; and, as Richard Topcliffe's attempts to discover Garnet's whereabouts from Henry Walpole revealed, no priest could divulge, even under the severest torture, information of which he was not in possession.* If, by now, the Government viewed the English Jesuit Superior as the linchpin of the Catholic mission, then the mission, itself, was doing everything in its power to protect him. Jesuit Oswald Tesimond, writing of his experiences on the mission, explained: 'the place where [Garnet] lived was known to very few persons, who could be thoroughly trusted'. It was only through these tried and trusted Catholics that a newly arrived priest, whether Jesuit or seminarian, could gain an introduction to Garnet and, through him, placement in a safe house out of town.†[38]

Even after a missionary had been stationed with a Catholic family it appears this practice of withholding all potentially damaging information from him continued. The unnamed priest's report on life in hiding quoted earlier revealed that 'except when the Superior visited them, [priests] scarcely ever saw one of the Society, or any other Priest'. Isolation, both from the family they

* It is unclear how or when the Government first learned Garnet was the Jesuit Superior in England, but Henry Walpole, during his interrogation in York, noted, 'The President [of the Council of the North] inquired of me who was the Superior of our Society in this Kingdom? Whether it was this, or the other, or who it was? Topcliffe answered, He knew who it was, and named him.'

† In a letter to Robert Persons of 9 April 1598, Garnet explained that every priest and layman captured was immediately 'asked for Henry'.

served and from each other, was now the missionary's main guarantee of safety. Even Garnet subscribed to this rule. In August 1587 he wrote to Aquaviva describing how 'yesterday I accidentally met our Robert [Southwell] in the street . . . I am altogether unable to tell you what joy this sight gave me'. The more a priest was in contact with his fellows and the more he knew where each of those fellows was stationed, then the more his capture and torture threatened to topple the entire operation.[39]

Loneliness was a cruelly ignominious fate for all those now pouring into England, heaven-bent on martyrdom, but it was on these lonely priests in their lonely attics that the hopes of every English Catholic rested. Loneliness was keeping the faith alive.

There are, though, exceptions to every rule. At the beginning of the second week in October 1591, from all corners of England, separately and in pairs, a party of men converged on Warwickshire, to a small manor house near Knowle, hidden from its neighbours by a dense belt of trees. From Essex rode John Gerard, joining, en route from London, Robert Southwell. From Hindlip Hall, just eighteen miles away, rode Edward Oldcorne and Thomas Lister, and from Yorkshire rode Richard Holtby. In all some dozen or so priests – all the Jesuits then at liberty in England plus representatives of the seminaries of Reims and Rome – made this journey, through woodlands of ochre and amber, through the first wraith-like mists of early autumn.* The timing was no accident. 'Our adversaries', wrote Garnet, 'are engrossed in a general election throughout the kingdom and with devising new methods of persecution': it was the ideal time to meet in secret.[40]

* In addition to those named, and to Henry Garnet the Superior, there were three other Jesuits then at work in England: John Curry, Holtby's travelling companion, and Thomas Stanney and John Nelson, seminarians who, like Lister, had joined the Society while in England.

These secret meetings were a biannual occurrence and their purpose was twofold: practical and spiritual. They were forums in which to discuss business and strategy, to plan for the future, to counsel lay Catholics, discipline the wayward and disseminate information picked up piecemeal from the farthest corners of the kingdom. Consequently, they were vital to the mission's success. As Garnet explained to Aquaviva, it was there that 'we forged new weapons for new battles'.[41]

They were also vital to each individual attending them. None of the Jesuits then in England had been a member of the Society more than fifteen years, none had yet taken his final vows, most were little more than novices. For a religious order founded on rigid training and discipline this was an unsatisfactory state of affairs, particularly given the adverse conditions in which its members now found themselves. The meetings provided an opportunity to redress this balance, allowing participants to renew their spiritual vows, to make confession and, above all, to experience a brief sense of community amid so much isolation. As John Gerard explained, 'I never found anything that did me more good. It braced my soul.' So twice a year, conditions permitting, Henry Garnet summoned the Jesuits together.[42]

It seems certain the meeting of 1591 was held at Baddesley Clinton, a medieval manor built of local honey-grey Warwickshire sandstone, moated and fortified, and belonging, since 1517, to the Catholic Ferrers family. Its current owner was Henry Ferrers the lawyer, antiquarian and diarist, described by contemporary historian William Camden as 'a man both for parentage and for knowledge of antiquity very commendable'. For four years now, though, Ferrers had leased the house to two sisters, the twenty-five-year old Anne Vaux and the widowed Eleanor Brooksby, two years her senior, daughters of Lord Vaux, the Catholic grandee who had already done so much to fund and assist the mission. The sisters had first begun harbouring priests at the family's Shoby estate in Leicestershire in the early 1580s; it was there that Henry Garnet was sent on his return to England, while Robert Southwell was stationed at the family's town house in Hackney, northeast of the city of London. Since then the two women had become an essential part of Garnet's team of lay assistants and it was in order to provide the Jesuit Superior with a base in the heart of England that Eleanor had rented Baddesley Clinton.[43]

On the evening of 14 October 1591 the meeting began. There is no record of the business discussed, but high on the agenda would have been the rumours then circulating the country of a new Spanish-led Armada against England and of the measures that would assuredly be taken against the English Catholics should that Armada sail. And certainly Garnet would have wished to outline his fears of more trouble ahead for the mission after a summer of comparative calm. The 'peace which we enjoy here from time to time is not due to any easement of the laws or to greater freedom in the practice of our religion', he would explain, 'but to a respite that will usher in a period of ever greater harshness'. Each priest then met with Garnet privately and on the final day, 18 October, after High Mass, they renewed their spiritual vows.[44]

Until then Garnet had remained calm, reassuring those who spoke of the dangers involved in congregating under one roof with the words 'Yes, we ought not to meet all at the same time now that our numbers are growing every day. But we are gathered for God's glory. Until we have renewed our vows the responsibility is mine; after that it is yours'. Now that those vows had been renewed, and as the priests sat down to eat, he grew agitated, warning 'us all', reported John Gerard, 'to look to ourselves and not to stay on without very good reason'. Garnet, himself, wrote afterwards, 'I know not what inspiration made me address them as follows: saying that, though up to now I had taken on myself all responsibility, I was no longer willing to guarantee them their safety, when dinner was over.' Immediately they had finished their meal a number of the party saddled their horses and rode away. At five the following morning Baddesley Clinton was raided.[45]

'I was making my meditation,' wrote John Gerard, 'Father Southwell was beginning Mass and the rest were at prayer, when suddenly I heard a great uproar outside the main door. Then I heard a voice shouting and swearing at a servant who was refusing them entrance.' The house was surrounded, with guards posted on every track leading up to it. The surprise was complete.[46]

It was Southwell who reacted first. 'He guessed what it was all about,' recorded Gerard, 'and slipped off his vestments and stripped the altar bare.' The others gathered up their possessions and hid them away. The 'beds presented a problem: as they were still warm and merely covered in the usual way preparatory to being made, some of us went off and turned [them] and put the cold side up to delude anyone who put his hand in to feel them'.[47]

Downstairs, the servants were holding the pursuivants back, explaining from behind the bolted and barred front door that 'the mistress of the house, a widow, was not yet up, but was coming down at once to answer them'. Eleanor Brooksby was spirited away to a hiding place at the top of the house; she 'was somewhat timid', noted Garnet, 'and unable to face with calm the threatening grimaces of the officer's men'. Anne Vaux now assumed Eleanor's role and still in her nightgown came to the door to interview the search party. She played for time. 'Does it seem right and proper that you should be admitted into a widow's house before either she or her maids or her children have risen?' For as long as she was able, Anne kept the men in conversation on the doorstep: each second now was precious. Finally, she could hold them back no longer and the ransack began.[48]

'They tore madly through the whole house,' wrote John Gerard, 'searched everywhere, pried with candles into the darkest corners.' Anne Vaux likened them to children playing blind man's buff, 'covering their eyes and then trying to touch and grasp' everything about them. 'You should have seen them', she told Garnet afterwards: 'here was a searcher pounding the walls in unbelievable fury, there another shifting side-tables, turning over beds.' After a while Anne offered them breakfast, presiding over the table with deliberate courtesy while her servants combed the house for anything incriminating. Then the search continued.[49]

Beneath the servants' quarters, huddled in a narrow sewage channel running the length of the southwest wing of the house, the hunters' quarry – five Jesuits, two seminarians, plus two or three Jesuit servants – listened in silence to the destruction taking place above their heads.* The channel was airless and dank and the water from the moat beyond lapped at their ankles. At its highest point the channel measured only four feet, so the men crouched or bent double in the darkness, straining to identify the noises echoing down the former garderobe shaft through which they had entered. Among them stood Nicholas Owen, the man responsible for converting the sewer into a hiding place; between his companions and capture lay a camouflaged trapdoor, covering the top of the garderobe shaft, which itself was concealed within the thickness of the wall. As the search overhead grew more frantic still, as furniture was overturned and panelling sounded for space behind, this trapdoor must have seemed like scant protection.[50]

It took a bribe of twelve gold pieces before the pursuivants were finally happy to leave the house.† Still the priests waited in the tunnel, though, allowing the men to go 'a good long way, so that there was no danger of their turning back suddenly, as they sometimes do', explained John Gerard. Only then did they emerge from the hide. They had been there over four hours.[51]

The raid at Baddesley Clinton had brought home two important points to those involved in it. First, that under the guidance of Nicholas Owen the policy of hide-building was working. In the lethal game of cat-and-mouse being played out across the country by pursuivants and priests, the mouse now stood at least a sporting chance of survival. Second, that the Jesuit network was now too successful ever to be put in such jeopardy again. After the

* The five Jesuits were Henry Garnet, Robert Southwell, John Gerard, Edward Oldcorne and Thomas Stanney. The names of the seminarians are unknown.

† Gerard wrote indignantly, 'Yes, that is the pitiful lot of Catholics – when men come with a warrant . . . it is they, the Catholics, not the authorities who send them, who have to pay. As if it were not enough to suffer, they are charged for suffering.'

abortive years of the early 1580s, when effort after effort to establish a Jesuit base had failed, there were now nine Jesuits at work throughout England, in charge of a fully functioning country-wide mission; and more were eager to join their number. At a single stroke this network had almost been destroyed. It seems that the 1591 Baddesley Clinton conference was the last occasion that all the English Jesuits met together in one place for many years to come.

For Henry Garnet the raid would prompt a crisis of confidence and within hours of the pursuivants' departure he had written to Claudio Aquaviva offering his resignation. He would also admit, some while later, that he had known Baddesley Clinton was under surveillance. Just before the meeting the chief pursuivant, a man called Hodgkins, had called at the house. Unhappy at his reception he threatened to return again within ten days, 'bringing with him a party of men to break down the doors and demolish the very walls'. But word in the district was that Hodgkins was now occupied elsewhere; furthermore, Garnet had been sure 'he could not come back into the immediate neighbourhood without our friends letting us know of it at once'. It had been a calculated risk to continue with the meeting, pitting danger against necessity, and this time it had paid off, but for Garnet the strain of leadership was telling. He had never been convinced of his suitability as head of the mission. Perhaps he also believed he had just come perilously close to carelessness. Now he requested permission 'to hand over the torch to someone more expert than myself . . . and be allowed . . . to run, not by my own discretion, but under the guidance of others'. Aquaviva refused his assent.[52]

Garnet had sought for the mission men outstanding in their piety. In the wake of the raid on Baddesley Clinton – explaining his decision not to call the meeting off and evincing his own piety – he had written, 'we . . . had exceeding confidence in God, for whose glory we were assembled'. Confidence such as this was the lifeblood of the mission and for those who had escaped the pursuivants at Baddesley Clinton their ordeal could only add to the certainty that God was on their side; John Gerard likened their experience to that of Daniel in the lion's den. In spite of their hardships their cause must be just, for providence was showing them so. Piety left little room for doubt (despite the many examples of pious priests, Campion and the like, already caught and killed). Indeed, the history of the Christian Church told them piety was a weapon of unparalleled strength in battles of faith.[53]

The problem was their struggle with the English Government was not a battle of faith. Each priest executed for treason to the realm rather than for his religion proved it so, representing one more victory for an administration determined to keep tight control of the terms by which this conflict was fought. And as the survivors of Baddesley Clinton rode back to the districts in their charge, the Government was preparing yet another broadside to ensure that its battle with the mission would be played out on political terrain. Against politics, simple piety did not stand a chance.

Chapter Seven

'In the beginning was the Word, and the
Word was with God, and the Word was God.'
St John's Gospel (King James Bible)

On 18 October 1591, the day of the Baddesley Clinton raid, Queen Elizabeth I put her signature to a new proclamation, the latest salvo in the English Government's ongoing campaign against the mission. Its contents were a rich stew of fact, half-fact, paranoia and spin, liberally seasoned with a colourful invective.* Its aim was to drive a wedge even further into the seam exposed three years before, during the Spanish Armada, dividing Englishmen more firmly than ever into two immutable and mutually exclusive camps: those loyal to Queen and country and those who were Catholic. And its timing – it was published a month later, in November – was guaranteed to earn it a sympathetic reading from anyone viewing the current hostilities between Europe's twin leviathans, Spain and France, with sweaty-palmed apprehension.[1]

These hostilities and, more particularly, the arrival of so many new Catholic priests into the kingdom were, according to the

* Many of the facts – including details of the missionaries' secret landing places and their use of aliases – were supplied by the seminary priest-turned-informer John Cecil. Cecil had returned to England early in 1591, along with James Younger and the Jesuit Richard Blount, disguised as a returning prisoner of war.

proclamation, simply the prelude to yet another attempted 'Invasion of this Realm' forecast for the following year. All those who regarded themselves as honest Englishmen were charged to stand to the defence of 'their Wives, Families, Children, Lands, Goods, Liberties, and their Posterities against ravening Strangers, wilful Destroyers of their Native Country, and monstrous Traitors'. The best way they could carry out this duty, advised the proclamation, was to make enquiry of every newcomer in their household, and, indeed, in the surrounding area, as to 'where he [had] spent his Time for the space of one whole Year before'. Anyone who proved unable satisfactorily to account for his movements was to be handed over to a specially appointed district commissioner for further investigation.[2]

Having stirred into action a neighbourhood watch of unprecedented size, suspicious of everyone, and having sounded a warning to all Catholic families in possession of a priest, the Government now turned its attention to the more pressing concern of mustering troops to set against this new Armada. The threat of invasion was real. Ever since the remnants of his first fleet had staggered home in disarray in 1588, Philip II of Spain had been marshalling his forces for a second attack. And recent events in France – events of which Spain was hungrily taking advantage – had offered England ample proof, if proof were needed, that there were among the Catholic Church those just as willing to kill for their faith as to die for it. In the summer of 1589 the French King Henri III was assassinated by Father Jacques Clément, a Dominican monk.

Clément was a member of the Catholic League, a fundamentalist organization financed by the influential Guise family, dedicated to eradicating heresy in France and powerful enough to challenge what it saw as Henri's over-conciliatory stance towards the French Huguenots. In 1588 Henri had attempted to clip the League's wings by ordering the murder of the Duc de Guise. This, in turn, had led to his own assassination. Now the Leaguers – with Spanish backing – were engaged in preventing the new French King, the Protestant

Henri de Navarre, from taking his throne. Were they to succeed there would be little to stop them joining forces with Philip II and crossing the Channel to invade England. 'The state of the world is marvellously changed', noted William Cecil drily, 'when we true Englishmen have cause for our own quietness to wish good success to a French king.'[3]

But if this was the state of the world, then the state of the nation was worse. For the England now crawling towards the century's close was an England haemorrhaging money on defensive strategies at home and counter-offensive measures abroad, taxed to the hilt and tired of starting at its own shadow. It would have been a rare government that failed, at such a moment, to play the treachery card, invoking the nascent creed, nationalism, over the ancient creed, religion; rarer still when there existed enough evidence to show that faith led men to murder just as easily as martyrdom. Elizabeth's Government was not so rare.

So the 1591 proclamation painted the very portrait of treachery. The Catholic missionaries, it explained, were 'dissolute young Men', criminals, fugitives and rebels, schooled in sedition, 'pretending to promise Heaven . . . threatening Damnation' and 'Undermining our good Subjects . . . to train them to their Treasons'. This was tabloid language and sentiment, fanning the flames of national paranoia, fuelling the panic, smoking out all those still standing undecided in the midst of England's religious divide. It was effective, low-punching, hard-hitting propaganda – cheap and cheerfully brutal. It deserved an answer. And almost as soon as the ink from the presses was dry, Robert Southwell, Jesuit poet, took up his pen to make it one. 'Most mighty and most merciful, most feared and best beloved Princess,' he began to Elizabeth. In the ongoing war of words between Catholics and Protestants a champion had just entered the lists.[4]

Even before the Reformation fractured the polished unity of Roman Catholic Christendom, dissenting voices had complained of the Church's stranglehold on the language of faith. The word of God, they argued, came to his flock in a tongue it did not speak, through the mangling maws of a parish priest schooled only so well as the latest educational policies permitted. The word of God was all too often unintelligible gibberish.

By the mid-fourteenth century a blunt-talking theologian, Yorkshireman John Wycliffe, had decided to take a stand against this and other disputed aspects of papal authority in England, launching a full-scale attack on what he regarded as Church corruption, culminating in the first translation into English of the Bible, *c.*1382. Rome's monopoly on the word of God had been broken.

The backlash was quick and severe. England gained new heresy laws in line with the rest of Europe permitting burning at the stake and several of Wycliffe's followers, who included a disproportionate number of Oxford University men, suffered accordingly.* Wycliffe, himself, was exhumed – he had died of a stroke in 1384 – and his body was burnt by order of the Pope. There could be few clearer indications of just how ugly the struggle to retain control of the written language might eventually become.

Then in 1450 a German goldsmith from Mainz called Johannes Gutenberg developed the first Western printing press, an invention brought to England twenty-six years later by William Caxton. Within the space of a generation the printed word was exploding about Europe with firecracker ferocity as the long and labour-intensive process of transcribing texts by hand was replaced with the rapid mechanical efficiency of metal typesetting and multiple copy-runs. The 'appearance and state of the world', as Sir Francis Bacon would later put it, had changed forever.[5]

* Lollardy – the movement that grew up in response to Wycliffe's teachings – was very popular among Oxford's academics (Wycliffe, himself, was a member of Merton College) and the university was heavily penalized as a result. Undoubtedly this contributed to Oxford's reluctance to embrace Protestantism in the sixteenth century.

The Church's critics were quick to identify the possible uses of this upstart industry. When the leading English Protestant William Tyndale fled abroad in 1524, it was his own new translation of the Bible that returned home in his place, secreted back from Antwerp and Cologne by the hundred-load. Draconian legislation was swiftly introduced in Parliament to try to stem the tide and the printing presses of England, previously unrestricted, now found themselves subject to stringent regulations, a practice Elizabeth, on her succession, saw no reason to discontinue. The Stationers' Company charter, as re-confirmed in 1559, read: 'no manner of person shall print any manner of book or paper of what sort, nature, or in what language soever it be, except the same be first licensed by her Majesty . . . or by six of her Privy Council or be perused and licensed by the Archbishops of Canterbury and York, the Bishop of London [and] the Chancellors of both Universities'. Such a formidable selection panel might be capable of muffling the voices of its domestic opponents; it was quite powerless against the presses of Europe.*[6]

Since the outlawing of Catholicism in England, book-running had become a popular pastime and the Continental printworks now produced a stream of Catholic literature to be smuggled back for the home market. As fast as the books entered the country so the Government hunted them down. Greater efforts still were made to keep such works from the eyes of susceptible university students; messengers flew between London and the heads of Oxford's colleges ordering searches of undergraduate rooms for suspect material. In the spring of 1577 pursuivants raided the house of Rowland Jenks, a stationer and bookbinder of Oxford. Jenks was charged with distributing banned Catholic texts. At the July assizes he 'was arraigned as a Catholic, found guilty, and being but one of the common people, was condemned to lose both his ears', as one commentator reported. In fact his sentence was even

* England was not alone in trying to regulate its printing industry post-Reformation; in 1535 François I of France – a country bordered to the east by Protestant states – would issue a ban on the printing of all books on pain of death.

more specific. His ears were to be nailed to the pillory and he was to be given the choice of cutting himself free or remaining there indefinitely.*[7]

It was Robert Persons, as part of the centralizing activities of the first Jesuit mission, who decided to set up a secret English press of his own. To his aid came William Brooksby and the printer Stephen Brinkley, both associates of the Jesuits' benefactor George Gilbert. Brooksby persuaded his father to lend them his house at Greenstreet near Barking, some six or seven miles outside London; Brinkley found seven workmen to join him there and by November 1580 the presses were rolling. The inaugural text was written by Persons himself and entitled *A Brief Discourse containing certain reasons why Catholics refuse to go to Church.*[8]

Great care was taken in distributing the new books. They were 'consigned to the priests in parcels of fifty or a hundred,' explained Persons, 'and sent at exactly the same time to different parts of the kingdom'. Then, to coincide with the predicted 'searches of Catholic houses . . . a number of young gentlemen [were] ready to distribute other copies at night in the dwellings of the heretics, in the workshops, as well as in the palaces of the nobles, in the court also and about the streets, so that the Catholics alone [could] not be charged with being in possession of them'. As a further protective measure the title page of each book testified to its having been printed in Douai. Little could camouflage their distinctive English typeface, though, and soon the government's net was closing in on Greenstreet. When one of his men was arrested, Brinkley finished off the print-run, dismantled his press and quietly departed the neighbourhood.[9]

For a while Brinkley relocated to Southwark near St Mary Overie (now Southwark Cathedral), to the house of Francis Browne, a staunch supporter of the Jesuit mission, but by April 1581 the press was on the move again, this time to the more discreet

* In 1581 Jenks went to work for Robert Persons in London. There, he was betrayed and arrested and by April of that year he was back in Oxford gaol. On his release he fled abroad to Douai, where he became baker to the English College. He died in 1610.

location of Stonor in Oxfordshire. Here, the propaganda war moved up a gear with Edmund Campion's *Decem Rationes.* Its full title was *Ten Reasons for the confidence with which Edmund Campion offered his adversaries to dispute on behalf of the Faith, set before the famous men of our Universities* and it contained an impassioned defence of the Catholic faith. It only needed Father William Hartley, Oxford's Catholic mole, to set the new book before those famous – and impressionable – young men of Oxford University to complete Campion's challenge. Hartley duly obliged. Those students attending Commencement at St Mary's Church on Oxford's High Street, on the morning of Tuesday, 27 June, found several hundred copies of the *Decem Rationes* waiting for them on their benches. Within weeks, though, Campion had been arrested at Lyford, Brinkley was a prisoner in the Tower and the Jesuit mission – and its secret press – lay in ruins.[10]

It took a former print worker and a budding poet to amend those fortunes. The pairing of Henry Garnet and Robert Southwell – so effective in building up a solid network of safe houses for the Catholic mission – was, in terms of the propaganda campaign still to be waged, inspirational. The driving force was Robert Southwell's. Six months into his new job it had become abundantly clear to Southwell how bleak the situation for English Catholicism was. Just the summer before, in August 1586, a small band of Catholics led by Anthony Babington had been rounded up, convicted of plotting to kill the Queen and executed with a brutality that shocked even those onlookers baying for their blood.* Now, as Parliament pressed for the death of Mary, Queen of Scots, bruited across the country as Babington's accomplice

* Elizabeth was keen to devise a new method of execution for Babington, even worse than hanging, drawing and quartering. However, William Cecil advised that the hangman should simply delay the conspirators' deaths for as long as possible, 'protracting of the same, both to the extremity of the pains in the action, and to the sight of the people to behold it'. This decision was amended after the first batch of executions when the Crown expressed revulsion at the savagery.

in the outrage, fresh details of the narrowly averted terror attack continued to spill from the popular presses, each one more sensational than the last. For England's Catholics, caught between the Scylla and Charybdis of Babington's folly and Philip of Spain's rumoured Armada, these were times to remain out of sight, away from the accusations of treason that continued to hound them because of their faith. Southwell's frustration at this unstoppable spew of propaganda was voiced in a letter to Claudio Aquaviva that January. The London publications reaching him were all of 'one cry', he railed: 'that traitors and assassins such as we should not be tolerated in the State, that we are plotting the ruin of the gospel and of all sacred things. And we are to conclude that our own ruin has come upon us because we are hateful to heaven and earth'. Bitterly he observed, 'It is an axiom of liars that what is won by lies must be kept and confirmed by lies.' When Henry Garnet met with Southwell that February, it seems Southwell put his case for confronting these lies head-on, in hard print. It remained for Garnet, with his technical expertise, to make this possible.[11]

It is likely the new Jesuit press was housed outside London, at Acton in Middlesex, in a small garden cottage belonging to Anne Howard, Countess of Arundel. Southwell's connection with the Howards had grown up out of a misunderstanding. Shortly after his arrival in England, while still based with the Vaux family in Hackney, he had received word that the Countess was searching for a Catholic priest. Southwell left Hackney for the Countess's town house on London's Strand and, after a few days there, began making discreet inquiries about the possibility of building a hiding place for this, his new residence. It was some years later – and only after Southwell had become a permanent fixture in her household – that the Countess revealed to him she had only been seeking a priest to visit on a temporary basis, not to stay. For the Jesuits, however, Southwell's mistake brought with it huge dividends. The Howards were England's foremost aristocratic family.[12]

On the execution of his father Thomas Howard, Duke of Norfolk, in 1572, Anne's husband Philip, Earl of Arundel had lost his dukedom, but gained, it seemed, a glamorous following at court, where he remained a popular favourite with the Queen.* Arundel, though, was no more a politician than his father was, and in the shifting sands of back stage life at Greenwich, Richmond and Westminster, wherever the royal party was in residence, he was soon stuck fast. By the early 1580s, having disastrously backed the latest marriage proposal to come from France over the Earl of Leicester's rival claims to Elizabeth's hand, Arundel withdrew from Court. Soon it was rumoured he was dabbling in Catholicism and in September 1584 Jesuit Father William Weston received him into the Roman Church. A year later Arundel attempted to flee the country, leaving behind him a letter to Elizabeth in which he accused her of countenancing 'mine adversaries in mine own sight of purpose to disgrace me'. He explained his departure by his refusal to be 'pointed at, as one whom your majesty did least favour'. As excuses go, it was hardly likely to endear him to the Queen. Even less so was his prominent conversion to the enemy faith and on his capture, mid-escape, Arundel was taken to the Tower of London. It seemed probable he would follow his father to the scaffold.[13]

While Arundel remained alive, though, his wife Anne could draw on the powerful Howard name for influence and from the plentiful Howard coffers – which even the fines exacted from the family for Arundel's Catholicism could not empty – for finance. Both were now at the Jesuits' disposal. From his new base at Arundel House on the River Thames, tucked precariously between the Earl of Leicester's London residence on one side and Somerset

* Norfolk – England's only duke and Elizabeth's cousin – had placed himself at the head of the anti-Cecil faction at Court. Though no committed Catholic himself, he allied himself with the northern rebels; he also planned to marry Mary, Queen of Scots. These facts, plus his connection with Roberto Ridolfi, a Florentine banker whose chief interest in life, apart from the many legitimate financial dealings which kept him in London, was in stirring up an invasion of England, were enough to guarantee his execution for treason.

House, belonging to the Queen, on the other, Robert Southwell set about rebuilding the fortunes of the Jesuit press.*[14]

His first publication was entitled *An Epistle of Comfort to the Reverend Priests, and to the Honourable, Worshipful, and other of the Lay sort, restrained in durance for the Catholic Faith.* Unlike the writings of Persons and Campion before him, this was largely a devotional work, containing arguments drawn from philosophy and theology and aimed at all those Catholics currently languishing in prison. Adversity was the 'livery and cognizance of Christ', he wrote, a 'royal garment' to be worn with pride by any disciple who aspired to be like his master. Mortal life was ever thus: 'Our infancy is but a dream, our youth but a madness, our manhood a combat, our age but a sickness, our life misery, our death horror'; it was the rewards of the world hereafter that would bring heart's ease. Put simply, Catholic suffering was not in vain. It was inspirational writing, hopeful, eloquent, charismatic, a powerful indicator of the uses to which the Jesuit thought to put his new press.[15]

At about the same time as Southwell was composing his *Epistle* he was also working on a poem, *Decease, Release,* commemorating the death of Mary, Queen of Scots, who, at Fotheringhay Castle on the morning of 7 February 1587, to a background noise of Parliamentary rejoicing and Elizabeth's anguish, placed her head on an executioner's block and so passed into legend. The last verse revealed a sentiment similar to that contained in the *Epistle,* but one drawn from a more personal agony:

Rue not my death, rejoice at my repose;
It was no death to me but to my woe;
The bud was opened to let out the rose,
The chain was loosed to let the captive go.[16]

* While Arundel House made for a highly dramatic location for Southwell's new base, it is also possible that he spent some time at another of the Countess's houses, near Spitalfields. Certainly, Spitalfields would have been a safer refuge, but sadly there is no evidence of precisely how he lived at this period.

To let the captive go: thoughts of death and of the escape that death brought with it had always haunted Southwell. While still in Rome he had written, 'Ah, dearest Jesus, help and rescue me . . . Canst thou not kill for pity one who longs to die?' A later poem contained the lines:

> In plaints I pass the length of lingering days;
> Free would my soul from mortal body fly
> And tread the track of death's desired ways.

On the eve of his departure to England he had headed his final letter 'from death's ante-room'. It was a fearful, almost hysterical letter. 'I know very well', he wrote, 'that sea and land are gaping wide for me; and lions, as well as wolves, go prowling in search of whom they may devour.' If it ever struck him as a bitter irony that these should be the words of a man returning willingly to the country he loved, he did not say.[17]

Even more so than Edmund Campion before him, Robert Southwell could have thrived in Elizabeth's England. Campion's had been a native wit and an ability to inspire, and the Queen and her circle had been more than ready to extend the ladder of social preferment to him. Southwell, no less keenly intelligent, was an insider from the start. Indeed, his was a fertile inheritance: the earth, blood and bone of old England – land, rank and wealth – coupled with the canny knack of navigating the new: Robert's grandfather Sir Richard Southwell had prospered through four different reigns, no mean achievement, and on his death in 1564 had left his family among the richest in England. Only enmity with the Howards, Robert's future benefactors, threatened the Southwells' continued success: Sir Richard was one of those to accuse Henry Howard, Earl of Surrey – Arundel's uncle – of treason in 1546. But if the Southwells were traditional gentry, with the added bite of a well-honed survival instinct, then Robert's mother's family, the Copleys, was peopled with coming men. It was through the Copleys that Robert Southwell was related to the Cecils and the Bacons, both families now well on the way to becoming dynastic power brokers. Through the Copleys, too, Southwell could expect personal recognition from the Queen. Bridget and Thomas Copley, Robert's mother and uncle, had stood by Elizabeth throughout her traumatic teenage years. Thomas had even gone so far as to convert to Protestantism in a show of loyalty to the princess. Now Elizabeth was godmother to Thomas' eldest son, no matter that Thomas had rejoined the Catholic Church and chosen voluntary exile over remaining in England. But perhaps of even more importance to the young Robert, through the Copleys the novice poet was related by marriage to the Wriothesleys. Henry Wriothesley, third Earl of Southampton, was William Shakespeare's patron.[18]

Under different circumstances Robert Southwell could and, likely, would have joined the swelling ranks of gentleman-writers peopling late sixteenth century England, another Sidney, another Ralegh, well heeled, well versed in the popular new style of chivalrous courtly love. It might still have been a cut-throat existence: both Sidney's and Ralegh's lives bore testament to the fact that even the best wits could flounder in a shifting world of capricious royal favouritism and jealous rivalry. It would not, though, have been a traitor's existence, stealing into England in disguise, with a price upon his head and the anticipation of an early execution.

So much of Southwell's life reads like an act of expiation. In 1588, as he described to Aquaviva the horrors taking place in England post-Armada, he confessed uncertainty as to 'whether it was better to confine to home my lament over our domestic calamity, or to impart to other nations the inward sorrow we here alone endure'. He feared 'lest the recital of [the Government's] impious conduct should bring more hatred on the English name'. For Southwell, the betrayal of English Catholicism had been a betrayal of everything English. It was a betrayal, avowed patriot that he was, that he felt deeply ashamed of broadcasting to the world and it was as though he now dedicated his life to atoning for it. Perhaps here were the roots of his bitter bouts of self-doubt, his self-disgust, his constant, almost clinical preoccupation with death: to take on the sins of England was a load far greater than he knew himself to be capable of bearing.[19]

But, if Southwell had wanted to take upon himself the sins of England, then he was already, from his earliest youth and on a familial level, familiar with the art. From the Southwells' manor house of Horsham St Faith near Norwich the family looked out over the fruits of old Sir Richard Southwell's religious pragmatism: the remnants of the Benedictine priory of St Faith, acquired during the dissolution. Nearby, a new house was rising from its rubble, promising an even statelier future. Yet for all those raised within sight of the many ruins such as these that scarred the English countryside, between their broken stones could also be seen the stumbling shapes of those monks, now reduced to vagabondage, who once had inhabited their walls. And about these spectral figures clung ghostly rumours, whispered among the country children and the superstitious, of a Monks' Curse:

> Of long ago hath been the common voice:
> In evil-gotten goods, the third shall not rejoice.

Robert Southwell, the third generation benefactor of the monks' eviction, the youngest son of a family of eight children, the acutely sensitive would-be poet, seems to have become the self-appointed sacrificial victim of his family's worldly success. And if for Southwell the mission meant martyrdom, then it was an end he anticipated with an eager terror.[20]

It was always unlikely Southwell could slip into England unnoticed. Sure enough, his first letter to Claudio Aquaviva, reporting his safe landing in July 1586, revealed that 'from the lips of the Queen's Council, my name has become known to certain persons'. Almost immediately the hunt was at his heels. That November the chief magistrate for Middlesex, Justice Richard Young, acting on information he had received from priest-turned-informer Anthony Tyrrell, led a posse of men against the Vaux house in Hackney.* The raid was set for the early hours when it was known Southwell would be at mass. The doorkeeper was overwhelmed and armed pursuivants dispersed about the house. Southwell had just time to conceal himself. Later he told Aquaviva, 'I heard them threatening and breaking woodwork and sounding the walls to find hiding places; yet, by God's goodness, after four hours' search they found me not, though separated from them only by a thin partition rather than a wall.' Young eventually withdrew, placing the house under surveillance and taking Lord Vaux's son Henry with him. Henry, brother of Eleanor and Anne Vaux and head of George Gilbert's relief operation since the latter's flight abroad, was caught with two of Southwell's letters addressed to Rome on him; he was led off to the Marshalsea prison. It was not until early December that Southwell was finally able to slip away from Hackney undetected to the country. There was to be no such escape for Henry Vaux. In prison he fell seriously ill. He was released in May 1587 and died that November, attended at the end by Henry Garnet.[21]

Vaux's death threw into painfully sharp relief Southwell's struggle. It was the Jesuit's duty to remain alive, sacrificing, if necessary, his lay helpers in the process, with all the attendant feelings of guilt and self-disgust to which he had always been prey. But the greater his sense of self-disgust, the greater this duty chafed at him, nagging at the raw nerve of his own craving for release.

Throughout this period Southwell remained pivotal to the mission's smooth running. He shared with Garnet the difficult task taken on by the Jesuits of coordinating the placement of newly arrived priests. He thrived at the even more difficult task of steadying the nerves of England's Catholics as the threat of the Spanish Armada drew nearer, then passed, leaving the spreading stain of

* Tyrrell, still incarcerated in the Clink prison, had had a hard time picking up any clues to Southwell's whereabouts. One misguided attempt saw him bringing evidence against Southwell's cousin, also Robert, recently knighted by Elizabeth. But on Friday, 4 November, he struck lucky, giving 'information of one Mr S——, a priest that for certain did lie at the Lord Vaux his house, by which means Justice Young went himself thither in the morning and made a search'.

Catholic blood in its wake. John Gerard would later say of him he 'excelled at this work. He was so wise and good, gentle and loveable'. Indeed, the two men quickly became close friends and Gerard offers a thumbnail sketch of Southwell struggling to learn the rudiments of falconry to serve him as a disguise. 'Frequently, as he was travelling about with me later,' wrote Gerard, 'he would ask me to tell him the correct [hunting] terms and worried because he could not remember and use them when need arose.'[22]

Throughout this period, too, Southwell continued composing poems. His style is lyrical, heavily influenced by Petrarch and by the English poets of earlier decades, Wyatt, Howard and Gascoigne. Still, it reveals a familiarity with the work of his contemporaries, in particular that of Sir Edward Dyer, the current Court favourite; twice, Southwell borrowed directly from Dyer, on one occasion adapting the latter's *Phancy*, a lover's lament, into his own *A Sinner's Complaint*. Elsewhere, he exerted a powerful influence of his own. Poet and dramatist Ben Jonson would later attest that had he 'written that piece of [Southwell's] *The Burning Babe*, he would have been content to destroy many of his [own poems]':

As I in hoary winters night stood shivering in the snow,
Surprised I was with sudden heat which made my heart to glow;
And lifting up a fearful eye to view what fire was near,
A pretty Babe all burning bright did in the air appear.[23]

The first line contained a chill echo of Southwell's recent exploits, as he described them to Aquaviva in a letter of December 1588. 'I have been', he wrote, 'on horseback round a great part of England in the bitterest time of the year, choosing bad roads and a foul sky for my pilgrimage, rather than waiting for the fair weather when all the Queen's messengers are on the prowl, much worse than any rainstorm or hurricane.' Then in 1591 came the Government's broadside against the mission and Southwell turned again from poet to polemicist. Some time in November of that year he sat down to write *An Humble Supplication to Her Majesty in Answer to the Late Proclamation.*[24]

'Most mighty and most merciful, most feared and best beloved Princess': Southwell framed his response around the assumption that Elizabeth, herself, had been kept ignorant of the iniquities being practised upon England's Catholics and that the proclamation was the work of her Government, in particular of William Cecil. That he truly believed this to be the case is unlikely – by now it was common knowledge among the missionaries that priest-hunter Richard Topcliffe was Elizabeth's man. Nonetheless, it was a literary conceit that served his purpose well, allowing him to make direct appeal to the Queen's oft-stated clemency. Elizabeth, he wrote, was 'the only sheet-anchor of our last hopes'. Then he proceeded, item by item, to refute the charges made in the proclamation. His arguments were logical, rational, a step away from the highly charged, often inflammatory outpourings that had characterized the war of words between Protestants and Catholics so far.* Bluntly, he addressed the unpalatable truth of Anthony Babington's treachery. Though he remained convinced the plot was as much a product of Francis Walsingham's invention as of Babington's own, he was quick to condemn the latter for his crime. And yet, he asked, was it right to judge all Catholics by Babington's actions? It 'were a hard course to reprove . . . all Protestants for one Wyatt', he pointed out, a fact of which Elizabeth, who, in her youth, had narrowly escaped execution for alleged complicity in Wyatt's rebellion against Queen Mary, was all too well aware. What reason, he asked, had English Catholics to kill the Queen when 'the death of your Majesty would be an Alarm to infinite uproars and likelier to breed all men a general calamity, than Catholics any Cause of

* The dangers of intemperate publishing were well illustrated by the case of *Leicester's Commonwealth*, a vitriolic book, printed in Paris, that accused the Earl of Leicester of every vice under the sun. Its likely authors were Charles Arundel (a member of the Howard family) and Lord Paget: both Catholics, but, more importantly, both rabid anti-Dudleyites. It is certain, though, that Robert Persons was aware of what they were up to: when Ralph Emerson, the Jesuits' assistant, travelled to England in 1584, he carried several copies of the book with him. On his arrest these copies were seized. Elizabeth was furious. She banned the book, saying 'none but the devil himself' could believe its lies. Soon the Jesuits and, in particular, Persons were being blamed for the work. It did little to further the English Catholic cause.

Comfort'? If the missionaries were rebels and assassins, then surely 'we should [have been] trained in Martial exercises, busied in politique and Civil affairs, hardened to the field, and made to the weapon; whereas a thousand eyes and ears are daily witnesses that our studies are nothing else but Philosophy and Divinity'? Here was a coolly reasoned attempt to engage with the charges of traitor and fifth columnist that had become so much a part of the mission's, and of English Catholic, daily life.[25]

There was still plenty for his opponents to find fault with. In response to the claim that Catholics would side with an invading enemy against their own countrymen, Southwell could only offer the non-committal assurance 'that what Army soever should come against you, we will rather yield our breasts to be broached by our Country's swords, than use our swords to th'effusion of our Country's blood'. This was scant comfort to a Government concerned with mobilizing the maximum number of troops to set against an attacking force. Likewise, dove-like among hawks, Southwell was quick to criticize what he saw as his Government's aggressive foreign policy.* Philip II, he argued, had been provoked into sending his Armada by a catalogue of English goadings: 'our surprising of the King's towns in Flanders, our invading his Countries in Spain and Portugal, our assisting his enemies against his daughter's right in Brittany, our continual and daily intercepting his treasure, warring with his Fleets, and annoying his Indies'. Assuredly, the English Government had been so provocative. As far back as 1568 it had impounded upwards of £85,000 from Spanish ships taking lawful refuge in English harbours, prompting a trade war between the two nations. Since then it had lost few opportunities to offer aid to anyone attempting to challenge Spanish hegemony in the Low Countries. But the origins of precisely who had thrown the first stone in this particular conflict were now long forgotten, if ever they were known. Besides, with a

* If the *Supplication*'s call for peace risked being unpopular, it was also an early acknowledgement by Southwell that the ending of the war with Spain was England's Catholics' best, perhaps only, hope of survival.

new invasion threatening, few Englishmen wished to be reminded that their side might have brought it upon themselves.[26]

But what made the *Supplication* so contentious was what also made it so unique. It conceded nothing by way of religious certainties, but it did contain the tacit recognition that if the English Catholic Church were to survive, then it could only do so in conjunction with the present English State, not in opposition to it. It was a plea for tolerance; relax your laws against us, Southwell promised the Queen, and you will see how loyal your Catholic subjects are. More radically still, it appeared to accept the right of every Christian of whatever sect to freedom of worship without fear of persecution; England's Catholics were now 'too well acquainted with the smart of our own punishments to wish any Christian to be partakers of our pains'. In light of the extreme penalties still being meted out to heretics in Catholic Europe, this was a startling concession by the Jesuit.* Essentially, the *Supplication* offered the possibility, faint still and barely formed, of a third way, a *via media*, in which English Catholics accepted the inevitability of their minority status in return for an end to their sufferings. Later, this possibility would divide the mission as surely

* Religious tolerance was rejected as an ideal at this period. The 1598 Edict of Nantes, which allowed French Huguenots to worship in public, was more an armed truce than an essay in freedom, recognition by the French King that France's religious wars must cease for the good of the State. It was condemned by the Vatican and by the Bishop of Geneva, who wrote to the Pope, 'At bottom, it leaves everybody free to think wrongly and act accordingly.' That same year the Jesuit Prefect of Studies at the English College in Rome, Father Henry Tichbourne, declared that England's Catholics must never accept religious toleration, even if offered to them by the Government. Toleration, he wrote, 'was so dangerous that what rigour of laws could not compass in so many years, this liberty and lenity will effectuate in twenty days'. William Allen regarded toleration as desirable only as the 'next best' thing to a full return to the Catholic Church. Robert Persons was a staunch opponent of toleration, reportedly commenting in 1597: 'It has been seen that in England in the first 12 years when the Queen did not persecute Catholics there remained practically none, and with persecution the faith has come to be enkindled. In Germany it has been seen that with liberty of conscience heresy has increased.' Only towards the end of his life did he shift his position, writing that Catholic princes might tolerate heretics 'when they are so multiplied, as they cannot be restrained without greater scandal [and] tumult'. For those prepared to entertain any idea of toleration, and they were in the minority, liberty of conscience was a necessary evil rather than a demonstrable good, entirely at odds with the charitable Christian desire to save the world from the sin of heresy.

as schism did England from Rome. For Robert Southwell now, though, unaware of the events he had just set in motion, there was to be little satisfaction in his achievement. It seems he completed the *Supplication* on the last day of December 1591 – this is the date appended to one of the few surviving copies. By January 1592 he was on the run again. And at his heels this time was Richard Topcliffe.[27]

Evidence of the vicious searches that took place that New Year as a follow up to the Government's proclamation comes in a short and atypically despairing sentence from Henry Garnet to Aquaviva. 'It is not worth the risk of sending [new missionaries]', he wrote wearily; not 'unless they are anxious to rush headlong into peril.' And, in his bleakest appraisal of the mission so far, he cried out, 'If God does not intervene, all things will reach the verge of ruin; there is no place that is safe.' Some time in January Topcliffe's men raided the house of George Cotton on Fleet Street, near St Bride's Church. Cotton was a cousin of Southwell's and for a time the Jesuit had taken his name as an alias. To Topcliffe's disgust, though, the house was empty: on Garnet's orders Southwell had already left London for Sussex. The chase continued.[28]

In Sussex Topcliffe picked up Southwell's trail at Horsham, staying with his Copley cousins. 'Young Anthony Copley, . . . and some others, be most familiar with Southwell,' Topcliffe told the Queen. Southwell moved on. January gave way to February, gave way to March. The countryside shook off its winter torpor and promised plenty: the last harvest had been good, this one would be better still. Southwell remained in Sussex. There are few details of his movements, but it seemed he used this time to put together his poems in publishable order. And as he circled Sussex, his *Supplication* began circling London. It is not known how or when a copy fell into Elizabeth's hands, but a memorial submitted to the Pope some time later revealed she had seen Southwell's plea; sadly it gives no indication of her response to it. Sir Francis Bacon wrote to his brother Anthony, in grudging praise of their disgraced relative, 'I send to you the *Supplication* which Mr Topcliffe lent me. It is curiously written, and worth the writing out; though the argument be bad.'*[29]

In April or May Southwell returned to London. So far his text had not been published – those versions passed around town up to now had been handwritten transcripts only. If he was to carry out the *Supplication*'s promise to Elizabeth, 'to divulge our petitions, and by many mouths to open unto your Highness our humble suits', then a decision had to be made whether or not to go to press. As always Henry Garnet, the Jesuit Superior, had the casting vote. Nothing is known of the conversation that took place between the two men, but it is characteristic of the mission's veiled and often abbreviated correspondence that the immediate fate of Southwell's *Supplication* should only be revealed in a single phrase written ten years after the event. In May 1602 Garnet would tell Robert Persons, 'Father Southwell wrote a very good answer to the Proclamation, but it could never be set forth.' Perhaps it was damage limitation, perhaps an attempt to cool off the search for his friend, but Henry Garnet forbade publication. The presses remained silent, Garnet returned to Warwickshire and Southwell resumed his London apostolate. And the most passionate defence of the mission's aims yet written was relegated to limbo, the only evidence of its existence a few copies passed between friends.[30]

On Saturday, 24 June 1592 Southwell recognized a young man named Thomas Bellamy in London's Fleet Street. According to his subsequent testimony Thomas had just come from a meeting with his sister Anne, at which Anne had been keen Southwell should visit the Bellamys at their house at Uxendon, near Harrow. Now Southwell – who, it seemed, had already been contacted by Anne privately – offered to ride up to Uxendon with Thomas. Thomas agreed. The following morning Thomas met Southwell at a prearranged rendezvous in Fleet Street and the two men turned their

* 'Curious' was popular slang for anything finely crafted. Shakespeare uses the word in *Venus and Adonis*: 'To cross the curious workmanship of Nature' [line 734].

horses towards Tyburn and the spur road leading out of London to the northwest. At about the same time a third man named Nicholas Jones left the city in some haste in the opposite direction, heading southeast to Greenwich. Here he would consult with Richard Topcliffe and soon the priest-hunter, too, would be on the move. Weeks of careful planning were now just twelve hours from fruition.[31]

Harrow-on-the-Hill, Sudbury Hill, Perivale, Greenford: their names tell a story of heights and valleys, of streams and former verdancy. It was here that the Bellamy family had its estates: the manor houses of Uxendon and, a little to the north, of Preston. In 1581 William Bellamy entertained Edmund Campion at Uxendon. Richard Bristow, academic and one of the original members of Douai's English College, took shelter here on his return to England, suffering from consumption. Bristow, with fellow ex-Oxford don Gregory Martin, was responsible for the Douai Bible, a translation into English of the New Testament, rivalling Elizabeth's authorized Protestant version (and an admission by the Catholic Church that the vernacular was superseding Latin in the battle for English souls).* Bristow would die in 1581 and Topcliffe would later accuse William's son Richard Bellamy of burying the priest in Harrow churchyard under the pseudonym Richard Spring. The family was well known to Robert Persons, who directed William Weston there on the latter's arrival in England in 1584, certain of his safe reception. Throughout the fledgling years of the Catholic mission the Bellamys of Uxendon would play a critically supportive role. Then, in 1586, the family gave food and clothing to those members of Anthony Babington's confederacy fleeing London and there the men were arrested. Babington, himself, was run to earth in nearby St John's Wood. Previously the family had been penalized for recusancy. Now it was to face far worse.

* In 1583 a Catholic commentator wrote of the Douai Bible: 'Every corner of [England] was searched for those books – the ports were laid for them, Paul's Cross is witness of burning many of them, the Prince's proclamation was procured against them; in the universities by sovereign authority colleges, chambers, studies, closets, coffers and desks were ransacked for them.'

Jerome Bellamy, William's son, described as 'a very clownish, blunt, wilful, and obstinate papist', was executed alongside the other conspirators. William's widow, Katherine, died a prisoner in the Tower of London. Another son, Bartholomew, is said to have died on the rack. Of all the victims of Babington's criminal lunacy, the Bellamys would suffer disproportionate agonies far above their deserts. They were agonies from which the family would never quite recover.[32]

In the post-Proclamation raids of January 1592 twenty-nine-year-old Anne Bellamy, eldest daughter of Richard Bellamy, was arrested and taken to the Gatehouse prison adjoining Westminster Abbey. In a matter of weeks she was pregnant. Her family, when it learned of her condition, was in no doubt who was responsible. In an appeal to the Privy Council Thomas Bellamy accused Richard Topcliffe of raping her.[33]

The events of that year come to us over a distance of four hundred years in piecemeal fashion – a statement here, a letter there – and the picture is far from complete. The reported facts are these. Within six weeks of her arrival at the Gatehouse, some time early in March, Anne Bellamy had become pregnant. Soon afterwards she was released on bail, with Topcliffe's instruction not to leave London. Anne took lodgings in Holborn, unwilling to confide in her family. At some unspecified time and in urgent need of money she agreed to Topcliffe's demands. His price was steep. In return for taking care of her he asked her to deliver Robert Southwell up to him.[34]

Initially, Anne appears to have contacted her parents and asked them that should Southwell ever visit Uxendon she should be told of it, as she was eager to meet the priest. Later, as she grew more desperate, she began searching for him herself. Her brother's chance encounter with Southwell merely hastened on an event already stamped with an awful inevitability. And as Nicholas Jones, the Gatehouse under-keeper, clattered into the courtyard of Greenwich Palace that Sunday afternoon, 25 June 1592, it was to deliver a message that had been anticipated now for some three weeks. Certainly that was the length of time Topcliffe's horses and men had been standing by.[35]

The raid of Uxendon Manor was carried out with a military efficiency. Anne had provided the pursuivants with a detailed map of the local area and in the still summer darkness the men had no difficulty in finding the house, or in quietly surrounding it. By midnight that evening the attack was set. When the hammering on the door came, accompanied by the familiar shouting and the familiar racing shadows, it dragged the household from sleep into a waking nightmare. This time there was no need to search for Southwell, no need to splinter the panelling or sound the walls. In his hand Richard Topcliffe held a piece of paper giving the exact location of the hiding place in which the priest lay concealed. Mrs Bellamy was permitted to make contact with her guest – Henry Garnet, who described the arrest to Aquaviva, was unclear how she did this – and Southwell was offered the choice of giving himself up, or being dragged out forcibly. Time slowed. Then, as the household and assembled search party watched, a slim man with auburn hair – such was an informer's description of him – made his way quietly downstairs and into the Great Hall. Priest and priest-hunter faced each other. Topcliffe demanded of Southwell who he was. Southwell answered, 'A gentleman.' Topcliffe replied, 'No, a priest, a traitor, a Jesuit.' Southwell asked Topcliffe to prove this assertion. Topcliffe drew his sword and ran at Southwell.[36]

It took a concerted effort to hold the priest-hunter back. All the while, wrote Garnet, Southwell watched impassively, seemingly unafraid of what was happening to him. Once Topcliffe had been subdued, the pursuivants turned their attention to ransacking the house for illegal books and massing equipment. This done, Southwell was bound, loaded into a cart and the cavalcade set off slowly back to London.[37]

Westminster was stirring as the party made its way towards the Abbey and to Topcliffe's house nearby. But although 'they passed through the least frequented streets,' wrote Garnet to Aquaviva, 'the report of [Southwell's] capture had spread already through the whole city'. Southwell was conducted to Topcliffe's private torture chamber and the door closed shut behind him.[38]

For Topcliffe, the next few days appear to have required careful stage-managing. The first letter he wrote was to Queen Elizabeth, informing her of Southwell's detention 'in my strong chamber in Westminster' and requesting permission to proceed with the Jesuit as he thought fit. The second letter was to Mrs Richard Bellamy. If there is any proof that the arrest of Southwell came about as a result of a secret deal between Topcliffe and Anne Bellamy then it is contained here, in this letter dated 30 June 1592. 'Mistress Bellamy,' wrote Topcliffe, 'It may be that I did leave you in fear the other night for the cause that fell out in your house.' Mrs Bellamy had due cause to be fearful: sheltering Southwell was a capital offence. So Topcliffe's reassurances to her are something of a surprise. 'And therefore', he wrote, 'take no care of yourself and for your husband, so as he come to me to say somewhat to him for his good. Your children are like to receive more favour, so as from henceforth they continue dutiful in heart and show. And although your daughter Anne have again fallen in some folly, there is no time past but she may win favour.' That Anne had just been returned to the Gatehouse for having invited Southwell to Uxendon was, it seemed, a carefully constructed smokescreen concealing the true nature of the transaction between her and Topcliffe. And that the Bellamys, themselves, had not already been arrested and incarcerated was, it must be assumed, the result of hard bargaining on their daughter's part.[39]

As for Robert Southwell, Topcliffe summed up the plans for his immediate future in a pungent phrase to Queen Elizabeth. The priest was to be made 'to stand against the wall, his feet standing upon the ground, and his hands but as high as he can reach against the wall, (like a trick at Trenchmore [a popular dance of the period])'. It sounded an innocuous fate; only the priest-hunter's contemptuous comments about the methods used in 'common prisons' to make victims talk – they 'hurteth not' – gave any indication of what was in store for Southwell. Topcliffe referred to a relatively new form of torture, the manacles: iron fetters fixed to the wall from which the victim could be suspended by the wrists for hours at a time. Their invention had been credited to Topcliffe himself. If so, it would explain his confidence in their application: the manacles, he assured Elizabeth, would 'enforce [Southwell] to tell all'.[40]

By the end of the first day of torture, though, this confidence had been shaken. Although Southwell had been strung up for several hours, yet still he had refused to answer any question put to him. Topcliffe sent urgently to the Queen, complaining about the Jesuit's obstinacy. 'The Queen', wrote Garnet, receiving his information from an unnamed source at Court, 'called Topcliffe a fool, and said she would put the matter in the hands of her Council who would soon finish it'. The following day two Clerks of the Council arrived to help Topcliffe with his interrogation. 'Yet still, they say, "the prisoner remains obstinate",' noted Garnet.[41]

By the end of the second day, it was clear that Topcliffe, even with assistance, was no nearer getting Southwell to confess to anything. The priest was removed next door to the Gatehouse prison, where the torture stepped up a gear. Now, he was left hanging for even longer periods, his legs bent back and his heels strapped up against his thighs. Members of the Council arrived to question him; Sir Robert Cecil came face to face with his cousin in a dank prison cell, its windows boarded up, the only light coming from a small pane of glass in the ceiling. But still Southwell refused to talk. Henry Garnet, desperate to know everything that was happening to his friend – from the conditions in which he was being kept to the comments made about him – reported one Councillor as saying, 'No wonder [Catholics] trust these Jesuits with their lives, when, from a man ten times tortured, not one word could be twisted that might lead others into danger.'[42]

Meanwhile, Southwell's father, Richard Southwell, was attempting to secure his son's release. Some time in July he petitioned Elizabeth directly. By now Robert was in a wretched state, lice-ridden, still dressed in the clothes in which he had been arrested, filthy, starved and tortured half to death. Such, anyway, were the reports being circled amongst his friends. His father's plea rang with indignant fury. 'That if his son had committed anything for which by the laws he had deserved death, he might suffer death. If not, as he was a gentleman, that her majesty might be pleased to order that he should be treated as such, even though he were a Jesuit. And that as his father, he might be permitted to send him what he needed to sustain life.' Elizabeth granted the petition.[43]

On 28 July the Privy Council wrote to the Lieutenant of the Tower of London with the following order: 'that her Majesty's pleasure is you shall receive into your custody and charge the person of Robert Southwell, a priest whom Mr Topcliffe shall deliver unto you, to be kept close prisoner so as no person be suffered to have access unto him'. Soon afterwards, with Topcliffe as his escort, Southwell was carried the few miles from the Gatehouse prison to the Tower of London. Here, he disappeared into the stronghold of the Lanthorn Tower, on the corner of the Queen's Privy Gardens overlooking Tower Wharf, and into a strictly enforced solitary confinement. For the next two and a half years he would remain there, hidden from view. His friends would struggle in vain to receive news of him. His enemies, it seemed, had virtually forgotten him. Denied all contact with the outside world, denied pen and ink to express himself, denied the glory of an inspirational martyrdom, he had simply passed into oblivion.[44]

But while Southwell's fate still hung in the balance, Richard Topcliffe, it seemed, was more immediately concerned with covering his own tracks. On 25 July, just days before Southwell's transfer to the Tower, Anne Bellamy was released from the Gatehouse prison. By now she was about four months pregnant. From Westminster she was taken to Greenwich, ostensibly to be examined by the Council, but in reality to be married off in secret. The man chosen to be her husband was Nicholas Jones.[45]

From Greenwich, Anne was taken to Somerby in Lincolnshire, Topcliffe's estate, where, towards Christmas, she gave birth. A month later, on 12 January 1593, Topcliffe wrote to the Bellamys denying all rumours of a hugger-mugger wedding. 'He very vehemently purgeth her from reports of slander, howsoever slanderous,' testified Thomas Bellamy later, adding that 'By a post-script [Topcliffe] writeth, that if any papist Catholic say she is with child, hold them knavish and false.' The priest-hunter signed his letter 'Your plain and known friend, Richard Topcliffe'.* In February Topcliffe again wrote to Anne's parents, defending her honour. The first news the Bellamys received of their daughter's condition came from Anne herself. In a letter dated 12 March 1593 Anne informed her mother of her hurried marriage to Jones, 'alleging many reasons thereof', wrote her brother, though none that satisfied her family's suspicions. She also confessed she had been prematurely delivered of a baby and she craved her mother's pardon.[46]

The Bellamys' grief did not end there. The following year Topcliffe contacted Richard Bellamy, requesting he make over his farm at Preston, worth one hundred marks a year, to Nicholas Jones. There was no need to put in writing the veiled threat behind this petition. And when Bellamy refused, retribution was swift to follow: Justice Young was detailed to ride out to Harrow and arrest the family. Richard Bellamy and his wife were taken to the Gatehouse, their two younger daughters were sent to the Clink prison and their two sons were sent to St Katherine's prison. Once again the family was suffering disproportionate agonies for someone else's actions.[47]

Richard Bellamy would remain in the Gatehouse for the next ten years, before departing England for the Low Countries. There, he would die in poverty. Mrs Bellamy's fate is unknown, though one report mentions she died in the Gatehouse. Audrey and Mary Bellamy, Anne's sisters, would remain loyal to the Catholic faith

* The extent of Topcliffe's duplicity is revealed in a letter of September 1592 to Lord Keeper Puckering. In it, Topcliffe suggested Puckering have Mrs Bellamy arrested. After a couple of days' imprisonment Topcliffe, himself, planned to play 'the part of a true man' and have her released; this act of charity would bring great, if unspecified, benefit to the State, he reckoned. However, he also warned Puckering not to mention the plan to anyone, 'Neither [to] Mr Young nor any other commissioner'.

and seem to have spent the rest of their lives in and out of prison; Thomas Bellamy and his brother Faith would eventually admit defeat and conform to the Anglican Church. Their Uxendon estate was sold off in 1603. As for Anne Bellamy, she was said to have continued at Somerby under Topcliffe's protection for the next three years, but from there the trail goes cold. Hers was a wretched outcome. In keeping with the attitudes of the period she was held responsible for her pregnancy and undeserving of sympathy. To her brother her troubles had come about as a direct result of 'her lewd behaviour'. It mattered little that he remained convinced to the end that Topcliffe had raped her. That rape, no less than her betrayal of Southwell, was accounted her own fault.[48]

On 18 February 1595, without warning, Robert Southwell was removed from the Tower of London and taken to Newgate prison, set in the city walls (on today's Newgate Street) and widely considered 'the most severe of the twelve London gaols'. There, he was led down into 'a subterranean cell of evil repute' called Limbo. The name was apt – particularly so, given Southwell's own limbo-like state for the last two and a half years – for the cell was regarded as a staging post to the scaffold, a waiting room for death. It seemed the Government had overcome its initial reluctance to kill him and was now about to proceed against him with an almost indecent haste. Speculation as to why this should have been was widespread among Southwell's friends, but the most likely explanation for it lies in the accusation levelled at him at his trial: Southwell would be charged with instructing Catholics in deceit, in teaching them how to lie.[49]

When Garnet had written to Claudio Aquaviva years earlier, describing the raid on Baddesley Clinton, he had included a telling sentence in his account. It came in reference to a young layman of the Vaux household who, on the pursuivants' arrival, had taken to his heels for nearby woods. When captured and interrogated later, he had sworn vehemently that he was no priest, an answer the pursuivants were happy to accept, 'believing at that time that none could deny that he was a priest without committing sin'. This simple statement reflected years of agonizing on the missionaries' part. The problem was this: as soon as a priest was run to ground the question would be put to him, 'Are you a priest?' On his next words hung his own life and that of the owner of the house in which he had been arrested. It might have been easy to make a swift and sure denial. It would also have been sacrilege.[50]

This was no new dilemma. The matter of the legitimacy of an untruth, if truth and justice were believed to be at odds, had perplexed philosophers and theologians for centuries. St Augustine had seemed to settle the matter definitively by ruling that any lie was intrinsically evil, regardless of circumstances, because it corrupted the essential function of language. But the Reformation had thrown this ruling into uncertainty. Suddenly, as Catholicism and Protestantism slugged it out for supremacy, the question of whether one could lie and deny one's faith had become a matter of some urgency. Martin Luther, ever the pragmatist, had ruled that an untruth, while always evil, was sometimes permissible. Catholic theologians were not so sure. For help, they looked to the law courts. In response, the theory of equivocation was born.[51]

According to this theory, if a man were unjustly questioned he might respond ambiguously, choosing words ambiguous in themselves or ambiguous in the context in which they were spoken. What constituted unjust questioning was not so subjective as it might appear (though it would have been unsurprising if the English Government had felt differently), for it was a privilege under common law that no man should be forced to incriminate himself; which action confessing to the priesthood (or to sheltering a priest) undoubtedly was. The use of equivocation was only permitted, though, in legitimate self-defence, not deliberately to mislead, and the user must have in mind a sense in which his words could be held to be true. By this reckoning, a Catholic householder, asked whether he was concealing a priest, might answer *No*, meaning, *No, you have no right to ask me that question.* If these qualifying safeguards seemed somewhat like hair-splitting, they reflected the deep conviction among most Catholics that lying was still irredeemably sinful, no matter the motive.* For the missionaries, though, the problem was much more than one simply of lying. Henry Garnet described his beliefs to Aquaviva thus: 'that Judas sinned by betraying Christ, but that Peter sinned also by denying Him'. To deny one's priesthood was an evil stretching back to the very first days of the faith.[52]

It is one of the many surprising points of the mission's history that, in between running for their lives, ministering to England's Catholics, organizing the building of hiding places and creating an underground nation wide network of priests, the missionaries, themselves, still found time to debate the morality of equivocation. A series of letters between Garnet and Aquaviva reveals just how uncomfortable most priests felt at the prospect of denying their calling.† As Garnet explained, 'their reason is that the canons of the apostles contain an instruction that anyone who denies his priesthood through fear is to be degraded; and they are uncertain whether this canon is merely human sanction or whether it asserts a divine principle'. For his part, Garnet was anxious that the use of 'a thousand ambiguities' risked diluting what were, to him, incontrovertible truths. Characteristically, Aquaviva ruled in favour of equivocation and Garnet responded obediently, 'now that we have your theologian's answer that this is lawful, many for the first

* Over the centuries the Catholic and Protestant view would divide even further, with Protestant writers promulgating the view that untruths were lawful when there was just cause, just cause being 'the preservation of life and property, defence of the law, the good of others'. Meanwhile Catholic writers continued preaching the virtues of ambiguity, still convinced that lying was evil.

† The case of Thomas Cottam illustrates the ethical constraints recognized by most priests. Arrested at Dover in June 1580, Cottam was entrusted into the safekeeping of a fellow traveller, to be taken to London and imprisoned. His escort, though, was a Catholic and, once outside Dover, he immediately gave Cottam his freedom. Unhappy at this breach of trust and concerned about what might happen to the man, Cottam eventually gave himself up to the authorities. He was arraigned alongside Edmund Campion and executed at Tyburn on 30 May 1582.

time are now acting on this opinion'. Southwell's promotion of this opinion would be his undoing.[53]

On 20 February 1595, just two days after Southwell's transfer from the Tower, a popular hanging took place at Tyburn. 'Almost all the city went out to see the execution,' wrote Garnet. At precisely the same time Southwell was taken by road the couple of miles from Newgate to Westminster Hall. In spite of this diversion word had leaked out and the courtroom was packed. Armed halberdiers guided Southwell's steps to the bar, where, with his hands freed, he 'put off his hat and made obeisance' to the men before him: Chief Justice Sir John Popham, Attorney General Sir Edward Coke and Richard Topcliffe. Then the Clerk of the Assizes read out the charges. 'Robert Southwell . . . You are indicted . . . for that you, since the first year of the Queen's Majesty's reign that now is, did pass without licence out of her highness' dominions beyond the seas, and there received order of priesthood from the pretended and usurped authority of the Bishop of Rome, and did return, and was found like a vile traitor at one Bellamy's house, nigh a place called Harrow Hill in Middlesex.' The court waited expectantly. Then Southwell replied, 'I confess I am a Catholic priest, and I thank God for it, but no traitor; neither can any law make it treason to be a priest.'[54]

The indictment was a formality: Southwell's ordination was not at issue, neither was his return to England. The question of whether this was carried out with treasonable intent could be – and was – debated long into the afternoon, but in the end it had little bearing on what was already a foregone conclusion. The real drama of the day was provided by the prosecution's surprise witness. The moment came after a long and angry denunciation of the Jesuit order by Sir Edward Coke. Turning to the jury Coke exclaimed, 'They pretend conscience; but you shall see how far they are from it' and into the courtroom walked Mrs Nicholas Jones.[55]

Anne Bellamy took the stand. Her testimony came with a devastating simplicity. Southwell, she said, 'had told her that if upon oath she were asked whether she had seen a priest or not, she might lawfully say no, though she had seen one, keeping this meaning in mind that she did not see any with intent to betray him'. It was damning evidence and the panel leapt at it. Did this mean the Jesuits advocated perjury? Were they teaching their flock to dissemble? Were all Catholics liars? There followed a debate in which Southwell attempted to explain the nature and value of equivocation and Richard Topcliffe shouted him down. 'Suppose,' argued Southwell, struggling to find some common ground between him and his accusers, 'that the French King should invade her Majesty, and that she (which God forefend) should by her enemies be enforced to fly to some private house for her safety, where none knew her being, but Mr. Attorney; and that Mr. Attorney's refusal to swear, being thereunto urged, should be a confession of her being in the house, . . . I say, Mr. Attorney were neither her Majesty's good subject nor friend.' It was an intelligent argument, but no one was prepared to rise to its challenge. Sir John Popham said that he should refuse to swear, but would debate no further, Sir Edward Coke that the cases were not sufficiently similar to merit a response. The jury was dismissed to consider the evidence.[56]

It was gone just fifteen minutes. When it returned, its verdict came as no surprise. Southwell responded, 'I pray God forgive all them that any way are accessory to my death'; his sentence was read, his hands were re-tied and he was ushered out of the hall. Briefly, it was discussed whether it were better to send him back to Newgate by river and avoid the crowds, but it was agreed Southwell 'would go quiet enough, and so he went joyfully with them through the streets where many of his friends and acquaintances awaited his coming'. Those who saw him testified they had never known 'him to look better or more cheerful'.[57]

The following morning, Friday, 21 February 1595, at first light, Southwell was led from his cell to the street outside Newgate prison. There he was bound to a hurdle, feet uppermost, his head level with the cobbles, and the execution party set off. Westwards towards Holborn it went, before skirting along the northern edge of Lincoln's Inn to St Giles in the Field. As a free man this had been Southwell's dominion, the extent of his ministry; now the streets were lined with friends and the curious, eager to see how the Jesuit would behave on this, his last journey. Midway, a kinswoman struggled forward to speak to him. Southwell thanked her, but begged her to be careful. For the rest of the time he prayed, 'holding up his hands and face as well as he could, towards heaven'.[58]

The horses' breath steamed in the raw morning air, their hooves quietened by the damp turf, while, behind them, the hurdle carved great gouges in the mud as the procession moved out into open countryside, along what today is Oxford Street, towards its final destination, Tyburn. Here, a large crowd had assembled and rising from among them was the gallows, three posts embedded in the ground, connected at the top by three crossbars from which a noose was suspended. Beneath it, a cart was standing ready and nearby was the hangman's table, the knives laid out upon it, and beside it the fire and the cauldron, offering fitful heat to those who had made it to the front of the throng. Southwell was untied from the hurdle and led to the end of the cart. Here, he wiped the mud from his face, scanning the crowd before throwing the handkerchief into its midst. The reports of his execution fail to say who caught the cloth, other than that it was 'one of the Society'. Perhaps Henry Garnet, who had made it a part of his office to witness the deaths of so many of his missionaries, was the face in the crowd Southwell searched for. If so, it was the last service he could offer his friend.[59]

Southwell was lifted into the cart. This was the moment he had waited for all his life. 'Take now your rest in the shade,' he had written nine years earlier, 'and open your mouths to draw in breath, so that when your hour comes, you too may go down into the sun-scorched arena.' Tyburn in February had become Southwell's sun-scorched arena now and he was ready. He commended into God's hands the Queen, his country and his soul. The hangman stripped him to his shirt and placed the noose about his neck. 'While we live we conquer, nor shall we be less victorious if we die,' he had written. Now he made the sign of the cross, murmuring, '*In manus tuas, Domine, commendo spiritum meum.*' And slowly the cart was drawn away.[60]

It took the hangman pulling on his legs before Southwell's body finally stopped moving, for the rope had been clumsily fixed about his neck. Three times one of the attending officers moved in to cut him down alive, but the crowd reacted angrily and from their number Lord Mountjoy stepped forward to stay the man. When the disembowelling was finished and Southwell's head was held aloft to the people there was silence; 'no one was heard to cry "Traitor, traitor!" as before times they were wont to do'. His head was set upon London Bridge and his four quarters upon four gates in the city walls. It remained to be seen what effect his death would have upon the mission, but for now Southwell had had his wish. The captive had been set free.

Chapter Eight

'. . . and since far greater is the fever of a woman once resolved to evil than the rage of a man, I humbly beseech your Lordship that that sex of women be not overlooked.'

attrib. Richard Topcliffe, 1592

WHILE ROBERT SOUTHWELL was writing his *Supplication* to Queen Elizabeth in the winter of 1591, John Gerard was continuing his East Anglian apostolate. 'Gerard doeth much good' was how Henry Walpole had described the newcomer's success and Gerard's own account of this period supports the view. He seemed happily to be fulfilling the promise of his first few hours ashore. Indeed, he was quickly developing a robust style of ministry all his own. It was a style that suited his disguise of sporting squire to the hilt of the silvered rapier he reportedly carried with him, at opposite ends of the spectrum to the agonized sufferings of so many of his colleagues. A story he recounted from the eve of the Baddesley Clinton conference illustrates this well. It had become Gerard's practice to journey north on occasion, visiting family and friends.* One such expedition saw him joining the hunting party of a Staffordshire cousin, eager for him to speak to and, if possible, convert the

* Before long he would write of his travels, 'I had so many friends on my route and so close to one another that I hardly ever had to put up at a tavern in a journey of a hundred and fifty miles.' On the assumption that he rode an average distance of about twenty to thirty miles a day (avoiding main roads), this would give some indication of Catholic density at the time.

husband of a relative of theirs. 'All day', wrote Gerard, 'I rode alongside him – the huntsman whom I was hunting down myself. Whenever the pack was at fault and stopped giving tongue, I used the pause to follow up my own little chase and gave tongue myself in real earnest.' It took four days of pursuit before the man finally caved in and became a Catholic. He also agreed from then on to maintain a priest himself. Slowly, cautiously, incrementally: this was how the Catholic faith would survive.[1]

It was not just Gerard's abilities in the field that were standing him in good stead. An unexpected facility at gambling was also proving ingenious cover for him, though frequently it seemed to cause some confusion as to his real identity. On one occasion a fellow card-player began sounding him out as a suitable match for his sister. On another, a woman wishing to be reconciled to the Catholic Church flatly refused to believe he was a priest at all, protesting, 'Why the man lives like a courtier. Haven't you watched him playing cards with my husband – and the way he plays, he must have been at the game for a long time.' Gerard, himself, took pains to justify his activities. 'I should explain', he wrote, 'that whenever I was with Catholics and we had to stage a [card] game [as a disguise], we had an understanding that everybody got his money back at the end and that the loser said an *Ave Maria* for every counter returned.'[2]

This, then, was the carefully constructed scene in play when George Abbot, Oxford's Doctor of Divinity, future Archbishop of Canterbury and 'a well-known persecutor of Catholics', called unexpectedly at a house where Gerard was staying for the night. According to Gerard, as the party sat around the gaming table, there ensued a heated discussion between the two men concerning the morality of suicide, which ended with Dr Abbot's stern admonition that 'Gentlemen should not dispute on theological questions.' Gerard cheerfully agreed, to his Catholic hostess's mirth, saying 'our profession is to play cards'. It was this willingness of Gerard's to revel in the frequent absurdities of his situation – and to rise to their challenge – that is the striking feature of his writing at this period. His seemed an uncomplicated faith, no less fervent than that of his contemporaries, no less compassed by the desire for martyrdom (frequently he bemoaned his 'unworthiness' to die for the cause), but augmented by a practical resolve that was soon reaping huge benefits for the mission.[3]

About seven months into his stay with the Yelvertons at Grimston, Gerard was approached by the Suffolk Catholic Henry Drury, heir to Lawshall Manor, six miles to the southeast of Bury St Edmunds. Like neighbouring Norfolk, Suffolk had been one of the first ports of call for the Protestant teachings of Martin Luther, an ideological cargo brought in by trading ship from the progressive Hanseatic cities. In contrast to the situation in Norfolk, these teachings seemed to have spread further than the coastal town houses of the merchant classes, to cover the whole county. Spanish assessment of English Catholic strength in preparation for the Armada listed Suffolk – a potential landing site – as being 'full of heretics'. Yet still there were those among the gentry (many of whom owned estates in both Norfolk and Suffolk) who resisted the changes, including Henry Drury.[4]

In April 1586, alongside John Bedingfeld and Edward Rookwood (the unfortunate victim of Elizabeth's 1578 East Anglian progress), Drury was listed as one of Suffolk's 'most obstinate' recusants. In 1587 he was imprisoned for his faith. Two years on, though, he was begging Gerard to move to Suffolk and renew his apostolate from there. Despite Drury being known to the authorities as a recusant and maintainer of priests, the advantages to the Jesuit of such a move were becoming increasingly obvious. As Gerard put it: 'the danger of recognition grew as I came to know more people [in Norfolk]'. So, with Henry Garnet's permission and having first secured a priest to take over his ministry at Grimston, Gerard crossed the county border to Lawshall Manor.[5]

Entering Lawshall was like stepping back in time. The altar furnishings were 'old and worn' (in Gerard's description) and spoke of decades of uninterrupted service in pursuance of the Catholic faith. Even the family chaplain – probably the seminary priest William Hanse – had been at Lawshall for many years. Here was a connection with a Catholic heritage that some half a century of religious changes had been unable to sever. 'In this new place', wrote Gerard, 'my life was much more quiet and congenial. Almost everyone in the house was a Catholic and it was easier to live the life of a Jesuit, even in the external details of dress.'[6] Although information about Gerard's time at Lawshall is sparse, other accounts from the period help flesh out a picture of the kind of pastoral life he was leading. One in particular, from the Yorkshire based priest James Pollard, bears quoting in some detail.

> In the house where I lived we were continually two priests, one to serve and order the house at home, the other to help those who were abroad, who especially in any sickness or fear of death would continually send to us for help, that they might die in the estate of God's Church . . . On the Sundays we locked up the doors, and all came to Mass . . . On the work days we had for the most part two Masses; the one for the servants at six of the clock in the morning, at which the gentlemen every one of them without fail, and the ladies if they were not sick, would even in the midst of winter of their own accord be present; and the other we had at eight of the clock for those who were absent from the first. In the afternoon at four o'clock we had evensong, and after that matins, at which all the knights and their ladies, except extraordinary occasions did hinder them, would be present and stay at their prayers all the time the priests were at evensong and matins.* The most of them used daily some meditation and mental prayer, and all at the least every fourteen days and great feast did confess and communicate; and after supper every night at nine of the clock we had all together litanies, and so immediately to bed.[7]

* Matins was traditionally a midnight office that might also be recited at daybreak. Within the Church of England the term now refers specifically to Morning Prayer.

Henry Garnet described it thus to Claudio Aquaviva: 'There is almost daily a large flow of people to these places just as if they were churches.' Amid fears of arrest, amid recusancy fines and the insidious implication that they were unnatural Englishmen, three-quarters of the way into Elizabeth's reign there were still English Catholics attempting to live a religious life. Dr William Allen's frustration at the level of compromising by Catholics in the 1560s, a frustration that had led to his founding the mission, had been replaced, some twenty, thirty years on, by feelings of relief and euphoria among his missionaries. As Robert Southwell exclaimed in the year after his return to England, 'the faith is still alive! The Church exults! The families are not falling away'.[8]

Political defiance or spiritual necessity? To most in the Government it must have seemed the former. Even as the Archbishop of Canterbury, John Whitgift, was curbing the spread of English Puritanism, successfully preserving the broad Church insisted upon by Elizabeth to keep the country together, English Catholics remained mulishly intransigent, an intransigence that appeared exponentially greater each time the Anglicans demonstrated their centrist reasonableness.* Worse, in the face of international Catholic aggression, their dissension was, at best, a fracturing of the united front required at such times, at worst, an act of outright treachery. To the Government, then, it was political defiance. But were the Government's concerns really the same as those of most ordinary English Catholics? Free from the crippling perception that their faith automatically rendered them enemy agents, free from the burden of rule, the latter were also free to look at the situation from a somewhat longer perspective. And in so doing, two things must have been immediately obvious. First, that there was still no clear guarantee that Elizabeth's Church of England

* In 1583 Whitgift published his Twenty-Four Articles, the fiercest challenge to Puritanism yet, reinforcing episcopal authority and declaring that the Book of Common Prayer contained nothing contrary to the word of God: both were the antithesis of Puritan belief. Although unpopular with Privy Councillors, many of whom had Puritan leanings, Whitgift acted with the complete support of Elizabeth.

would survive her: within living memory was proof that a change in ruler could bring about a change in religion. So, the Northern Rebellion had discovered altar-stones and holy-water stoups hidden away in quarries for just such an eventuality and the Protestant Reverend Alfield had advised his son Thomas to work for the Jesuits – he became Robert Persons' servant – to help him when that change in religion came. If the new Church was an uncertain entity, then the old faith was a far safer certainty to cling to. And this was where the second consideration came into play. For not only had Elizabeth's Church of England been a sectarian fudge, it had also been a theological one. Amongst all the doubts that still existed over what did and did not constitute the means to salvation, there was all too much scope, as the century drew to a close, for the unwitting layman to stray from the path of righteousness and plunge headlong into the fires of hell. And the only thing he could be sure of as he plummeted was that his fall would be accompanied by the sound of those Anglican divines appointed to sort out the confusion bickering the while. A quick death at the hands of an invading Spanish army or an Elizabethan hangman was one thing; eternal damnation because the country's new bishops could not agree between them what was sacred was a fate somewhat harder to stomach.[9]

Birth to death had always been a hazardous journey. Famine, fire and sword, plague and pestilence, sins mortal and venial: they lay in wait at every turn, to tempt the weak and carry off the unwary. Religious certainty provided a bulwark against these forces. Disease was rife, so a Christian burial dispatched the newly departed to a far better pain-free place. Frequent confession along the way – and extreme unction at the end – helped bring the sinner back on track for his final journey into the afterlife. And if these simple precautions were ignored, then the consequences were no less certain. According to the medieval Church, the child who died without baptism was destined for limbo, for the very borders of hell. So to prevent this, the newborn child was exorcized: 'Go out of him unclean spirit and give place to the Holy Ghost the comforter'. Then, with his head bound tight with a white cloth called a chrisom (his shroud should he die in infancy), he was anointed with consecrated oil and balsam. The child who cried at his christening was expelling the devil.[10]

But these certainties had been thrust aside at the Reformation as so much conjuring and necromancy, the exorcism of the new-born being dismissed with the breezy cry 'For, if ye may make at your pleasure such things to drive devils away . . . what need have ye of Christ?' Under Edward the baptism ceremony had been purged of its drama, its anointings and chrisoms, in the flurry of back-to-basics religious reform sweeping the country. But new broom though it was, the Protestant ascendancy had failed to provide a satisfactory answer as to the fate of those infants who remained unbaptized. The 1549 Prayer Book, Edward's first, continued to stress the vital importance of baptism within the first few days of life; Elizabeth's 1559 version recognized cases of urgency by permitting christenings to take place on days other than Sundays and holy days; and the current hot topic amongst the Queen's theologians was whether the baptism sacrament was absolutely necessary for salvation, or just formally necessary.* It was a fine distinction and made for good debate, but can have been of little comfort to bereaved parents concerned their unbaptized offspring were heading straight to hell.[11]

This was where the mission came into its own. For where there was controversy it brought assurances, where there were vagaries it brought absolutes, where there was strife it brought the guarantee of eternal peace. In a world of rapid change, accelerated ascendancies and breath-taking falls from favour, in a world of war, torment and terror, the mission offered a lifeline back to an earlier age – rendered golden by the present turmoils – in which certainty reigned supreme. God was in His heaven and you, too, could join Him there, if only you followed these simple guidelines.

* As late as 1569 the vicar of Ashford in Kent declared that those children who died before baptism were the 'firebrands of Hell'.

On the way you might risk arrest, imprisonment, poverty, exile and execution, but at least the life hereafter was assured.

Birth to death: as the new underground English Catholic Church readjusted to its changed and vastly straitened circumstances and the mission entrenched further, families and their priests found themselves taking extraordinary measures to ensure that this most momentous of journeys was hedged with the necessary ceremony. The dangers were great, as a letter from Henry Garnet to Aquaviva described:

> 'It is impossible, save at the greatest risk, to baptise infants, celebrate marriages, give the sacraments or offer the sacrifice according to the Catholic rite. Therefore expectant mothers travel to remote parts for the birth of their infants in order that they may not be asked questions later about the christening of their offspring. When they marry, they ride to some distant place for the ceremony and then return home, to avoid questioning on the celebration of their marriage.'[12]

These evasions were as much a legal necessity as a spiritual one. As Garnet explained, 'It is a crime punishable at law for a mother to give birth to a child and not to have it baptised, or for her to move about in public before she has been childed.'* Indeed, if a woman could not show that her child had been christened or that she, herself, had been purified after the birth, she would be branded a harlot and her offspring a bastard. There are many stories detailing the lengths to which Catholics went to avoid the law. 'A certain woman with child,' wrote Garnet to Aquaviva 'when her time of delivery drew near, travelled to another county where she might

* The churching of women after childbirth (referred to here by Garnet as childing) was a medieval practice continued after the Reformation by the Anglican Church. The ceremony was one of thanksgiving for a safe deliverance and represented society's recognition of the woman as a mother. However, the ritual was ringed with superstitious beliefs (frequently enforced by the more orthodox Anglican clergy), in particular that the unchurched woman was unclean and should not leave her house until she had been purified.

have her child . . . So by chance it happened that this woman, after a short labour, gave birth in an open field by the road, without any other woman present; and then she carried her infant son at the breast to the house of a neighbouring [Catholic] lady.'[13]

In death, more trickery abounded. A case brought before the Star Chamber revealed one commonplace method of cheating the churchwardens. 'The Archbishop [of Canterbury, Richard Bancroft] in his speech delivered that it is the secret practice of the Papists to wrap their dead bodies in two sheets, and in one of them they strew earth that they themselves have hallowed and so bury them they care not where, for they say they are thus buried in consecrated earth.'*[14]

Countrywide, English Catholics, with the connivance of their priests, were finding ways to outwit the law and keep the faith. It was a gamble, pitting the danger to their lives and livelihoods of arrest and prosecution against the danger to their souls of failing to observe Catholic teachings, and some seem to have played this gamble to its limits. Francis Swetnam, baker to the Vaux family, reluctantly agreed to attend his local church after some two years of avoidance (in which time his fines totalled about £520), 'for that he had rather adventure his own soul than lose his five children'. With recusancy fast becoming a rich man's sport, it was all too often poorer Catholics, like Swetnam, who ended up conforming to the Anglican Church to avoid bankruptcy, hoping that a last minute deathbed reconciliation would be enough to appease their God, as much as a lifetime's obedience had appeased the State.[15]

In Oxford, Nicholas Owen's father Walter had remarried and, despite the fact that all the sons from his previous marriage had joined the mission, was raising his second family in strict accordance with the law. On 2 October 1585, just a week after their

* Any priest who died on the mission had to be disposed of with great care, for it was dangerous to bring out a corpse for burial that could not legally be accounted for. When the London-based Jesuit John Curry died of an illness in 1596 'he was buried in a secret place,' wrote Gerard, 'for all priests who live in hiding on the mission are also buried in hiding'. Whether the words 'secret place' refer to a priest-hole is unclear, but it seems that some priests were buried in the houses in which they had died.

half-brother John Owen had been banished from England for the first time for his priesthood, the twins Robert and Elizabeth were baptized with Protestant rites at the Owens' parish church of St Peter-le-Bailey.* Whether Walter arranged for a Catholic minister to perform an alternative ceremony any time before or after is unknown, but the extent of his particular gamble is reinforced by a glance at the church's register of burials: Elizabeth Owen was dead within days of her christening. She was interred at St Peter's on 10 October. Over the years a succession of Owens would be baptized, married and buried at St Peter's, yet still Walter retained his Catholic faith: in July 1604 he was indicted for recusancy along with his second wife Agnes and his daughter Dorothy. Only now, it seemed, with his children of an age to be self-sufficient (his youngest daughter Katherine was born in the spring of 1590), did he feel it safe openly to declare his allegiance to the Catholic Church.[16]

Birth to death: perhaps the dangers inherent in the first had brought the second much closer to home for one half of the population, because in spite of the very masculine nature of the mission, its disguises, deceptions and daring midnight escapades, Catholic women would come to play a vital role in the religious resistance. The Vauxes, Anne and Eleanor, and Anne, Countess of Arundel were just three among countless wives, daughters, sisters and spinsters who defied the Government and kept the English Catholic Church alive. Their legacy far outweighed their actual position in society.[17]

Contemporary opinions of women kept little back. As a sex they were held to be 'light of credit, lusty of stomach, unpatient, full of

* John Owen's first taste of the mission was disastrous. He landed at Rye in October 1584, but was arrested in Winchester just four months later, on 28 February 1585. At first he gave his name as John Gardiner, until he was brought before a member of the cathedral staff who had known him at Oxford, at which point he confessed his true identity. He was banished around Michaelmas that year, but by the end of January 1586 he was back in England. Within two months he had been re-arrested, at Battle, in Sussex. This led to his eventual conviction and recantation at the Chichester assizes in 1588.

words, apt to lie, flatter and weep, all in extremes, without mean, either loving dearly, or hating deadly, desirous rather to rule than to be ruled, despising naturally that is offered to them'. Other commentators opined that women were 'full of tongue and much babbling'. These were views against which Elizabeth, herself, no less than her female subjects had had to struggle. In 1569, eleven years into her reign, as England lurched unsteadily towards an earlier conflict with Spain, Privy Councillor Sir Francis Knollys drafted a letter to the Queen begging her to hand decision-making over to the men of her Council and blaming the country's current woes on her 'mild [female] disposition'. Privately he complained to William Cecil (to whom he posted the letter, asking if it were wise to send it to Elizabeth) that 'if she will needs be the ruler or half ruler herself, my hope of success is clean gone'. Edmund Campion, meanwhile, called upon to debate extempore before Elizabeth at Woodstock in the wake of his 1566 Oxford triumph, was understood to have been overawed at the prospect, 'until', or so the Spanish ambassador revealed, 'after a few moments . . . he remembered that she was but a woman, and he a man, which is the better sex'. And for opponents to the break with Rome, hope had always burnt bright, right up to the death of the Duc d'Alençon in 1584, that Elizabeth (by then aged fifty-one) might wed one of the many Catholic suitors paraded before her and succumb to the civilizing influence of a husband.[18]

Over the years Elizabeth would learn to play her gender to every advantage. Her rallying call to the troops at Tilbury in August 1588, Armada year, was the high point in a lifetime of speechmaking in which the 'weak and feeble woman' who occupied the throne was shown to have a will of iron. Indeed, the king's 'heart and stomach' of which she boasted seemed all the more formidable for the obvious disadvantages Elizabeth laboured under as a female. The Catholic women of England would become no less adept at manipulating the male-dominated world around them.

The mission had efficiently transformed Catholic houses into the religion's new churches, complete with priests, chapels, vestments and vessels. However, in place of an administrative hierarchy to run them, these new churches now depended for their organization as much on the cooperation of the supporting household as on the coordinating efforts of the Jesuits – and in this domestic sphere women held sway. While the Elizabethan husband presented to the world the public face of his estate, it was the Elizabethan wife who was responsible for its below-stairs, behind-closed-doors private face. And since the missionaries stationed in a house perforce were occupiers of its most private recesses, by default it was with Catholic women that priests had the most contact.

This public/private division quickly became an important weapon in the English Catholic armoury of law evasion. The Montagues of Cowdray Park in Sussex were perhaps its best exemplars. Anthony Browne, Lord Montague, was one of the many Catholics to have benefited from the redistribution of monastic lands under Henry VIII. Under Mary his career flourished further, but when, in 1559, he became the only temporal peer to speak out against Elizabeth's Act of Supremacy, it looked certain his stellar ascension was about to come to an abrupt halt. Elizabeth, however, continued to favour him. In 1560 she sent him to Spain as her new ambassador, in 1565 to Flanders in the same capacity and from 1569 to 1585 Montague shared the role of Lord Lieutenant of Sussex with the Protestant Lord Buckhurst, only being removed from office on the outbreak of war with Spain. Even then, he continued to serve his country, acting as one of the commissioners at the trial of Mary Stuart and equipping an army of some two hundred horsemen to ride against the Spanish in 1588. This was all the more remarkable when it was discovered one of his brothers had sailed with the Spanish aboard the *San Mateo* (he was killed either during the gun battle in which the ship was holed, or when the English overran her).* Montague's career offered startling proof that it was still

* This information came from William Borlas of the English fleet, who reported to Sir Francis Walsingham that, on the *San Mateo*'s capture, 'I was the means that the best sort [including the ship's commander, Don Diego Pimentel] were saved, the rest were cast overboard and slain at the entry.'

possible to succeed in England as a Catholic as long as you retained the Queen's favour (always fickle) and were prepared to offer a token conformity to her religious settlement. In Montague's case, this was achieved by making occasional appearances at his local Anglican church. To do so was, of course, a grave sin – the ruling had come from the Pope to confirm it and any missionary could daily remind him of the fact – but Montague's tokenism bore rich fruit. His estates at Cowdray and at Battle Abbey were soon established among the mission's most important support centres, both enjoying comparative immunity from the random searches frequently suffered by other Catholic estates. And, it seemed, both were managed for the mission by Montague's wife.[19]

Soon after Montague's death in October 1592, his wife's chaplain, Richard Smith, wrote an account of life at Cowdray, as run by the formidable Magdalen, Lady Montague.*

> She built a chapel in her house . . . and there placed a very fair altar of stone, whereto she made an ascent with steps and enclosed it with rails, and, to have everything conformable, she built a choir for singers and set up a pulpit for the priests, which perhaps is not to be seen in all England besides. Here almost every week was a sermon made, and on solemn feasts the Sacrifice of the Mass was celebrated with singing and musical instruments, and sometimes also with deacon and sub-deacon. And such was the concourse and resort of Catholics, that sometimes there were 120 together, and 60 communicants at a time had the benefit of the Blessed Sacrament. And such was the number of Catholics resident in her house and the multitude and note of such as repaired thither, that even the heretics, to the eternal glory of the name of the Lady Magdalen, gave it the title of Little Rome.[20]

* With Lord Montague's death Lady Montague's seeming immunity from investigation suffered slightly. In the summer of 1593 Richard Topcliffe was ordered to conduct three separate searches of her houses, for eight 'divers and dangerous persons', including two named priests. It was emphasized, however, that the searches were to 'be done with regard to the quality of the lady'. The outcome is not recorded.

To Cowdray's Little Rome flocked numerous members of the mission, including Robert Southwell, who was probably there some time in the summer of 1590. Later, as Southwell lay in prison, Richard Topcliffe combed the district for evidence against the 'Father Robert' who had stayed with the Montagues 'the summer before the Queen's Majesty came to Cowdray'. It is likely John Owen visited Battle Abbey (he was arrested nearby in the spring of 1586); earlier he had travelled to Guildford 'in the company of one of the Lord Montague's men'. Montague's pragmatism coupled with his wife's zeal provided English Catholicism with as near to a stable base in the south of England as it was ever likely to achieve in those precarious times. Public conformity to the state and private observance of conscience, the male and female faces of Catholic resistance, appeared a rational solution to an impossible situation. Indeed, such behaviour was firmly in accordance with Elizabeth's own views on religion, as stated and restated throughout her reign: that so long as her subjects obeyed her laws she had no intention of making enquiry into their innermost beliefs. Many Catholic families would adopt this Janus-like attitude over the years.*[21]

Elsewhere, though, some Catholic women were being more

* From this was born the concept of the Church Papist, despised by everyone of unwavering religious conviction, but seized on by anyone with half an eye to surviving the immediate future (with any degree of comfort) as the only practical option available to them. A seventeenth-century writer defined a Church Papist as 'one that parts religion between his conscience and his purse, and comes to church not to serve God, but the King. The fear of the Law makes him wear the mark of the Gospel, which he useth, not as a means to save his soul, but his charges. He loves Popery well, but is loath to lose by it, and though he be something scared by the Bulls of Rome, yet he is struck with more terror at the apparitor [an officer of the ecclesiastical court]. Once a month, he presents himself at the church to keep off the churchwardens, and brings in his body to save his bail; kneels with the congregation, but prays by himself and asks God's forgiveness for coming thither. If he be forced to stay out a sermon, he puts his hat over his eyes and frowns out the hour; and when he comes home, he thinks to make amends for his fault by abusing the preacher. His main subtlety is to shift off the Communion, for which he is never unfurnished of a quarrel, and will be sure always to be out of charity at Easter. He would make a bad martyr, and a good traveller, for his conscience is so large he could never wander from it, and in Constantinople would be circumcised with a mental reservation. His wife is more zealous in her devotion, and therefore more costly, and he bates her in tyres what she stands him in religion.' The fine polish of this attack goes some way towards indicating just how common the public/private male/female face of Catholic observance had become by this period.

private still, not to say devious. Agnes, Lady Wenman of Thame Park, twelve miles to the east of Oxford, approached John Gerard (through a kinswoman of hers) with what appeared a common request. Her husband, she said, was a Protestant 'and though she was very anxious to do it, she could not keep a priest in her house'. Instead, Gerard arranged for her 'to support a priest who could visit her regularly during her husband's absences'. 'I found', wrote Gerard later, that 'she never omitted her hour's meditation or her daily examination of conscience, except on one occasion when her husband insisted on her staying with her guests. Yet she had a large household to keep busy, and she was seldom without people staying with her.' Lady Wenman's example, and that of women like her, helped fuel a shift in training practices on the Continent, where seminary priests were soon being taught that it was perfectly reasonable for wives to spend their conforming husbands' money on the mission, since they did so for their husbands' spiritual welfare.[22]

Grace, Lady Fortescue was cut from similar cloth. Her father-in-law was Sir John Fortescue, Chancellor of the Exchequer and cousin to the Queen. Her husband, Sir Francis, 'was a schismatic (that is, a Catholic by conviction),' wrote Gerard, 'but there was no hope of converting him. He was content with wanting to be a Catholic, and refused to go beyond for fear of offending his father'. Grace, herself, was a Protestant; however, she was understood by friends to have expressed a strong interest in Catholicism, so on several occasions Gerard journeyed out to speak with her. Again, his card-playing huntsman's demeanour proved impenetrable cover, for when, in his own words, 'I . . . brought the subject round to the state of her soul . . . she looked at me in astonishment – I was the last person in the world she expected to speak in this fashion.' Gerard quickly abandoned his disguise and explained his purpose. The pair set aside a period of time in which Gerard undertook to satisfy her on all points of the Catholic faith; then he received her into the Church. Next, he began persuading her to transform the family estate at Salden, in Buckinghamshire, into a Catholic centre. 'It would be a great pity, I pointed out, if there were no priest in a house like this . . . [for] I had never come across a house in the whole of England where a priest could live so conveniently in secret.' Brushing aside whatever objections her husband might have had to this plan, Lady Fortescue agreed to Gerard's request and the Jesuit introduced her to Father Anthony Hoskins, newly arrived from the Continent and in need of shelter.*[23]

If these stories told only of single instances of Catholic women defying their menfolk (and only of aristocratic women at that), then the Government's response to the problem of obstinate female recusants told of a countrywide phenomenon, spanning the classes. In November 1576 fifty-one Yorkshirewomen were brought before the Northern Court of High Commission to answer questions about why they still refused to come to church. In a number of instances their husbands gave evidence against them. Christopher Kinchingman attested that he had dragged his wife to church by force, George Hall that he had beaten his wife to make her obedient. Many husbands agreed to pay their wives' fines, but said it would be impossible to make them change their minds. And Lord Mayor Dinley of York was roundly condemned for the fact that, employed to govern a city, he was unable even to govern his household. His wife would prove one of the most obstinate recusants in the area.[24]

Meanwhile, in Norfolk, Anne Howlet was being released from prison on a bond of forty pounds, on the understanding that her husband was conformable to the Church and was prepared to persuade her to the like obedience. Whether Mr Howlet succeeded in his task is unclear, but a paper drawn up in 1586 by Sir Francis

* Hoskins was born in Herefordshire in 1568. He became a Jesuit in 1593 and served the English mission for six years, before being recalled to Brussels. He died in Spain in 1615. Sir John Fortescue built Salden House towards the end of the sixteenth century. Queen Elizabeth may have visited it in 1602; James I certainly did in 1603. Only a fragment of the house remains, converted into a farmhouse; sadly nothing is left of the original interior. Thanks to Francis Fortescue's refusal to convert, the family never suffered for its faith and local tradition reports the Fortescues living at Salden in great splendour with as many as sixty servants.

Knollys revealed the extent to which husbands were held responsible for their wives' actions: 'for no man can deny but that the law giveth to every man so much power over his wife, that he may constrain his wife to come to church, and there to remain quietly for the service time'. This being the case, advised Knollys, 'her majesty should show herself offended with such as do pretend to be good subjects, and yet do suffer their wives to be open recusants'. But it seemed many women were beyond constraint. In Hampshire, when Mrs Pitts of Alton (the 'most obstinate' sister of papal envoy Nicholas Sanders) was imprisoned, Bishop Cowper of Winchester wrote to the Council, begging she might be held there indefinitely. Her return to the diocese, he felt, 'would do more harm than ten sermons do good'. For emphasis he added the opinion that 'no man whose wife is a recusant is sound himself'.[25]

Then, in 1588, the Spanish Armada set sail for England and attitudes towards this monstrous regiment of recusant women began to harden. In February that year Francis Cromwell, Sheriff of Cambridge, wrote to the Council seeking advice on how to proceed against Catholic women, 'whom he durst not presume to apprehend without further direction'; so, too, did the Sheriff of Leicester and the Earl of Kent. Four years later a memorandum, attributed to Richard Topcliffe, appeared to sum up official feelings in regard to recusant women during wartime: that they were 'needful to be shut up . . . as much as men . . . [because] . . . although they cannot go to the field and lie in camp (for their sex and shame) yet they want no deceit nor malice'. And in March 1594, as the country braced itself for a new Spanish invasion, William Cecil drew up a list of defensive 'Considerations', one of which commanded 'all men's wives, being recusants, to come to church' on penalty of their husbands being interned. That this was, in itself, against the law is a clear indication of the level of Government concern about recusant women.[26]

The difficulty was that no matter how much the responsibility of enforcing a wife's behaviour lay at her husband's door, justice itself was toothless when it came to proceeding against married women recusants. Since no married woman could own property, she could not be indicted and fined for her recusancy. Neither could a husband be punished for his wife's criminal acts – fining or arresting him in lieu of his wife was illegal.* The only option left to the courts was imprisoning recusant wives as excommunicates (under the writ of *de excommunicato capiendo*) until they proved conformable, unpopular since it allowed too many women to assume the role of martyr to the faith, particularly if they should die in prison.† It was far easier to proceed against widows and spinsters: they were entitled to own property and could therefore be indicted and fined for their recusancy with impunity. With fining still the Government's weapon of choice against Catholics, it is significant that the record books of the period are full of suggestions of how to make husbands financially liable for their wives' misdemeanours. A 1586 list of Bedfordshire recusant women even went so far as to provide precise details of their non-recusant husbands' annual incomes.[27]

By 1590 William Cecil had set out amongst his new religious priorities a series of specific measures whereby wives could be indicted for recusancy and their husbands compelled to pay their fines, and when Parliament met in February 1593, high on its agenda was the female problem. Immediately, a bill was introduced into the Commons that left little doubt recusant women had become the Government's latest targets (with husbands seen merely as the cash cows capable of funding their wives' law-breaking). Recusant heiresses were to lose two-thirds of their inheritances; recusant wives were to lose their dowries and jointures. In addition, all children aged seven and over were to be removed from their

* If a husband died before his wife, two-thirds of her jointure could be seized in payment of her outstanding recusancy fines. The Exchequer Roll of 1593–94 reveals sixty such instances.

† Traditionally bishops were able to call down secular punishment on an excommunicated person by applying to the Court of Chancery. Chancery would then issue a writ to the county sheriff, ordering the arrest of the excommunicate until such time as he/she conformed. In 1563 a Parliamentary Act [5 Eliz. c.23] stipulated that a regular return of these writs should be made to the Court of the Queen's Bench to enforce the practice.

recusant parents' care and given to selected families to raise – at their parents' expense (such was the pernicious influence of recusant mothers on the next generation). Finally, in a clause left deliberately vague, householders were to be fined ten pounds a month for every member of their household refusing to attend church.[28]

On 14 March 1593, as the bill was batted back and forth between Commons and Council for amendment, Thomas Barnes wrote to fellow exiled Catholic Charles Paget, informing him how little these proposals were liked. Francis Craddock, MP for Stafford, had objected to the generalized clause against householders; Mr Wroth, MP for Liverpool, had identified that husbands were the real victims of this law. One parliamentary diarist even noted that many MPs kept 'special eye . . . [during the bill's progress] . . . that no such thing might be inserted which might wind them with such a penance'. Few men, it seemed, were willing to legislate against themselves. Nonetheless, when the completed 'Act against Popish Recusants' was passed that April, although all its other clauses had been stripped away, the remaining clause stipulating householders could be fined for their recusant household remained with all its old ambiguity intact. Indeed, courtesy of the Law Lords, it now contained an added twist. From this time onwards the courts were able to sue a husband jointly with his wife for her recusancy and a host of such prosecutions swiftly followed.*[29]

Whatever the complications of proceeding against women for their recusancy, the law was unequivocal about what would happen

* Curiously, the Commons played a moderating role in this Parliament, reserving their real energy to debate whether the bill applied to all recusants or just Catholic recusants; the Government insisted on the former (an indication of its increasing animosity towards the Puritans), the largely Puritan Commons insisted on the latter. The Government won, but whether the Commons' comparative leniency towards Catholics was a result of growing identification with their fellow persecuted or growing self-confidence is unclear. An additional clause to this legislation, poached from a bill introduced in the Lords, prevented all Catholics over the age of sixteen from travelling five miles from their place of residence without a licence, on penalty of exile. It was specifically designed to prevent Catholics from evading indictment for recusancy by moving about.

to those of either sex caught harbouring Catholic priests. Yet, in practice, women seem to have been treated leniently in the courts. Out of the thirty people who would eventually be executed under the 1585 statute against aiding and abetting Jesuits and seminarians, only three of them were women. But of those three, the death of Margaret Clitherow, in 1586, provided the benchmark against which all other Catholic women considering supporting a priest could measure their actions.

Clitherow was the daughter of Thomas Middleton, a former Sheriff of York. Her stepfather was Henry Maye, Lord Mayor of York in the year of her death. In 1571 she married John Clitherow, a successful butcher and city chamberlain living in York's Shambles (the name comes from the Saxon, *Fleshammels*, meaning the street of butchers). Hers was a solid, respectable background – marred only by her faith. Some time *c.*1574 Margaret Clitherow, then aged about eighteen, converted to Catholicism. Two years later she was arrested and imprisoned for recusancy. There followed a string of similar convictions, but all the while Clitherow continued maintaining priests, two at a time: one in a concealed chamber adjoining her house, the other stationed across the city. It was a high-risk strategy and several of her priests would be captured; she, herself, made no secret of her desire for martyrdom. On 10 March 1586 she was arrested for harbouring.[30]

On 14 March Margaret Clitherow was brought before the judges at the York Assizes and asked to enter her plea: guilty or not guilty as charged. Clitherow declined to do so, saying, 'I know no offence whereof I should confess myself guilty.' Asked once more by the judges to enter a plea, she responded, 'Having made no offence, I need no trial.' The law was clear about the fate of those felons who failed to plead in court (their refusal was felt to be a shameless spiking of the wheels of justice). Accordingly, the York judges invoked the fourteenth century penalty of *peine forte et dure.** 'You

* The punishment was finally removed from the statute book in 1827; the last recorded case of its being invoked came eighty-six years earlier in 1741. From then on all those refusing to plead were adjudged to have entered a plea of not guilty. Convicted felons

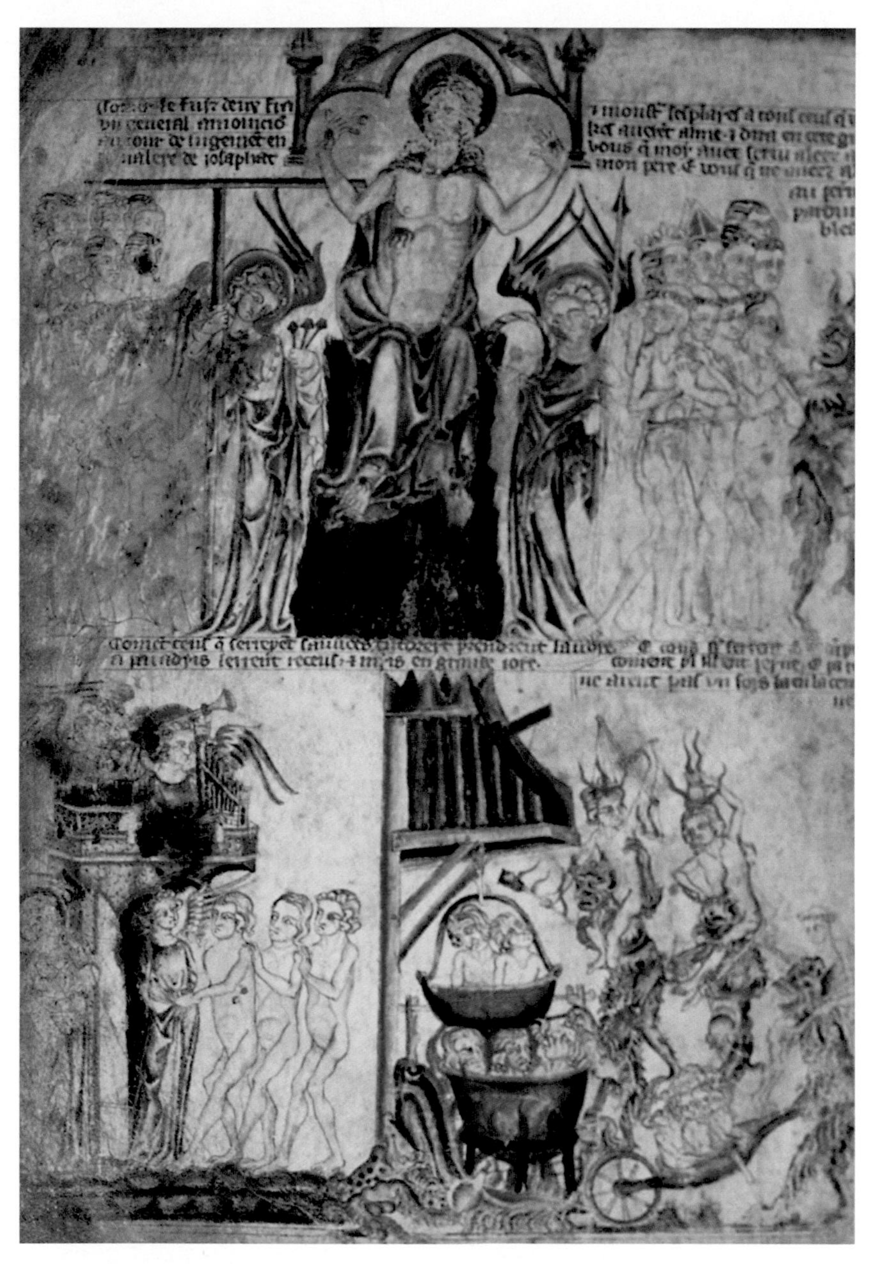

A fourteenth-century English depiction of the Last Judgement, showing the seated Christ, flanked by angels, welcoming the blessed into heaven. Below, an angel greets three more of the saved, while, to the right, demons carry the damned away to hell. Two boiling cauldrons stand in hell's mouth, full of damned souls. Images such as this would have been familiar to every pre-Reformation Englishman and woman, providing a dreadful warning of the perils of defying the Church. When English Catholics spoke of the danger to their mortal souls of attending 'heretic' services, this was what they had in mind.

Queen Elizabeth I, shown standing on a map of England. The England that the 25-year-old Elizabeth inherited was a country fractured by religious conflict.

Above King Philip II of Spain, painted by Rubens in martial pose. In fact, according to those English Catholic émigrés trying to persuade him to invade England, Philip was terrified of war.

Right Sir William Cecil, Elizabeth's principal political and financial advisor. He was made 1st Baron Burghley in 1571. English Catholics, reluctant to believe that Elizabeth was in any way responsible for the laws against them, blamed Cecil squarely for their troubles.

Oxonia Antiqua Instaurata, Sive Urbis & Academiæ Oxoniensis Topograp
Huic in Margine accessit è Cod. MS. Bodl. Collegiorum & Scholarum
COLLEGIVM MERTONENSE
COLLEGIVM NOVVM
ECCLESIA CHRISTI

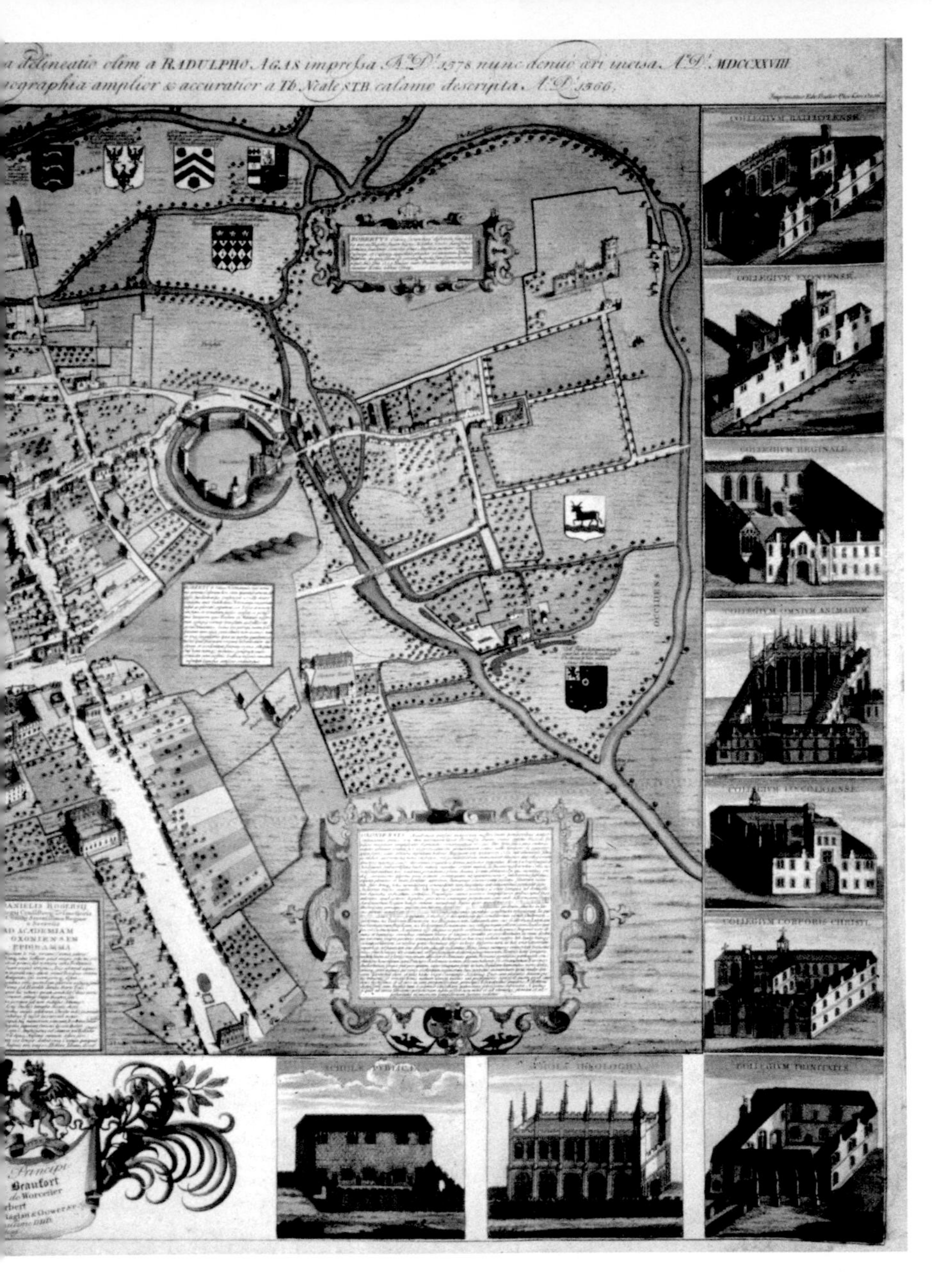

The city of Oxford in the sixteenth century, showing images of the University's halls and colleges, with Oxford Castle to the right. Castle Street, where Nicholas Owen's family lived, extends away to the left of the castle. Many Oxford students joined the mission, including almost half of those priests executed under Elizabeth.

Left William Allen, the mission's founder. Young English Catholics flocked to Allen's seminaries in France, Italy and Spain, to be schooled in argument, imbued with a hatred of heresy and prepared for martyrdom back home. Allen was concerned to distance his students from political matters, stating repeatedly that theirs was a purely religious endeavour. His own dabbling in politics made a collision between the students and the English Government inevitable.

Above Pope Pius V, responsible for excommunicating Elizabeth in 1570, and releasing her subjects from their allegiance to her. This act, more than any other single event, defined the English Catholic dilemma of divided loyalty.

Left Pope Gregory XIII, who sanctioned the first Jesuit mission to England in 1580.

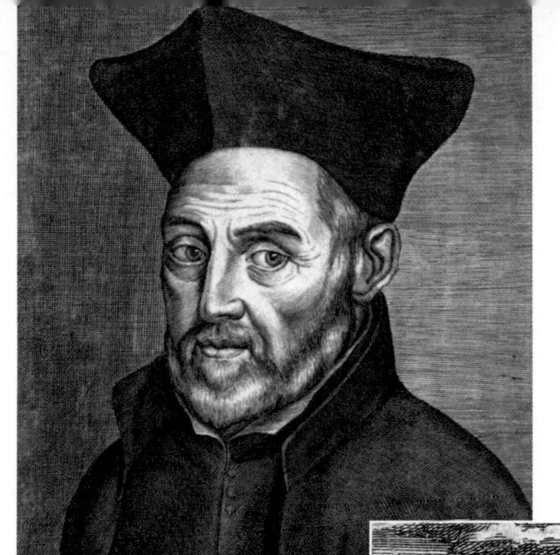

Left Robert Persons, who led the first Jesuit mission. His reputation remains controversial: he is widely regarded as an excellent example of that dangerous species, the politicized Churchman.

Right Edmund Campion, 'one of the diamonds of England', according to Sir William Cecil.

Overleaf The opening page of a 1st edition copy of Edmund Campion's *Decem Rationes*. . ., printed at Stonor, near Henley-on-Thames, in 1581. Copies of the book were distributed in secret among the students at Oxford University. Campion's skill at rhetoric, which, as a young man, had earned him the patronage of the Earl of Leicester, became compelling reason for the Government to silence him. Just weeks after *Decem Rationes* was published, Campion was arrested.

¶ *Edmundus Campianus,*
Doctissimis Academicis, Oxonii
florentibus & Cantabrigiæ,
S.

ANNO pr[illegible]m ex instituto [illegible] in hanc In[illegible] sem, claris[illegible] di sanè fluctus haud pa[illegible] Anglicano litori quàm [illegible] Oce-ano Britannic[illegible] s à tergo reli-queram. Mo[illegible] em in Angli-am vbi penet[illegible] nihil [illegible] mi-liarius, quàm in[illegible] hil certius, quam in[illegible] ile-gi me vt po[illegible] mor temporum [illegible] ne prius for[illegible] corri-perer, quàm auditus à quopiam fuis-sem, scripto protinus mandarem offici-um meum, qui venissem, quid quæ-rem, quod bellum, & quibus, in[illegible] cogitarem. Autographô apud [illegible] bus, vt mecum, si caperer, caperentur must return from whence you came,' read Clitherow's sentence, 'and there, in the lowest part of the prison, be stripped naked, laid down, your back upon the ground, and as much weight laid upon you as you are able to bear, and so to continue three days without meat or drink, except a little barley bread and puddle water, and the third day to be pressed to death, your hands and feet tied to posts, and a sharp stone under your back.'[31]

There followed eleven days in which a stream of prison visitors attempted to persuade Clitherow to enter a plea and she refused them all. Perhaps she was unwilling to subject her family to the ordeal of testifying against her; perhaps she was just unprepared to accept that harbouring priests was a crime. Finally, on 25 March she was led out to the Tollbooth, some six or seven yards from the castle gaol, stripped of her clothes and stretched out on the ground, a wooden panel balanced on top of her. In an act of charity the judges appear to have accelerated the process that came after: 'She was in dying one quarter of an hour,' wrote John Mush, her biographer (and one of the priests she had maintained). 'A sharp stone, as much as a man's fist, [was] put under her back; upon her was laid to the quantity of seven or eight hundred weight [eight hundred and ninety-six pounds] at the least, which, breaking her ribs, caused them to burst forth of the skin.'[32]

Clitherow's execution passed quickly into Catholic legend. Here was a martyr whose refusal to speak had emphatically given the lie to the belief that all women were 'full of tongue and much babbling'. Her stubbornness served to remind both priests and government that the female sex was a force to be reckoned with. Over the years many missionaries would have cause to be grateful for this fact.

forfeited their estate to the Crown, whereas those who died as a result of *peine forte et dure* did not – this may have been one of the few compelling reasons to choose such a fate.

Two miles south of the parish church of All Saints in Wimbish, Essex stood Broadoaks Manor, high gabled, tall chimneyed, built round three sides of a courtyard, some time *c.*1560. The house was encircled by rich farmland; Wimbish, itself, lay just to the south of the prosperous market town of Saffron Walden. In Gerard's words, the owner of Broadoaks, William Wiseman, enjoyed the wealth and independence 'of a little prince'.[33]

It was Henry Drury, Gerard's host at Lawshall, who had first brought the Wiseman family to Gerard's attention, inviting Thomas and John Wiseman, William's younger brothers, to meet the Jesuit there. Gerard had drilled the pair in the *Spiritual Exercises* of Ignatius Loyola. This was the Society's *vade mecum*, a collection of documents (containing instructions and admonishments) and practical tasks (prayers and meditations) to help move the faithful to greater piety. (The full title of the work was *Spiritual Exercises to conquer oneself and regulate one's life, and to avoid coming to a determination through any inordinate affection*, which gave a reasonably clear indication of the Spanish ex-soldier's views on religious self-discipline.) Both men had been inspired by Gerard's teaching; both had decided to leave England and join the Society.* Before they left, though, Thomas Wiseman had persuaded his brother William to try the *Exercises* for himself.[34]

In Gerard's description, William Wiseman was a Catholic, 'but his thoughts were very far from Christian perfection'. Like so many others of his religion, he had kept his head down and his family clear of the law, preferring to safeguard his estate rather than assist the mission. His encounter with Gerard was to change that. Within days of beginning the *Spiritual Exercises* William Wiseman 'had resolved henceforth to devote the whole of his life to furthering God's greater glory'. This was no mere boasting on Gerard's part, for Wiseman immediately invited the Jesuit to

* The younger brother, John, died as a novice in Rome in 1592. Thomas died of consumption at St Omer in 1596. During their time abroad they took as aliases names used by Gerard during his stay in East Anglia: John became Starkey, Thomas, Standish, in tribute to their mentor.

move from Lawshall to Broadoaks and take up residence with him and his family. The man who up to now had avoided the company of seminary priests for fear of the harm they might do his reputation was now consorting with the even more harmful Society of Jesus.[35]

Wiseman's invitation came at a good time for Gerard. Henry Drury had recently decided to leave England and become a Jesuit lay brother; the seminarian William Hanse had declared himself happy to stay on at Lawshall and care for Drury's elderly mother; Henry Garnet was keen the Jesuits should open up another county for the mission.* So in the winter of 1591 Gerard shifted his centre of operations to Essex.[36]

Essex has been described as the headquarters of sixteenth century Puritanism. Like its neighbours Suffolk and Norfolk, it, too, had received an early dose of Protestant theology, courtesy of the east-coast trading vessels. In 1532 'the image of the crucifix was cast down and destroyed in the highway' near Coggeshall, to the west of Braintree, as the muscles of the new religion began to flex. Under Mary the Bishop of London was ordered 'to send into Essex certain discrete and learned preachers to reduce the people who hath been of late seduced by sundry lewd persons named ministers'. When the execution of heretics started, Essex suffered accordingly, with burnings at Colchester, Manningtree, Harwich and Saffron Walden among other sites. Under Elizabeth, and no doubt in part as a result of these widespread burnings, Protestantism in Essex had crystallized into a more hard-edged religious non-conformity, termed Puritanism for convenience's sake, though this implied a greater unity of belief than was in fact ever the case. In the 1580s Robert, Lord Rich of Rochford Hall, prominent within the Puritan party, was imprisoned for a time to prevent this non-conformity spreading further; Rich was considered 'the wealthiest lord' in Essex. None of this deterred John Gerard. From Broadoaks he journeyed northwest to Sawston

* Drury died in Antwerp on 10 September 1593 as a novice of the Society.

Hall, just over the county border in Cambridgeshire, to convert his host's brother-in-law Henry Huddlestone, and northeast to Bury St Edmunds in Suffolk to supervise the religious education of the Rookwood children of Coldham Hall. He was even in conference with Lady Rich, attempting to win her to the faith.* Unknown to him, though, the Wisemans had come under suspicion.[37]

Some time in December 1592 an Essex magistrate named Nicholls led a raid on Broadoaks. The pursuivants spread out across the house and immediately set about sounding the wooden panelling in search of concealed hides. Very soon they struck lucky. Cowering in 'a secret place between two walls' they discovered the elderly priest Robert Jackson. As a Marian priest, Jackson was no outlaw (hence Wiseman's long-standing support of him in favour of any seminarian): Elizabeth's new treason legislation had made a clear distinction between those priests ordained in the previous reign – when England was Catholic – and those who had deliberately gone abroad to receive holy orders after England's split with Rome. But when one of Wiseman's servants, Edward Harrington, confessed to having heard Jackson say mass for the household on two separate occasions the family was now recognized to be in contravention of the law. On 12 January 1593 the Council received notice of the indictments against William Wiseman, his wife, mother and sisters for being present at one or both of two masses celebrated by Robert Jackson at Wimbish. Of John Gerard's name there was no mention; neither does he, himself, record the event. Most likely he was absent at the time of the search.[38]

Clearly, though, it would now have been safest for Gerard to change base again. That he did not suggests either that he and Wiseman were happy the search was a one-off and would not be repeated, or that, for all his abundant confidence, Gerard

* Gerard was cheerfully honest about his failures as much as his successes. He spent many hours with Lady Rich, the sister of the Earl of Essex, but in the end she was persuaded against conversion by her lover, Lord Mountjoy. However, on her death in 1634, it appears she was reconciled to the Catholic Church by an unnamed Jesuit.

recognized few families in the area would be ready to welcome him as a permanent addition to their household. Instead, Gerard summoned Nicholas Owen to Essex. Within days of the Nicholls' raid, Owen was at Broadoaks strengthening the house's internal fortifications.[39]

The following year, 1593, appears to have passed without incident. No doubt the family paid the requisite fines for hearing Catholic mass, currently one hundred marks apiece (Robert Jackson would have been liable for a two hundred marks fine for saying mass); no doubt Gerard continued his work in Essex; there is evidence of neither. The focus of most Catholics would have been on the proceedings in Parliament that spring and on the fear of increased penalties against them. When those penalties failed to materialize (barring the new facility for fining householders and the restriction on Catholic travel), there can have been little sense of relief. Yes, for the first time Parliament had acted as a check on the Government, stifling harsher measures against Catholics. Yes, there seemed to be a growing sense among some Englishmen that enough was enough; that year Henry Garnet reported Lord Grey as saying, 'I was under the impression that our purpose hitherto was merely to keep the Papists humbled and in subjection so that they should cause no trouble. We have sucked them dry and reduced them to extreme poverty. Now we strive to harass them yet further. It is plain to me that we are persecuting religion.'* Yet few can have believed for an instant that their troubles were at an end. For the Wisemans, their troubles were just beginning.[40]

On Boxing Day 1593 pursuivants unexpectedly raided the North End house of the Widow Wiseman, William's mother. Mass was in preparation when the raid took place. The priest, a man named Brewster, was successfully spirited away up a chimney to hide, but evidence of his activities lay all around and magistrate

* This was probably Henry Grey of Enville in Staffordshire. Grey was married to Anne, daughter of William, Lord Windsor, who was known to Garnet from the Hurleyford conference. Though clearly no Catholic himself, his marriage makes it possible he was possessed of Catholic sympathies, hence his reported outburst.

Richard Young, informing Lord Keeper Puckering of the event, advised bringing Mrs Wiseman in for examination. North End he described as the chief place of resort 'for all these wicked persons'. When, some time the following year and after an unexplained pause, Mrs Wiseman was suddenly carried up to London and imprisoned in the Gatehouse gaol for receiving seminary priests, it may have appeared that this was something more than an ordinary case of bad luck.[41]

And still things got worse. On the evening of Friday 15 March 1594 Richard Topcliffe, apparently acting on his own authority, organized a raid of all known or suspected Catholic houses in London. 'The uproar was such that Hannibal himself might have been at the gates or the Spanish fleet in the river Thames,' wrote Garnet to Aquaviva in the months following. Local magistrates were called in to assist the priest-hunter and overnight the city's churches were drafted into use as holding pens to contain all those arrested in the raids.[42]

That same evening Gerard was visiting Garnet at the latter's new base in an undisclosed location some 'four or five miles from London'. He had been scheduled to continue on to London that night, to a house at the upper end of Golden Lane (running between Old Street and the Barbican) rented just recently for his own and friends' use. The house had been let from a neighbour, a Mr Tute, and the name on the lease was that of William Wiseman. But now Garnet begged Gerard not to go there. It was the second time the Jesuit Superior had had a premonition of danger. Gerard had experienced the first occasion for himself at Baddesley Clinton; accordingly, he remained with Garnet that night. 'Early next morning', he recorded afterwards, 'rumours reached us that papists had been seized in the house.' Garnet later wrote to Robert Persons that 'there [should have] been there Long John with the little beard [Gerard] . . . if I had not more importunately stayed him than ever before'. Those arrested during the raid on Golden Lane included Gerard's servant, Richard Fulwood, and the Wallis brothers, servants to William Wiseman, one of whom was vocal in his admiration for the Jesuits and declared himself glad to be arrested so that now he might suffer persecution for his faith.* They were taken away for interrogation. Under a barrage of questioning all of them were able to keep hidden from the authorities Gerard's connection with the house, but William Wiseman's own association was clearly documented and he was called to London.[43]

On 19 March 1594 Wiseman appeared before the Attorney General, Edward Coke, to assist in the inquiry into Golden Lane. He strongly denied that the house was a resort for priests and though he did admit that he had stood guarantor for a third party who was renting the property, he refused 'for charity's sake' to give that party's name. He was led off to the Counter prison in nearby Wood Street where he was placed in solitary confinement. Only his servant John Frank was permitted contact with him, and back at Broadoaks the family waited anxiously for Frank to bring news. Towards the end of the month Frank left London for Essex with a letter containing reassurances from Wiseman and precise details of the questions he had been asked and the answers he had given, necessary if the family's stories were to be consistent. Meanwhile, Gerard had returned to Wimbish to offer what comfort he could and vitally, now that the Wisemans were facing investigation, 'to get everything [at the house] hidden neatly away'. Thus it was that on 1 April 1594, Easter Monday, he was at Broadoaks when the pursuivants raided.[44]

It was first light and Gerard was preparing mass when he was halted by the sound of horses approaching. Within moments the house had been surrounded. Escape was now impossible. The servants barred the doors while Gerard stripped the altar, concealing his books and papers in one of the many caches with which the house had been provisioned. Then he began making his way downstairs from the attic chapel to one of Nicholas Owen's new

* The arresting officers, Robert Watson and Edward Vaughan, revealed to Sir Robert Cecil that when asked if he were a Jesuit, Wallis replied, 'no, I am not learned, and would to God I were worthy enough to carry their shoes'. They also revealed that when the suspects were questioned, their answers 'all seemed to vary, and not one could tell an even tale'.

hides near the dining room, stocked with basic supplies for just such an event. But here Mrs Wiseman intervened, drawing Gerard back towards a second Owen-built hide leading off the chapel. Speed was now of the essence and the Jesuit did not resist, crawling into this new hiding place with the altar furnishings crammed in around him. At the last moment Mrs Wiseman handed him a jar of quince jelly she happened to have with her and the hide was closed over Gerard's head.[45]

Downstairs the pursuivants fanned out across the house. Mrs Wiseman and her two daughters were locked away in her chamber, the family's Catholic servants were segregated in different parts of the building, and the search began. Candles were shone into dark corners. Long rods were brought out to measure walls, checking where the internal and external dimensions did not tally; any section that could not be accounted for was demolished. Panelling and floorboards were sounded for hollow spots then swiftly smashed through. This took two days.

By late Wednesday the justices leading the raid were ready to admit defeat. A detachment of pursuivants was left behind to escort Mrs Wiseman, her two children and the Catholic servants to London the next morning. The non-Catholic servants, of whom there was a handful, were detailed to watch the house; among their number was John Frank. That evening Mrs Wiseman got a message to Frank that as soon as the house was empty he was to call to Gerard from the room beneath the chapel, telling him the search was over (such was the secrecy of Broadoaks' hides that even now she was unwilling to divulge their precise location). This message Frank now conveyed to the pursuivants.

On Thursday, 4 April the justices returned to Broadoaks and the search resumed. Mrs Wiseman's instructions to Frank suggested the existence of a hiding place positioned somewhere within the interior masonry of the first floor in the north wing. That day the corresponding rooms were measured and sounded even more carefully as the pursuivants closed in on their prey. That night a watch was set. Two men now guarded the attic chapel, stopping off

Gerard's only way out of the house. As darkness fell so, too, did the temperature and the men decided to light a fire laid ready in the attic grate. From his hide beneath the fireplace, accessed by lifting a small section of floor under the grate, Gerard listened as the thin layer of bricks constituting the false hearth came loose with a rattle. Immediately above him the wooden floorboards, on which the bricks rested, began to smoulder and shift. He heard one of the guards comment on this peculiarity, but neither seemed inclined to pursue it further, putting off their investigation until the following day.

By Friday, 5 April Gerard's capture seemed a formality. All night the fire had burnt over him, the embers raining down on his head. Now, from where he crouched, he could look up through the charred remnants of the hide into the attic room beyond, just as clearly as anyone approaching the fireplace could look down on him. Yet no one looked. At first light his guards were called from their post to the rooms below, which now became the focus of even greater scrutiny as, systematically, the pursuivants began to strip the plaster from the walls. During this process they discovered the hiding place Gerard had initially intended using, before Mrs Wiseman dissuaded him; inside, untouched, lay his store of provisions. Meanwhile, the stripping of the plaster continued. By late afternoon the room beneath the chapel was entirely denuded, bar a small section of wall immediately surrounding the fireplace. Here a large and finely carved chimney-piece impeded the pursuivants' progress; this, plus the growing belief that Gerard had somehow managed to escape during the night, was enough to halt the proceedings. Immediately behind the chimney-piece lay Gerard's hide.

On the evening of Friday, 5 April the search was officially called off.* Mrs Wiseman and her daughters were released, along with the

* On 20 May 1594 William Cecil wrote to the justices in charge of the raid, Francis Barrington and Mr Frank (no relation to John Frank), asking for a report on the behaviour of two pursuivants – Worsley and Newell – during the search. It seemed Mrs Wiseman had complained about their 'bad demeanour'. Sadly, there is no record of the follow-up to this

Catholic servants, and, once the pursuivants had finally departed the house, John Gerard was lifted bodily from the hide in which he had crouched for the last four days. All this time he had eaten nothing but the quince jelly thrust upon him by his hostess as the raid began. In an extraordinary gesture of solidarity – part desire to share his suffering, part means of ascertaining how long he could survive without food – Mrs Wiseman had starved herself for the same period. 'When I came out,' wrote Gerard afterwards, 'I found her face so changed that she looked a different person; and had it not been for her voice and her dress I doubt whether I would have recognised her.' But any relief at the near miraculous nature of their escape was misplaced. No one yet suspected John Frank to be an informer.*[46]

As soon as he was rested enough to travel, Gerard left Broadoaks, lying low at a nearby friend's house for a fortnight to recuperate properly. Then he headed for London. William Wiseman was still being held prisoner in the Wood Street Counter and the Jesuit was keen to see if there was anything he could do for him, but with London's Catholics in disarray following Topcliffe's raids it was imperative Gerard first find a safe place of refuge. Nicholas Owen was drafted in to help him in this search; meanwhile, Gerard

---

complaint. The problem was not unusual, though. One Government informer wrote to Walsingham, warning him of the behaviour of Anthony Munday, employed by Topcliffe 'to guard and to take bonds of recusants'. According to the agent, Munday's 'dealing hath been very rigorous, and yet done very small good, but rather much hurt; for in one place, under pretence to seek for Agnus Deis and hallowed grains, he carried from a widow 40 [pounds], the which he took out of a chest. A few of these matters will either raise a rebellion, or cause your officers to be murdered'. This complaint rather gave the lie to Topcliffe's assertion, made to Lord Keeper Puckering, that Munday was 'a man who wants no wit'.

* Little is known about Frank. For a time he was employed by Thomas Wiseman, who recommended him to his brother on his departure abroad. He had a house in London, which Gerard used as a staging post for Catholics leaving for the Continent, and he had free access to Broadoaks, despite the fact that he was not a Catholic himself. His motives for this surprising betrayal of the Wisemans are unclear, as is his subsequent fate (though he later admitted to Gerard that he only came to suspect him as a priest because of the great respect in which he was held by the family).

shifted precariously between the Strand house of Anne, Countess of Arundel and a house belonging to the Wisemans somewhere in Holborn. It was there, at around ten o'clock on the evening of 23 April, that John Frank arrived with a letter from Mrs Wiseman that he said required Gerard's urgent attention. And it was there just two hours later that the pursuivants arrived, hammering on the door to wake Gerard and Owen, both asleep in the upstairs chamber.[47]

This time there was no way out: the chamber had only one exit; the house did not have a hiding place (Owen had just sufficient time to burn Mrs Wiseman's letter before armed men burst into the room). And this time there could be no pretence as to his real identity: Gerard recognized one of the pursuivants, probably from his time in the Marshalsea prison ten years before.* Gerard and Owen were ordered to get dressed. Their room was searched for incriminating evidence. Then both men were led away under escort.[48]

* Worsley and Newell, the two pursuivants who had so offended Mrs Wiseman during the Broadoaks search, led the Holborn arrests. It is not clear which of them recognized Gerard. Newell would later take part in a raid on the Derbyshire house of Gerard's sister, Mrs Jennison.

Chapter Nine

'As soon as thou comest before the gate of the prison, do but think thou art entering into Hell, and it will extenuate somewhat of thy misery.'

Geffray Mynshall, 1618

Royal Commissioners:*	Who sent you over here?
John Gerard:	The Superiors of the Society.
Commissioners:	Why?
Gerard:	To bring back wandering souls to their Maker.
Commissioners:	No, you were sent to seduce people from the Queen's allegiance to the Pope's, and to meddle in State business.

Day Two of Gerard's arrest and the questions – and accusations – were flowing.[1]

From Holborn, in the early hours of Wednesday morning, 24 April 1594, John Gerard and Nicholas Owen were conducted to the house of the pursuivant who had recognized Gerard. There they spent the remainder of that night. Soon after daybreak Owen was taken away for interrogation while Gerard was handcuffed and

* Leading the interrogation was Sir Thomas Egerton, who that year replaced Gerard's cousin Sir Gilbert Gerard as Master of the Rolls of the Court of Chancery. Egerton was a lapsed Catholic – he had been listed as a recusant in 1577. In 1596 he was appointed Lord Keeper and in 1609 he became Lord Chancellor.

locked in his room. Twenty-four hours later it was his turn to appear before the Commissioners.[2]

There are no transcripts of these first examinations other than Gerard's own, so neither the accuracy of his recall nor its impartiality can be verified. According to him, though, his first answer was a lie. When his inquisitors demanded his name, he gave an alias. It was a last-ditch response, automatic and not a little desperate, for it was unlikely the pursuivant known to Gerard had not passed this information on. Sure enough, the commissioners made it clear they now knew not only his real name, but also his profession. Gerard moved quickly to limit the damage: 'I said [to them] that I would be quite frank and give straight answers to all questions concerning myself, but added I would say nothing which would involve others.' For the rest of the session he talked openly about the Jesuits' injunction not to deal in political matters, but refused to give away any details of his landing in England, or where he had stayed for the last six years, so as not to implicate anyone else. Finally he was committed to the Counter in the Poultry, a prison four doors to the west of St Mildred's Church, in the maze of streets that made up London's Cheapside.[3]

The Counter in the Poultry 'hath been there . . . time out of mind', observed London chronicler John Stow. To the playwright Thomas Dekker it was one of London's 'thirteen strong houses of sorrow'.* In such a place, wrote Dekker, with an insider's experience, 'the prisoner hath his heart wasting away sometimes a whole 'prenticeship of years in cares . . . O what a deal of wretchedness can make shift to lie in a little room!' John Gerard's cell in the Counter was a very little room. 'It was a small garret which had only a bed in it

* The thirteen to which he referred were the two Counters, in Poultry and Wood Street, Newgate prison and Ludgate prison (all serving the City); the Bridewell and the Fleet; the Gatehouse and the Convict prison (both in Westminster); the Southwark Counter, the Marshalsea, the King's Bench, the Clink and the White Lion (all in Southwark). In addition to this, of course, there was the Tower.

and such a low ceiling that I could not stand upright except near the bed . . . The doorway was so low that I could not enter the room on my feet, and even when I crawled down on my knees I had to stoop to get through. But this proved an advantage, since it helped to keep out the smell of the privy next door.' It was the stench of an Elizabethan prison that wore the prisoners down, even more so than their confinement. James Younger, the priest-turned-informer (and an internee of the Poultry Counter just the year before Gerard), had complained to Lord Keeper Puckering, 'such is the corruption and unsavoury air of the place in which I remain that my body is not able any longer to sustain those annoyances, but fainteth continually'. Poet Edmund Gayton would put it even more succinctly, in the following century:

> . . . the Prison smelt of lice,
> Of urine, and of seige [excrement] and mice
> And rats' turds.[4]

Like all political prisoners Gerard was held in solitary confinement. In addition, his legs were shackled. This was standard punishment for the worst sorts of criminals, but one from which the payment of a specific fee – 'for easement of irons' – could usually buy relief. In Gerard's case, though, he was rumoured to have welcomed the additional suffering, as a letter from Henry Garnet to Claudio Aquaviva some years later revealed.

> 'When he was first taken and the gaoler put very heavy irons on his legs, he [Gerard] gave him some money. The following day the gaoler, thinking that if he took off the irons, he would doubtless give him more, took them off but got nothing. After some days he came to put them on again and received a reward; and then taking them off did not get a farthing. They went on playing thus with one another several times, but at last the gaoler, seeing that he did not give him anything for taking off his irons, left him for a long time in confinement, so that the great toe of one foot was for almost two years [afterwards] in great danger of mortification.'

Gerard, himself, only noted, 'I did not feel the least bit sorry for myself. Quite the contrary. I became very happy – so good is God to the least of his servants.' With each new misery he endured, he was moving a step nearer martyrdom.[5]

Three or four days into his imprisonment, Gerard was led out once again for interrogation. This time he faced more practised adversaries than the Royal Commissioners: Richard Young, Chief Magistrate for Middlesex, and Richard Topcliffe. Young, so prominent in the search for Southwell and in the recent arrest of Mrs Wiseman, began with a question about the places where Gerard had lived since his return to England and about the Catholics he knew. Gerard refused to answer. Now Topcliffe entered the fray; sixty-two-year-old priest-hunter faced twenty-nine-year-old priest. The scene, as reported afterwards by Gerard, was a bravura display of bluster by the older man, bravado by the younger: 'Topcliffe looked up at me and glared. "You know who I am? I am Topcliffe. No doubt you have often heard people talk about me?" He said this to scare me. And to heighten the effect he slapped his sword on the table close to his hand as though he intended to use it, if occasion arose. But his acting was lost on me. I was not in the least frightened.' Hereafter, wrote Gerard, 'I was deliberately rude to him.' The meeting ended with Topcliffe laying out the charges against Gerard.[6]

In a paper addressed to the Privy Council Topcliffe wrote: 'The examinee was sent to England by the Pope and by the Jesuit, Persons, on a political mission to pervert the Queen's loyal subjects and to seduce them from the Queen's allegiance.' Then Gerard was allowed to write his defence: 'I am forbidden to meddle in State affairs and I have never done and never will. My endeavour has been to bring back souls to the knowledge and love of their Creator, to make them live in due obedience to God's laws and man's, and I hold this last to be a matter of conscience.' The lines had been drawn for the conflict ahead.[7]

* * *

Gerard spent three months in the Poultry Counter. During this time Nicholas Owen and Gerard's servant Richard Fulwood (captured during the raid on Golden Lane) were pumped for information. Again, there are no official records of these examinations, nor is it clear where Nicholas Owen was being held, but Fulwood was now a prisoner in that former royal residence, turned notorious house of correction, the Bridewell.[8]

The Bridewell, sprawled between the eastern end of Fleet Street and the Thames, was unique among London's lock-ups. In 1553 Edward VI had given the palace to the City as an experimental workhouse for vagrants, to be run according to theories of poor relief coming from the Continent. Inmates were to be taught new skills such as weaving and spinning, and subjected to violent punishment so as to prevent them falling 'into that filthy puddle of idleness which [is] the mother . . . [of] all mischief'. With regular floggings, it was felt, the vagabond, drunkard and prostitute would learn again to 'walk in that fresh field of exercise which is the guider and begetter of all wealth, virtue and honesty'. So infamous had the prison become in its short life span that soon prisoners were being sent to the Bridewell simply to force confessions from them. Fulwood's internment there revealed the extent to which the Government was eager to proceed swiftly against Gerard, an eagerness that would explain why – according to the Jesuit – both Owen and Fulwood were now tortured. 'For three hours on end (I think)', wrote Gerard, they 'were hung up with their arms pinned in irons rings and their bodies suspended in the air.' Neither man revealed a single name – either of the places Gerard had stayed in, or of the people who had assisted him.[9]

John Frank, though, was busy revealing everything he knew. From Gerard's betrayer came the information that Fulwood was Gerard's servant and had written to him from prison. About Gerard, Frank knew that he was a Jesuit who went by the aliases Tanfield and Staunton, that he was instrumental in smuggling Catholics over to the Continent (namely Jane and Bridget Wiseman, William Wiseman's sisters) and that he had been present at Broadoaks during the search that April. Against the Wisemans, Frank offered still more damaging testimony. For the first time the Council now learned the full details of the North End raid on the Widow Wiseman's house: that the priest, Father Brewster, had been hidden in the chimney during the search and then spirited away to Broadoaks by one of William Wiseman's servants. William Wiseman, himself, was frequently seen by Frank in Gerard's company, notably at Frank's own London house. As for Nicholas Owen, he had been at Broadoaks over Christmas 1592, he was known as Little John and he wore a cloak belonging to Wiseman. This cloak was of 'sad green cloth with sleeves,' added Frank, 'caped with tawny velvet and little gold strips'. It was on such detail that the balance of a man's life might hang: at Gerard's next interrogation he was asked to try on a suit of clothes discovered at Broadoaks and said (correctly) to be his. A positive identification would have been sufficient to indict William Wiseman with harbouring. Gerard vehemently disowned the suit.[10]

Though Frank had placed Nicholas Owen at Broadoaks a year and a half before his arrest, and though Owen was clearly familiar with Gerard, neither amounted to a capital offence. Under torture Owen had revealed nothing incriminatory. Probably he had come across as a minor player on the mission: part carpenter, part servant, in no ways vital to the infrastructure of Catholic resistance. Certainly, no one was yet aware of his status. Had Frank known that the Broadoaks hides were Owen-built he would have said so, among the welter of other detail he provided. Consequently, Owen was bound over in prison until, at an unspecified date, bail was purchased on his behalf. If the Government had no use for him, then the Catholic families of England assuredly did. Therefore, wrote Gerard, some 'gentlemen . . . paid down a sum of money and had him released'. Owen returned to Henry Garnet's service. Gerard returned to the interrogation room.[11]

By the summer of 1594 it was clear his inquisitors were making little headway. In Gerard's words, 'only my priesthood could be proved against me'. Now his friends stepped in to try to improve the conditions in which he was being held. They 'achieved this', wrote Gerard, 'by bribing no less person than [Richard] Young himself'. On 6 July Gerard was led out from the Poultry Counter, over London Bridge to the teeming streets and narrow lanes of Southwark, to the Clink prison on Clink Street, tucked in between the Bishop of Winchester's palace and a row of former brothels. 'I looked on this change to the Clink as a translation from Purgatory to Paradise,' Gerard recorded.[12]

Sixteenth century prisons – the Bridewell excepted – were places of detention not reform, packed with a rich cross-section of society, some post-trial, some pre-trial, some unlikely ever to see trial; all too many prisoners succumbed to disease in the cramped conditions. And sixteenth century prisons were more cramped than most. Enclosure – the practice, popular during the first half of the century, whereby estate owners consolidated large stretches of open field for extensive agricultural use – had chased the poor from the land and into vagrancy. The dissolution of the monasteries had chased a host of monks, clerks and charity-seekers out to join them. And one solution to this growing problem was to clap them all in prison. Later they were joined there by all those whom the extravagance of Gloriana's golden age had driven into mounting debt and all those whose religious beliefs had placed them in opposition to the Established Church. Vagabonds, debtors and Catholics now made up the majority of the prison population.*[13]

The Clink, Gerard's new home, was a Catholic hothouse. After just a few months, he wrote, 'we had, by God's grace, everything so arranged that I was able to perform there all the tasks of a Jesuit

* In 1583 Robert Persons wrote to the Rector of the English College in Rome, highlighting Government concern at the numbers of priests it was holding in prison. '[O]ur opponents are not so anxious now to capture priests', he noted, '. . . and so nowadays the heretics are usually annoyed if any priests give themselves away easily. This was the case lately with Lomax, one of your men, who was secretly scolded by the Magistrates because, when seized at port, he guilelessly confessed that he was a priest at the first word from the Magistrates, and so they were forced by his needless confession to send him to prison.' Father James Lomax was arrested in the summer of 1583. He died in prison early in 1584.

priest, and provided only I could have stayed on in this prison, I should never have wanted to have my liberty again'. Through a hole in his cell wall, covered over with a picture, Gerard was passed paper, pen and ink, with which to write to Henry Garnet, informing the Superior of all that had taken place during his interrogations. Through this same hole he was also able to make confession and receive the holy sacrament. His solitary confinement was not long lasting. Soon some Catholics in the prison had fashioned a key for his cell door and Gerard was free to roam. From nine o'clock at night, when prisoners were locked into their cells, until the following morning, when the warders returned to check no one had escaped, Gerard had the run of the gaol, joining all those other Catholic prisoners who had also made keys to their cells. His next step was to win over his gaoler.[14]

It is helpful to think of an Elizabethan gaoler as something akin to an Elizabethan innkeeper. Both ran a boarding house, both sought a profit, neither received any subsidy from the State: as you passed through the portals of either's establishment you signed your name in the entry book and reached for your purse. How much you paid, in prison no less than at an inn, depended on what you could afford. Gaols were divided into 'Sides', subdivided into wards and further subdivided into individual rooms. Inmates chose their Side according to their social status and their ability to pay: the better the Side, the better the rooms, the more expensive the rates.* Of course, better was a comparative term: Gerard's room in the Poultry Counter appears to have been situated on the best Side. Bed and bed linen, a light to see by, meals at the canteen (which ranged from the dire to the digestible, again according to your ability to pay) and a discharge fee at the end of your stay: these were just the basic expenses you might incur in prison;

* In most prisons there appear to have been three ranks of paid accommodation, the Master's Side, the Knight's Ward and the Twopenny Ward. The penniless were cast into the Hole, a beggar's ward whose inmates were dependent on alms from the rich for their survival; William Cecil was a long-standing benefactor of the residents of Newgate's Hole. In 1606 official prison rates were set: according to these, a bed cost 4d for a single occupant, or 6d a night shared between several occupants. Mattresses were 1d extra.

and pocketing all of them was the prison gaoler, one eye on his immediate posting, one eye on his pension. In an Elizabethan prison, money talked. And the head gaoler at the Clink seemed disposed to listen. 'With bribes and a little coaxing', wrote Gerard, 'I induced him not to pry too closely into our doings, and to come to me only when I called him.'*[15]

From his prison cell Gerard now took up where he had been forced to leave off in Essex. Hearing confession was straightforward pastoral work and soon, wrote Gerard, 'so many Catholics came to visit me that there were often as many as six or eight people at a time waiting their turn to see me' in the next door room. Some were fellow prisoners in the Clink; many were from the outside, all quietly taking advantage of the fact that, as Gerard put it, 'my whereabouts were known and never changed and I could be found without difficulty'.† Meanwhile, with the gaoler continuing to turn a blind eye on Gerard's activities, a chapel was created in the cell of a fellow Catholic, from which the Jesuit could conduct divine service and instruct those who were interested in the Society's *Spiritual Exercises.* By January 1596 Henry Garnet could write to Aquaviva: 'The work John does in prison is so profitable that it is hardly possible to believe it.' Crucially, Gerard was now able to take over Robert Southwell's work as coordinator of the London end of the mission.[16]

Southwell's capture had left a gaping hole in the mission's chain of command. The majority of newly arrived priests were still

* In 1596 Sir Robert Cecil noted that, of all London's gaols, the Clink had been particularly corrupted by its Catholic prisoners and that it might be necessary to dismiss the gaoler.

† In March 1583 William Allen wrote: 'In one of the prisons called the Marshalsea there are, besides the other Catholics, twenty-four priests who live there . . . Both in this and the other prisons many masses are said every day, with the leave or connivance of the gaolers, who are either bribed or favourable to religion; people from without are admitted from time to time for conference, confession or communion; and more than this, the priests are allowed to go out everyday to different parts of the city and attend to the spiritual needs of the Catholics, on condition that they return to prison for the night.' On 25 February 1600 forty-eight people were caught attending a Catholic sermon at the Marshalsea. Most of the congregation, as well as the priest (a Scottish Capuchin friar), were visitors.

pouring into London. The Jesuits were still responsible for dispatching them to safe houses in the shires. Henry Garnet, shuttling between the capital and the surrounding countryside, was stretched to breaking point; and the March raids on London's Catholics suggested that, for the time being at least, the city was too dangerous a place for him, or any priest, to remain in for any length of time. It seemed John Gerard's incarceration in the Clink was heaven sent. For just as Catholics in search of absolution were guaranteed a meeting with the Jesuit, so too were newly landed missionaries, all of them profiting from the unparalleled freedom with which visitors, for a fee, could come and go in prison. The 'majority of priests coming from the seminaries over here', he noted, 'were instructed to get in touch with me, so that I could introduce them to their Superior and give them other help they might need'.[17]

Next, Gerard solved the problem of what to do with these newly arrived missionaries while a safe house was being made ready for them. Through an unnamed third party he 'rented a house with a garden of its own in a suitable [but unspecified] district' of London. There he sent all those priests who called on him, 'until, with the help of my friends, I was able to get them . . . clothing and other things they needed, or find them a residence'. He put in charge of running this house a widow, Anne Line, who for some time had been living as a guest of the Wisemans. Line had already tasted persecution on account of her faith: her Protestant father had disinherited her, her Catholic husband had died in exile. She was the perfect choice to serve the mission now: as practical as she was willing 'to die for Christ'. For Gerard, she was 'able to manage the finances, do all the housekeeping, look after the guests, and deal with the inquiries of strangers'; moreover, she was 'very discreet'. With a priest permanently stationed with her – to undertake the personal calls that Gerard, himself, was unable to make – and with Gerard overseeing the entire operation safely from his prison cell, Line's boarding house effectively secured the London end of the mission for Henry Garnet. New priests now had a fixed first contact, as well as fixed lodgings and a fixed source of necessary supplies; and no Jesuit need risk his liberty to provide them.*[18]

In late December 1594 Gerard was led out from the Clink, back across the Thames to the City, to face his sternest test yet, at London's Guildhall before a panel of commissioners led by Richard Topcliffe.[19]

The question put to him there, the so-called Bloody Question, went as follows. 'What would you do if the Pope were to send over an army and declare that his only object was to bring the kingdom back to its Catholic allegiance? And if he stated at the same time that there was no other way of re-establishing the Catholic faith; and commanded everyone by his apostolic authority to support him? Whose side would you be on then – the Pope's or the Queen's?'[20]

Whose side would you be on then – the Pope's or the Queen's? This question summed up the English Government's hostility towards the mission. Moreover, it illustrated the fundamental dishonesty at the heart of the English Catholic position, a dishonesty that the missionaries themselves had done everything in their power – indeed had been trained – to ignore. Because, as things stood in this conflict, there were only two sides, Pope's and Queen's; and England's Catholics were attempting – for dear life no less than for their dearer faith – to pretend there were not. If you were Catholic then Elizabeth was illegitimate; if Elizabeth were illegitimate she had no right to occupy the throne, she was a usurper; moreover, the Pope had confirmed her illegitimacy and had sought to depose her from that throne. Conclusion: if you were Catholic then you must support that deposition. You must support

* Line ran the Jesuits' boarding house for three years, until her notoriety made it unsafe for her to continue. Soon afterwards she took lodgings in another building, from where she continued sheltering priests. She was arrested on 2 February 1601, charged with harbouring, and executed at Tyburn on 27 February, the last woman to suffer under the 1585 act.

the Pope. Except for two factors, which every English Catholic, missionary and lay person with half a mind on survival now clung to with a fervour born of desperation.

First, the full extent of the Pope's powers had always been a matter of dispute. It was an eleventh century document, the *Dictatus Papae*, attributed to Pope Gregory VII, that laid claim, on the Pope's behalf, to supreme legislative power over all Christendom, including the right to depose monarchs. It made sense for this document to have sprung from Gregory's pontificate: at the time he was engaged in a long-running and bitter campaign to stop the papacy becoming a sinecure in the gift of the Holy Roman Emperor. But when, three centuries later, Pope Clement V decamped from Rome to Avignon as a personal favour to an ambitious King of France, it gave the lie to the belief that the Pope was the impartial Father of all Europe.* The history of Christendom was littered with examples of wilful monarchs dictating policy and compliant popes obeying. Indeed, in the hands of a powerful king, a pliable pope could be the deciding factor in international disputes and personal rivalries, brought in to add a biblical seal of approval to a wholly political conflict. In 1521, before thoughts of schism had come to haunt him, Henry VIII had won the title Defender of the Faith from a grateful Pope Leo X, for his attack on Protestant heresy. But arguably, Henry could have secured his divorce and avoided schism altogether had Leo's successor, Clement VII, not lived in fear of the then Holy Roman Emperor, Charles V. Charles, Catherine of Aragon's nephew, had an unfortunate tendency to sack Rome if the Pope displeased him. Under these circumstances it was highly unlikely Clement would ever sanction the divorce, disgrace and ruin of Charles' aunt.

To add further uncertainty to the issue of papal power, the history of Christendom was similarly littered with examples of popes claiming authority over monarchs and monarchs choosing

* In another example of papal impartiality, in 1155 Pope Adrian IV (the former Englishman Nicholas Breakspear) granted to King Henry II of England all temporal rights over (and possession of) Ireland.

to turn a deaf ear. Even Mary Tudor – 'a person not a little devoted to the Roman religion' in Sir William Cecil's masterly understatement – was known to have disobeyed the Pope when he attempted to dismiss her favourite, Cardinal Pole. So though advocates of papal supremacy might scour the Bible for texts with which to support their claims – Dr William Allen was a practised exponent of the art – there was just enough doubt surrounding the subject to admit the flicker of hope offered by the all-important second factor. And this came stamped with the authority of Christ, himself.[21]

'Render unto Caesar the things which be Caesar's, and unto God the things which be God's': Christ's response to a situation of potential dual loyalty similar to that faced by English Catholics now. Crucially this response suggested that there *was* a distinction between spiritual and secular jurisdiction and that Caesar had as much claim on a subject's obedience as God, so long as each kept to his own sphere of influence. The problem was, with the Pope claiming powers of deposition and with English religion a matter of Parliamentary legislation, those two spheres did not just overlap, they were almost indistinguishable; in Elizabeth's England there was nothing so helpful as Caesar's image on the back of a coin to indicate where loyalty lay at any given moment. The Bloody Question was the government's attempt to impose from the outside, by means of a single interrogatory, a forced separation of the secular from the spiritual. Unfortunately for English Catholics the Bloody Question was unanswerable.

Edmund Campion had been asked a version of the question back in 1581 – did he 'acknowledge her majesty to be a true and lawful queen, or a pretended queen, and deprived [of her throne]?' – and had had the good sense to fudge the answer: 'this question depends upon the fact of Pius [V], whereof [I am] not judge'. Alexander Briant, executed alongside Campion, chose the path of theological uncertainty and told his interrogators that 'whether the pope have authority to withdraw [subjects] from obedience to her majesty, [I] know not'. By May the following year the question had crystallized into its more recognizable form and been put to those priests arraigned with Campion, but still not yet executed. Thomas Ford and William Filby both deferred making an answer 'until that case should happen'. Robert Johnson gamely attempted to plead the secular/spiritual divide, saying, 'if such . . . invasion, should be made for temporal matters, [I] would take part with her majesty; but, if it were for any matter of [my] faith . . . [I] were then bound to take part with the Pope'. All three were swiftly dispatched to Tyburn.[22]

Campion, Briant et al. had been justified, perhaps, in treating the question as entirely hypothetical. After all, at the beginning of the 1580s England was not yet at war. Furthermore, if centuries' worth of theologians had failed to reach any agreement on the matter, who were they to venture an answer now? But as the Spanish Armada set sail, weighed down by the Pope's blessing and Philip II's dynastic ambitions, for both England's Government and England's Catholics the Bloody Question acquired sharp new teeth. The concern for the Government was whether it could trust English Catholics not to support the invasion; for English Catholics it was whether, should those invasion forces land and order them to fight, they would be fighting – and dying – for God or a foreigner. The one was acceptable, the other markedly less so for most Englishmen.

In the event, England's Catholics were never forced to make their choice, but some time in the spring or summer of that year the Privy Council approached the Crown's lawyers with what seems to have been a highly controversial request. The request built upon a suggestion to Elizabeth made by Sir William Cecil in 1583, in a paper entitled *Advice to Queen Elizabeth, in Matters of Religion and State.* This suggestion, part of a general summary of the Catholic position at the time, was characteristically pragmatic. Instead of asking Catholics to take the Oath of Supremacy (which might be accounted a religious oath), wrote Cecil, they should be asked the Bloody Question. Those who swore that they would not take up arms against the Pope could then immediately, and confidently, be accounted traitors – and no one could accuse the Government of religious persecution. Interestingly, Cecil, himself, was convinced that most Catholics, however devout, would chose country over conscience at this point. The question remained, though, what to do with those Catholics, like Campion, Filby or Ford, who refused to answer, or who fudged the issue. Could their silence, or fudging, be taken as an indication of their guilt? This appears to have been the proposition put to the Crown's lawyers. And while the Crown's lawyers met to consider the case, Henry Garnet and Robert Southwell seem to have called an emergency meeting of their own to decide how best to respond to this latest threat to their co-religionists.[23]

Whether or not the Jesuits consulted a team of Catholic lawyers is unclear, but their agreed new position, as Southwell explained it in his letter to Claudio Aquaviva that summer, was eminently legalistic. Since 'it was not a question of faith', explained Southwell, '. . . it was thought more prudent to use language that was truthful and yet would not irritate the magistrates'. Priests should therefore swear 'that as priests it was unlawful for them to bear arms'. Furthermore, they should declare that they were praying to God for Him 'to favour the side on which His cause and that of justice stood': so God could choose who to fight for, Elizabeth or the Pope. Laymen were advised to ask for the chance 'to prove their loyalty to sovereign and country by defending both against . . . unjust aggression', neatly concealing, behind a flurry of promised activity, the fact that some might actually view a papal invasion as just aggression.[24]

It was the best counsel the Jesuits could give under the circumstances, but it was the Crown's lawyers who saved the day for the mission. When their answer came back on 20 July it marked a conclusive victory for the forces of justice over the forces of fear. No, the Government could not take silence as a statement of guilt. Nor were those Catholics guilty who declared themselves 'unlearned and ignorant and so not able to answer', nor those who pointed out the illegality of being questioned on what they *might* do some time in an unspecified future. Certainly those who answered in this fashion were 'dangerous persons', but 'unless some other action drawing them in danger of the law may be proved against them', they were not traitors and could not be proceeded against as such. With this polite but firm admonishment to the Government for attempting to enshrine thought-crime in the statute books – and removing the right to silence into the bargain – the Crown's lawyers rendered Cecil's Bloody Question legally toothless. Morally, though, it still retained its bite.[25]

John Gerard had now to come up with an answer that would protect him from accusations of disloyalty to the State, while not compromising his loyalty to the Pope as head of the Church. Nor must he provide the kind of definitive judgement on the matter that might be used against other Catholics in the future. All this, while retaining his attitude of studied insolence in the face of Richard Topcliffe's bullying. Gerard, himself, was clear about the difficulties: Topcliffe 'had so framed his question', he wrote, 'that whatever I answered I would be sure to suffer for it, either in body or in soul'. His answer was a considered one: 'I am a loyal Catholic and I am a loyal subject of the Queen. If this were to happen, and I do not think it likely, I would behave as a loyal Catholic and as a loyal subject.' It was a brilliant attempt to square the unsquarable, to assert that English Catholics were bound by a dual loyalty to Rome and Elizabeth, and to redefine that loyalty in terms of individual conscience rather than blind, objective allegiance, but it utterly failed to address the issue at stake. For Rome had ruled that loyal English Catholics could not be loyal Elizabethan subjects, not while Elizabeth remained under sentence of deposition. It was hardly surprising that Topcliffe should regard this as casuistry at work, slippery and not to be trusted. Gerard was returned to his cell while his accusers returned to pondering his immediate fate.[26]

On 13 April 1597 the Privy Council drafted the following letter.

> You shall understand that one Gerard, a Jesuit, by her Majesty's commandment is of late committed to the Tower of London for that it hath been discovered to her Majesty he very lately did receive a packet of letters out of the Low Countries which are supposed to come out of Spain, being noted to be a great intelligencer and to hold correspondence with Persons the Jesuit and other traitors, beyond the seas. These shall be therefore to require you to examine him strictly upon such interrogatories as shall be fit to be ministered unto him and he ought to answer to manifest the truth in that behalf and other things that may concern her Majesty and the State, wherein if you shall find him obstinate, undutiful or unwilling to declare and reveal the truth as he ought to do by his duty and allegiance, you shall by virtue hereof cause him to be put to the manacles and such other torture as is used in that place, that he may be forced to utter directly and truly his uttermost knowledge in all these things that may any way concern her Majesty and the State and are meet to be known.

After almost three years of an extraordinarily liberal captivity the situation for John Gerard had just taken a sharp turn for the worse.

The chief question raised by this letter is, if Gerard were such a 'great intelligencer' – i.e. an intriguer – why was he permitted to remain in the Clink for so long, under so little supervision and without any form of action being attempted against him? To put his captivity into perspective, he was arrested on 23 April 1594 and transferred to the Tower of London on 12 April 1597, during which time he was subjected to an unspecified number of examinations, none of which appears officially to have been recorded. Over the same period a total of thirteen other Catholics were executed, of whom nine were priests; of those nine, the two with most bearing on Gerard's condition were Robert Southwell and Henry Walpole, both executed in 1595, on 21 February and 7 April respectively.

Southwell, too, had experienced an unaccountable delay

(between arrest and trial), but he had spent his incarceration in the Tower, in strict solitary confinement, whereas Gerard was lodged in an openly Catholic prison known for its comparative laxity. Walpole, on the other hand, had been fast-tracked from York, to the Tower and then back again to York for execution, a victim of circumstances beyond his control. His case makes for a strange and tortuously convoluted hunt through the State Papers, but it is worth outlining the salient facts in order to indicate how the Government's mind was working against the Jesuits in 1594.

In the December of Walpole's return to England – 1593 – an adventurer named William Polwhele voluntarily confessed – on the grounds of an over-burdened conscience – to a plot to murder the Queen. He, Polwhele, was to be the assassin and he suggested the Council investigate a man named John Annias to discover more.[28]

Annias, under questioning, provided two further names: those of Patrick Cullen and John Daniel, both Irishmen.[29]

Cullen, in turn, pointed the finger at Father William Holt, an English Jesuit connected with the Scottish Court back in its pro-Catholic/pro-France days of the early 1580s and now known to be living on the Continent, consorting with Catholic exiles, in particular the northern rebels. Holt, said Cullen, was the man behind the plot, but actually he, Cullen, was to be the assassin.[30]

Daniel, when it came to him, pointed the finger wider still. The first name he mentioned was that of Hugh Cahill, yet another Irishman (and yet another who voluntarily admitted to being the assassin-designate). Daniel – supported by Cahill – said that those behind the plot were Holt, Hugh Owen and Sir William Stanley.[31]

Owen was an English Catholic, living in Flanders and widely regarded as one of the leaders of the discontented exiles. Stanley was a former officer in Elizabeth's Army who, sent to the Netherlands to fight against the Spanish, had surrendered to Spain first the besieged town of Deventer and then his entire regiment; now he fought alongside Philip's Catholic forces, stamping out heresy. Holt, Owen and Stanley and a plethora of Irishmen: three names, and an entire nation, long suspected by the English Government as a source of Catholic unrest.* On 17 February 1594, as the confessions flowed, the Council appointed officers to every English port to search, interrogate and, if need be, detain anyone entering the country.[32]

But it was the very last name Daniel and Cahill mentioned that is of interest here. As he waited in Calais to board a boat for England, said Cahill, he was met by a short, tanned, well-set man, 'with black hair, very like a Spaniard, about 33 or 34 years old'. The man's name was Henry Walpole. He was accompanied by a Jesuit named Archer (who specifically urged Cahill on in his assassination attempt). Both Archer and Walpole advised Cahill to travel in secret, a statement confirmed by Daniel. If true, this was sound advice from Walpole to any Catholic Irishman wishing to enter Protestant England, but the implication was clear: Walpole was aware of Cahill's mission and was therefore a potential accessory to murder.[33]

In the event Walpole appears to have satisfied his interrogators that he was entirely innocent of conspiring the Queen's death: witnesses to his trial make no mention of any assassination plot. Under examination he admitted meeting both Cahill and Cullen in Calais, but since all three of them were engaged in the same search for transport across the Channel this was not in itself suspicious. He also admitted that he had heard rumours of a planned assassination attempt while still in Valladolid and that, in hindsight, he now wished 'he had taken more intelligence thereof, but withdrew for fear of entangling himself'. He believed that 'if any member of his society had been known to deal in such

* Government informers were employed to tag the English exiles, but a letter, dated 2 January 1592, from W. Sterrel, an agent provocateur, to Thomas Phelippes, the Government's code-master, revealed that the process was not foolproof. Sterrel writes about an informer, Cloudesley, a former servant to Hugh Owen. Cloudesley had just returned from the Continent, but had succeeded in messing up his mission, delivering Sterrel's letters to the wrong people: 'giving Owen's to Westmoreland [Charles Neville, the northern rebel], and Westmoreland's to Holt, which may do hurt, they being of divers factions'. Sadly Sterrel does not reveal the contents or purpose of these letters, but he does suggest Cloudesley should instead be employed 'about the prisons', to hunt out information there.

a horrible enterprise, the General [Aquaviva] would cast him out'. It was not enough to save his life, but it was enough to clear his name of misprision of murder.[34]

But the Jesuits as a whole remained severely tarnished by these accusations. In their absence, Holt and the unknown Archer were considered guilty as charged. This purportedly Catholic-led, Irish-backed mishmash of a plot, for which Patrick Cullen would eventually hang, may have seemed farcical in so many of its details, but it confirmed the government's suspicion that the Society was growing ever more inimical to English interests.* For if, as had been claimed, its members were prepared to promote Elizabeth's assassination in order to save English Catholicism, then the Bloody Question had just been provided with an equally, and unequivocally, bloody answer. So why then was John Gerard, a known Jesuit, allowed to remain so long unpunished?[35]

One cannot escape the conclusion that this was a regime that still actively did not want to kill. Campion and Southwell had both had distinct and inspirational voices that had needed silencing, and quickly. Walpole, however unwittingly, had been linked to the Cullen conspiracy. But William Weston was now entering his tenth year of captivity, despite being the one-time head of the mission. And, significantly, since Southwell's death, not a single Catholic had been executed in London. If this was a reaction to the outcry at his execution, then one legacy of those Tyburn scenes seems to have been to make the Government loth to proceed indiscriminately against Gerard now. So although Richard Topcliffe might write to Robert Cecil that the Jesuit was 'very desperate and dangerous',

* The credibility of this conspiracy is in doubt, as is that of the so-called Lopez plot, discovered that same month. Sceptics look to the fact that the Earl of Essex, who had taken onto his payroll many of Francis Walsingham's spies on the latter's death in 1590, was keen to prove himself indispensable to the ageing Queen after she over-ruled his choice of candidate for the post of Attorney General. Suspicion of the case grows in direct proportion to the number of people apparently so willing – voluntarily – to claim the assassin's role. Neither is it helped by Daniel's testimony that when he and Cahill took their leave of the Jesuits they were told 'to use all haste, as there were an Englishman and a Scotchman appointed for a similar purpose'. Earlier, Cahill had said that his recruiters particularly 'wanted to employ a tall, resolute, and desperate Irishman' for the job.

clearly, in the absence of any pressing reason to proceed against Gerard, the Council was prepared to play a waiting game.* Meanwhile, they kept him in prison and under surveillance.[36]

Clues about how long Gerard was watched in the Clink, and by which of his fellow prisoners, are few. The only fact that is clear is that the informer was a Catholic missionary, for on 2 October 1596 the Secretary to the Privy Council, Sir William Waad, wrote to Sir Robert Cecil, revealing the latest information to come from his 'priest in the Clink'. The priest had told Waad 'of a very tall handsome man . . . come from beyond the seas, apparelled all in black, a black satin doublet, velvet gascons, a long cloak with buttons. He was thrice in one week at the Clink, but being warned by Garnet, cometh no more'. Clearly Waad meant Gerard here, rather than Garnet, because in his next sentence he advised removing Garnet to a more secure prison (impossible with Garnet still at liberty) on the grounds that 'he giveth advertisement [information] beyond the seas'. But this confusion aside, both Waad and Robert Cecil were now aware that their prisoner was receiving visitors and communiqués from the Continent. It took until spring the following year, some six and a half months later, for the Government to act on the knowledge.[37]

Now Gerard takes up the story. For some time he had known that a fellow priest in the Clink 'was a little unsteady and seemed rather too anxious to be free again'. Gerard had therefore been 'careful not to confide in him'.† However, at the beginning of April

* This did not stop the rumours flying on the Continent, though. On 9 May 1595 Charles Paget, in Brussels, wrote to Thomas Throgmorton, in Rome, saying, 'In England, executions are out for Fathers Edmonds [Weston], Walpole, and Gerard.' By this point, of course, Walpole was already dead.

† Unwilling to give his name, so as not to compromise him, Gerard says of the priest that he was someone 'whom I had occasion to help many times. On his arrival in England I had arranged for him to live in a fine house with some of my best friends'. From this clue it seems likely the priest was Robert Barwise, or Barrows (*alias* Johnson and Walgrave). Barwise was a Londoner, born in 1563 and educated at Reims. He returned to England in 1589 and during Lent 1593 he stayed at Broadoaks. On 10 March 1594 he was arrested 'in the Queen's highway by Newell and Worsley' and committed to the Clink by Richard Young. Topcliffe mistakenly lists him as a Jesuit and describes him as 'very dangerous', but it seems the greatest danger Barwise posed was to his fellow Catholics. While at Broadoaks he

1597 this unsteady missionary (who evidently had not yet been set free) went to the authorities to inform against Gerard. 'He said he had been standing next to me', recorded Gerard, 'when I handed a packet of letters from Rome and Brussels to Father Garnet's servant "Little John" [Nicholas Owen] . . . and that I was in the habit of receiving letters from priests abroad addressed both to me and my Superior'. Gerard, himself, neither confirms nor denies these accusations, but given his other activities in the Clink it seems highly likely he had taken on the role of sorting-office for the mission, and highly unlikely he had done so in front of a man he mistrusted. Nonetheless, the accusation provided sufficient excuse for the Government to turn against him. On the night of 11 April 1597 John Gerard was unceremoniously removed to the Tower of London.[38]

He was taken to a 'tall tower, three storeys high with lock-ups in each storey'. This was the Salt Tower, standing at the south eastern corner of the Inner Ward, the first circle of the palace's defensive battlements. He was assigned a room one floor up. Next morning he was able to examine this new cell. One window had been blocked up, reducing what light there was to a trickle, but on either side of its stone surrounds Gerard could make out the names, in chalk, of all the orders of the angels. Above the window were the names of the cherubim and seraphim, above them the names of Mary and Jesus and, at the very top, God's name, written in Latin, Greek and Hebrew. Close by, chiselled into the wall, was the name of the author of this private oratory: Henry Walpole.[39]

Gerard remained in this cell just a single day, drawing comfort from 'a place sanctified by this great and holy martyr'. The

---

'rode a gelding of Wiseman's and wore his cloak, though Wiseman told him it was very dangerous for him to go to his house, because of the often watches and searches'. In 1594 he provided Richard Young with information about a Catholic book called *News from Spain and Holland*. In 1602 he provided undisclosed evidence to Robert Pooley (which Pooley duly forwarded to Cecil), as well as details of 'the secret in and out passages of the Jesuits, the conveyance of their closest affairs, and in what places they remain'. Interestingly, at his arrest Barwise admitted knowing John Annias and also mentioned a vague plot to kill Elizabeth and Cecil.

following day he was moved upstairs to the second floor, to another cell which was 'large and, by prison standards, fairly comfortable'.* His warder now offered to fetch him a bed, if his friends could provide him with one, and Gerard directed him to the Clink, from where he returned shortly, with a mattress, a coat and fresh linen, and the promise of a retainer from the Clink Catholics if he treated Gerard well.[40]

Next day Gerard was led out for interrogation. Waiting for him were Sir Richard Berkeley, the Lieutenant of the Tower, Sir Edward Coke, the Attorney General, Sir Thomas Fleming, the Solicitor-General, Sir Francis Bacon, then still at the start of his legal career, and Sir William Waad, representing the Privy Council. The examination began. And this time there is an official transcript.

The first question was whether or not Gerard had received any letters from beyond the seas. He admitted he had. Now his inquisitors began firing questions at him: whose letters were they, where were they going to, who had delivered them, what did they contain? Only once did Gerard seem rattled, confessing at first that he had burnt the letters, before admitting that, in truth, he had forwarded them on. But for the rest of the session he refused to name the senders, conveyors, or intended recipients of the letters. As to their contents, he had only read two or three of them and they had dealt solely with matters concerning the maintenance of scholars overseas; these he had sent on to parties more concerned than himself with financial affairs. Again and again, his examiners were forced to report that 'he refuseth to disclose' names. Towards the end of the examination Gerard was permitted to set down the reasons for this obstinacy. The writing is small and neat, each letter printed separately, very different from the extravagantly cursive secretary script surrounding it. 'I refuse not for any disloyal mind I protest as I look to be saved, but for that I take these things not to have concerned any matter of state with which I would not have dealt, nor any other but matters of devotion as before.' Questioned

* Carved into the wall of this second storey cell is a heart pierced by an arrow, symbol of one of the five wounds of Christ, and what look like the initials JG.

about why he had written the above in a feigned hand, he replied 'that he would bring no man to trouble and for that he will not acknowledge his own hand'. Two matters remained. First, why had he recently attempted to escape from prison (a fact of which Gerard, himself, makes no previous mention)? He had done so in order to have the 'more opportunity to save more souls'. Last, and with the only name they had at their disposal, his interrogators asked him about Henry Garnet. Were the letters for Garnet? Where was Garnet? Gerard refused to answer. Now his examiners produced the Privy Council's torture warrant. 'I do not think', Robert Persons had written confidently, and incorrectly, in 1584, 'that they will be so ready again to torture the priests they seize; so that those of us who are now sent will run much less danger of suffering than those who went before.' He continued: 'This is a great point, because truly to be hanged is child's play in comparison with being tortured.'[41]

'We went to the torture-room in a kind of solemn procession,' Gerard recalled, 'the attendants walking ahead with lighted candles.' It is likely that the room in question was in the basement of the White Tower: underground, cavernous and very dark, though in the candlelight the instruments of torture were clearly visible.* Gerard's interrogators showed him the devices and asked him to confess. Gerard refused.[42]

He was led to an upright wooden post, one of the roof supports for the chamber, into which a number of iron staples had been driven. Metal gauntlets were attached to his wrists and he was commanded to climb up a set of wicker steps at the foot of the post. His arms were raised above his head and an iron bar was thrust through the rings of one of the gauntlets, then through the highest staple and then through the rings of the second gauntlet.

* There seems to have been no one single torture-chamber in the Tower and frequently prisoners were tortured in their cells, but Gerard's account does suggest he was taken to the White Tower. There is rumoured to have been an underground passage between the Lieutenant's lodgings, where his initial examination took place, and the White Tower vaults, and from Gerard's description of the attendants' torches it is possible he used this passage now.

The bar was then fastened with an iron pin to prevent it slipping and one by one the wicker steps were removed from under him. Still, though, his toes touched the floor and his captors were forced to scrape away the earth beneath his feet until they hung clear of the ground. Again he was asked to confess.

Gerard opened his mouth to respond, but, as he recalled later, 'I could hardly utter the words, such a gripping pain came over me. It was worst in my chest and belly, my hands and arms. All the blood in my body seemed to rush up into my arms and hands and I thought that blood was oozing out from the ends of my fingers and the pores of my skin. But it was only a sensation caused by my flesh swelling above the irons holding them.' For a while Gerard was tempted to answer any question put to him and it was only (he wrote afterwards) thanks to God's mercy that he was able to resist this urge. He consoled himself with the thought that 'the utmost and worst they can do to you is to kill you, and you have often wanted to give your life for your Lord God'. Seeing his continued refusal to speak, his examiners now left Gerard in the care of the gaolers.*[43]

Some time in the early afternoon Gerard fainted. The gaolers replaced the wicker steps beneath his feet, supporting his body until he came to. Then, when they heard him beginning to pray, they removed the steps again. This continued some eight or nine times over the course of the afternoon until the Tower bell rang at five o'clock. Soon afterwards Gerard was released and taken back to his cell. The next day he was returned to the torture-chamber.

By now his wrists were so swollen that the gauntlets could barely be fastened and the pain was intense. This time when he fainted it took the gaolers so long to resuscitate him they sent for Sir Richard Berkeley in case he was dying. Gerard came to, sitting on a bench, supported by gaolers on either side, with warm water being

* On 7 May Garnet wrote to Aquaviva, 'The inquisitors say that [Gerard] is very obstinate, as they cannot draw the least word out of his mouth, except that in torment he cries "Jesus".'

trickled down his throat. Berkeley appeared reluctant to continue with the session, but with Gerard still refusing to talk he had little choice; this time, though, the ordeal lasted only an hour before Berkeley called a halt. Two months later Berkeley would resign his position as Lieutenant of the Tower. It was reported to Gerard afterwards that he had done so 'because he no longer wished to be an instrument of torture of innocent men'. Whether or not Berkeley believed Gerard to be innocent is unknown – this might just have been the partisan view of Gerard's source. Nonetheless, it certainly seemed Berkeley was unhappy implementing a process that had no judicial basis in English law.[44]

In 1565 the English ambassador to France, Sir Thomas Smith, had written in disgust at Continental practices: 'Torment . . . which is used by the order of civil law and custom of other countries . . . is not used in England, it is taken for servile.' This suitably jingoistic condemnation of foreign practices – torture was legally permitted in most European countries – neatly masked the fact that, though it might never appear in English statute books, torture did occur in England and in Elizabethan England more so than at any other time.* Of the eighty-one recorded cases of torture (of which Gerard's is one) that took place between 1540, when the Privy Council registers begin, and 1640, after which no further mention is

* Pre-1215, most criminal cases were tried by ordeal, the ultimate judge being God, himself. The person who floated during the water ordeal was guilty; the person whose hand remained infection-free after a hot iron had been applied to it was innocent, and so on. In 1215, though, the fourth Lateran Council [Pope Innocent III's ecumenical conference] banned the ordeals and countries had to find other methods of trial. In England the jury system was already in use in exceptional circumstances, namely when the accused wished to avoid the ordeals, and this system was now adopted for all cases. In mainland Europe the system chosen was that of the Roman-canon law of proof, which laid down strict criteria for conviction: to return a guilty verdict the court needed two eye witness reports or the accused's own confession; circumstantial evidence was not permitted. However, by a complicated system of grading (a point for one eye witness report, a point for substantial circumstantial evidence, etc.), if the accused was felt to have half-proof against him, and was therefore highly likely to be guilty, he could legally be tortured to see if any further proof could be obtained. The practice was strictly controlled: the accused could not be asked leading questions and all information had to be verified, the aim being to extract evidence that no innocent person could know. The use of judicial torture was finally abolished in the eighteenth century.

made of the practice, fifty-three of those cases took place in Elizabeth's reign, mostly for crimes against the State. Furthermore, not all instances of torture appear to have been recorded – there is no torture warrant for Southwell, for example, despite Topcliffe's letter clearly detailing what he had in mind for the Jesuit. Probably some warrants are simply missing from the registers; possibly the Catholic chroniclers of the period have talked up the amount of torture that took place; or there was sufficient Governmental queasiness surrounding the practice that some torture cases went deliberately unrecorded. But certainly the Government *was* made deeply queasy by the practice, as a report of 1583 revealed. This report, denying the popular and widespread rumours that Campion and his fellows had been barbarously racked in the Tower, stressed again and again that only those known to be guilty were ever tortured. No 'innocent was at any time tormented; and the Rack was never used to wring out Confessions at Adventure upon Uncertainties'. Further, 'the Queen's Servants, the Wardens, whose office and Act it is to handle the Rack, were ever, by those that attended the Examinations, specially charged to use it in as charitable manner, as such a Thing might be'.[45]

Charity and torture are not words that usually go hand in hand, but evidently Richard Berkeley had, in the case of John Gerard, elected to be charitable. Gerard's torture now ceased. In one final act of charity before he resigned his job, Berkeley wrote to Robert Cecil, saying: 'Gerard, a prisoner in the Tower, being ill and weak, hath importuned me to signify his petition to be allowed to take the air on a wall near his prison. I am told to advertise you of this, being their mouth, as they term me. The man needs physic.'[46]

For the next few months Gerard remained in his warder's care. The effects of his torture were such that he was unable to move his hands for several days and he was entirely dependent on the warder to feed him. 'He had to do everything for me,' Gerard recalled afterwards. Nonetheless, the authorities made sure to remove Gerard's knife, scissors and razors when he was alone: standard Tower procedure for a prisoner facing torture.[47]

'At the end of three weeks, as near as I can remember, I was able to move my fingers, and to hold a knife and help myself.' Now Gerard was able to send to his friends in the Clink for some money with which to bribe his warder. The first items he bought with this money were oranges, which he presented to the warder as a token of friendship. Now the warder was happy to bring him more oranges. 'Each day', wrote Gerard, 'I did exercises with my hands after dinner ... cutting up the orange peel into small crosses; then I stitched the crosses together in pairs and strung them on to a silk thread, making them into rosaries. All the time I stored the juice from the oranges in a small jar.' Then Gerard asked the warder if he might send his crosses and rosaries to friends in the Clink. The warder agreed. Gerard asked if he might have paper to wrap them in. The warder agreed. Gerard asked if he might write a few lines in charcoal to his friends. The warder agreed, so long as he could check the words himself. Gerard agreed. Then, with the warder out of his cell, Gerard scribbled a second note to his friends, this time in his secretly hoarded orange juice. It was an old trick, familiar to every Catholic in the Clink, and the recipients of Gerard's crosses knew exactly what to do with the paper, holding it up to the fireside until their hidden message became visible. Gerard had successfully restored communications with the outside world.* And the outside world brought happy news. From Henry Garnet, Gerard now received word to prepare himself for trial and execution.[48]

On 13 May 1597 Gerard was led out again to meet his examiners (only Francis Bacon was absent this time). Attorney General Coke laid out the case for the prosecution. He started with Gerard's priesthood and his coming to England as a Jesuit, a fact Gerard did not deny. Then he asked whether Gerard had ever withdrawn

* Gerard's account of this correspondence gives careful advice on the merits of orange juice and lemon juice as invisible inks. The advantages of the latter are that it can be read just as well with water as with heat, and that if the paper is once dried out, the writing disappears again. Orange juice writing can only be read with heat and once the heat has brought it out, it stays out. The advantage of this is that the recipient of an orange juice letter can always tell if his message has been intercepted and read.

the Queen's subjects from their allegiance to the English Church. To this charge Gerard also pleaded guilty. This was sufficient to convict him of treason. In a supplementary question Coke now asked Gerard how he could hope to bring about the conversion of England without meddling in politics, this surely being the best means to achieve his ends. Gerard picked his words with care, recording them afterwards with equal care:

> 'I would want the whole of England to return to Rome and the Catholic faith: the Queen, her Council, and yourselves also, and all the magistrates of this realm; yet so, my Lords, that neither the Queen, nor you, nor any officer of state forfeit the honour or right he now enjoys; so that not a single hair of your head perish; but simply that you may be happy both in this present life and in the life to come . . . I am not at enmity with the Queen nor with you, nor have I ever been.'

For all the sophistication of the missionary training, the schooling in rhetoric and debate, this was a staggeringly simple statement, defining an ambition of almost childlike innocence. And it seems impossible to take it at anything but face value. To Gerard, politics equalled intrigue and possibly bloodshed; his mission – a religious mission – was to achieve a bloodless coup, a quasi-miraculous return of England to the Catholic Church without any loss of life, any loss of office, or any loss of any freedom currently enjoyed by any person in the land. It was little wonder Coke, as Gerard recorded, 'was at a loss for an answer'.[49]

Now the examination moved onto the vexed issue of equivocation, used by Gerard again and again throughout his examinations to avoid naming names and incriminating others. To Coke, equivocation 'countenanced lying and undermined social intercourse between men'. To Gerard, 'the intention was not to deceive, which was the essence of a lie, but simply to withhold the truth in cases where the questioned party is not bound to reveal it': to deny a man that to which he had no claim could not be accounted deception. Significantly, Gerard ruled that equivocation could only be used in cases of religion, not in criminal cases under lawful interrogation. It was a statement that left his examiners in little doubt as to his views on the legal basis of the Church of England. He summed up his position thus: 'In general, equivocation is unlawful save when a person is asked a question, either directly or indirectly, which the questioner has no right to put, and where a straight answer would injure the questioned party.' All this Coke noted carefully, promising to use it against Gerard at his trial.[50]

The examination over, Gerard was sent back to his cell with the expectation that this promised trial would take place in a matter of weeks, at the commencement of Trinity Term, the summer sitting of the law courts. But then inexplicably the whole process stalled. The assizes passed with no trial date being set and suddenly it seemed Gerard was about to enter the same kind of limbo as Southwell, forgotten to the world. By the end of July he had determined that this being the case he should at least earn himself the consolation of saying mass again.

At the opposite corner of the Queen's Privy Garden from the Salt Tower was the Cradle Tower, home to fellow Catholic prisoner John Arden. Arden had been arrested in January 1587 on suspicion of complicity in the Babington Plot and had been condemned to death, but so far sentence had not been carried out and now he was permitted to receive visitors again. Gerard began the complicated procedure of secretly making contact with him. 'I signed to the gentleman to watch the gestures I was going to make . . . He watched me as I took a pen and paper and pretended to write; next, I placed the letter over the coal fire and held it up in my hands as though I were reading it; then I wrapped up one of my crosses in it, and went through the motions of despatching it to him. He seemed to follow what I was trying to indicate.'[51]

Arden had not followed what Gerard was trying to indicate. When he received the cross from Gerard's warder, specially bribed for the occasion, he carefully threw the wrapping paper in the fire. It took another three days, and another painful charade in which Gerard mimed the making of orange juice ink, before Arden realized what was required of him and finally read Gerard's proposal. The pair began to plan for their forbidden mass. Friends from the outside agreed to provide the necessary equipment. A date was set: 8 September, the feast of the birth of the Virgin Mary, the day after Queen Elizabeth's sixty-fourth birthday. The hardest part for Gerard was persuading his warder to let him visit Arden's cell and a significant amount of money appears to have changed hands. On the evening of 7 September Gerard was duly escorted to Arden's cell, ready to say mass early the following morning before the Tower was properly awake. It was only then that he realized precisely where Arden's cell was situated: 'while we were passing the time of day together, it struck me how close this tower was to the moat encircling the outer fortifications'. Now Gerard voiced a new proposal to Arden: escape.[52]

Days later Henry Garnet received a secret communiqué from Gerard asking for permission to attempt a break-out of the Tower of London. The Jesuit Superior had long been used to fielding such enquiries. While Walpole was still a prisoner in York, word had reached Garnet via his Jesuit colleague in the north, Richard Holtby, that there was a chance Walpole might be able to escape. Holtby asked for counsel. Garnet deferred to Holtby's knowledge of local conditions and Holtby advised against it. His reasons were twofold. Even in prison Walpole was proving inspirational to northern Catholics. More practically, were he to escape it was likely that 'a general search . . . would follow and that, for one priest at liberty, several might be captured, along with many laymen, some of them perhaps less well prepared to endure torture and death'. Walpole, himself, agreed: 'Similar considerations occurred to me. I proposed it only to satisfy others.'*[53]

But this did not mean that no one ever attempted escape. In 1596

* Early in William Weston's imprisonment friends had offered to buy his liberty in return for his exile. Weston had begged Southwell to dissuade them on the grounds that 'it was a despicable method of liberation' and Garnet had banned the escapade. He explained to Aquaviva, 'Since he [Weston] has, as it were, gained an eminent and illustrious position for himself, he is exposed to the observation of all and it would look shameful for the shepherd to fly from his flock at such a time as this.'

Richard Fulwood had fled the Bridewell, accompanied by the Jesuit Father John Percy, two seminary priests and six laymen. Garnet wrote to Aquaviva:

> 'Your Lordship should not be astonished that Catholics sometimes break out of prison. While there are certain prisons in which they are treated with humanity and give a pledge that they will not attempt escape, this prison was utterly barbarous and reserved chiefly for whores and vagabonds. Catholics were forbidden access to it and there was no communication permitted between prisoners. Here no pledge was given, for no one had the humanity to ask it. This escape, therefore, caused no scandal at all'.

The crucial factor determining whether Garnet permitted an attempt or not was the nature of any accompanying repercussions, and in the case of Gerard's proposal it seemed Garnet could find no reason to deny him. He only begged him 'not to risk [his] neck in the descent'.

Now the preparations intensified. Since his early orange juice correspondence with the Clink Catholics, Gerard had succeeded in bribing his warder to carry more letters in and out of the Tower for him. Through this means he made contact with his old servant Fulwood and with John Lillie, a former Catholic prisoner from the Clink and the occupant of the cell used as a chapel. It was Lillie who, since Gerard had arranged his release from prison, had undertaken the Jesuit's business. Between them they formed a plan.[55]

On the evening of 3 October Gerard was conducted to the Cradle Tower by his ever-compliant warder. Once again he had begged permission to spend the night in Arden's cell so that the two of them might celebrate mass early the following morning. As soon as the warder had barred the cell door behind him, Gerard and Arden set to work, chipping away the bolt on a second door that led up to the roof. By midnight they were standing on the battlements, looking out over the moat to Tower Wharf and, beyond that, to the River Thames. There they could just make out a rowing boat, with Fulwood and Lillie at the oars, and at the tiller Gerard's old warder from the Clink. Slowly the three men brought the boat alongside the wharf.

Then the plan began to unravel. From the rooftop Gerard and Arden watched as a man came out onto the wharf and hailed the rescue party, taking them for fishermen. For some time he stood there talking. Eventually he moved away, heading back towards his cottage on the wharf, while Fulwood and Lillie stood off from the river bank, making to row upstream. Time was ticking fast. The escape had been planned for the slack water at low tide, but as the rescue party paddled up and down waiting for the man to go to sleep the tide began to turn, gathering force by the moment. In frustration Gerard and Arden saw the boat now alter course and row away. The attempt had failed. Worse, the incoming tide was sweeping the boat towards London Bridge, pinning it against the piles driven into the river bed to break the flow of the current.* As the water level rose so each new wave threatened to capsize the boat.[56]

In the darkness Gerard could hear the men's shouts for help and the answering cries of people on the river bank. Lights were moving and soon the water was alive with boats circling the trapped vessel, but not daring to close in for fear of being swept onto the piles themselves. More lights were lowered from the bridge and in their glow Gerard could see a kind of basket swinging from the end of a rope as shadowy figures attempted to pull the three men from their craft. It took a sea going ship to brave the tide and move near enough to the stricken rowing boat to haul the men to safety. Finally the river was quiet again and Gerard and Arden returned to their cell. It still remained for them to repair the broken lock before the warder arrived in the morning.

Next day Gerard received a letter from John Lillie. It began, 'It

* Old London Bridge had nineteen small arches spanning the river, severely restricting the flow of water and making the current there dangerously strong. An old proverb stated that the bridge was for wise men to go over and for fools to go under.

was not God's design that we should succeed last night, but He mercifully snatched us from our peril – He has only postponed the day. With God's help we will be back tonight.' So, at midnight on 4 October, having once more bribed the warder, Gerard and Arden stood again on the battlements of the Cradle Tower, overlooking the river. This time the rowing boat drew up alongside the wharf without incident. The gaoler from the Clink stayed behind while Lillie and Fulwood came to the edge of the moat, fastening one end of the rope they carried to a stake on the bank. They listened carefully for the chink of metal as Gerard tossed a small ball of iron down to them from the rooftop. Tied to the ball was a length of cord, which the two men now fastened to the free end of their rope, and Gerard began hauling it up to the battlements. The rope had been doubled at Henry Garnet's insistence, to prevent it from snapping, and its passage aloft proved alarmingly noisy. Then another difficulty presented itself. As Gerard recalled, 'The distance between the tower at one end and the stake at the other was very great and the rope, instead of sloping down, stretched almost horizontally between the two points.' At this point John Arden began to waver. Gerard remembered, 'he had always said it would be the simplest thing in the world to slide down. Now he saw the hazards of it'. But there could be no going back, not without more noise and more risk of being spotted.[57]

Arden went first, pulling himself bodily along the rope until he reached the wharf. Then it was Gerard's turn: 'I gripped the rope with my right hand, and took it in my left. To prevent myself falling I twisted my legs round the rope leaving it free to slide between my shins.' But now Gerard realized that Arden's crossing had slackened the rope and where once it had stretched taut across the moat, now it hung loosely. 'I had gone three or four yards face downwards when suddenly my body swung round with its own weight and I nearly fell. I was still very weak, and with the slack rope and my body hanging underneath I could make practically no progress.' It took huge effort for Gerard to reach the centre of the rope and there he stuck. 'My strength was failing and my breath, which was short before I started, seemed altogether spent.' Three and a half years of imprisonment and two bouts of savage torture had taken their toll.

It was 'the help of the saints and . . . the power of [his] friends' prayers' that finally, Gerard reckoned, got him to the wall on the far side of the moat, but here his strength gave out altogether, leaving him dangling, his feet just touching the top of the wall. John Lillie seized his legs and pulled him up over the wall and down onto the wharf and Gerard was half carried, half dragged to the waiting boat.

They rowed a good distance from the Tower before putting into shore. From there, Lillie escorted John Arden to Anne Line's house, while Fulwood took Gerard to Garnet's latest London base at Spitalfields, to the east of the city, where Nicholas Owen was waiting with horses. As dawn broke, Gerard and Owen were heading west towards the village of Uxbridge, some dozen miles from London on the Oxford road. On the outskirts of the village stood Morecrofts, a house newly leased by the Jesuits. Here, Garnet was expecting them. 'The rejoicing was great,' remembered Gerard afterwards.

It just remained to put the final part of Gerard's plan into operation. Before leaving his cell Gerard had written three letters. The first was to his warder, giving his reasons for escaping; the second was to the new Lieutenant of the Tower, assuring him that the warder had not been privy to the escape; the third was to the Privy Council, justifying his actions and assuring them that neither the warder nor the Lieutenant was in any way culpable. But concern for what might befall the warder led Gerard to go further still and a messenger was dispatched to intercept the man on his way to work that morning. Through this messenger Gerard now promised the warder an annuity of two hundred florins for life and the guarantee of a safe refuge if he left London immediately. The warder accepted and Richard Fulwood accompanied him to a house some hundred miles from London, belonging to friends of Gerard. In time he was joined there by his family, setting up his own house with the money Gerard sent him. He also converted to the Catholic faith.[58]

On the morning of 5 October the new Lieutenant of the Tower, Sir John Peyton, sat down to write a letter of his own – to the Privy Council. 'This night there are escaped two prisoners out of the Tower, viz., John Arden and John Gerard. Their escape was made very little before day, for on going to Arden's chamber in the morning, I found the ink in his pen very fresh . . . [The warder] is also gone this morning at the opening of the gates . . . I have sent hue and cry to Gravesend, and to the Mayor of London for a search to be made in London and all the liberties [outlying districts].'[59]

Strangely, though, Peyton's call for a search received little support from the Council. The prevailing view was that if Gerard had friends prepared to help him escape in this manner, then he also had friends prepared to find him a secure hiding place. One Councillor was even reported to have told a gentleman in attendance he was glad Gerard had got away – this was the news that filtered back to Gerard. Garnet, writing to Robert Persons three days later, noted, 'There is no great enquiry after him.' Gerard, himself, somewhat coolly recorded, 'A search was made in one or two places. As far as I could discover, nobody of note was taken.' This was not arrogance, so much as a realistic appraisal of his worth to the mission. With Elizabeth entering her sixty-fifth year the promise of change was now palpable in the air and after his own enforced absence from the field Gerard was poised to seize whatever chance that change might bring. Few could doubt that the events of the immediate years to come would be vital to the continued existence of English Catholicism. Fewer still could doubt that Gerard had earned the right to be at the centre of those events.[60]

Chapter Ten

'I know the inconstancy of the people of England,
how they ever mislike the government and have
their eyes fixed upon that person that is next to succeed.'

Queen Elizabeth I

London, 28 April 1603

LEADING THE WAY was the Knight Marshal's man. Behind him walked two hundred and forty poor women, in lines of four, the servants of the country's assorted knights, esquires, and gentlemen, all in rank, two porters, four trumpeters, the Rose Pursuivant at Arms, two Sergeant at Arms, the first of the royal standard-bearers and two equerries leading a riderless horse. This was just the beginning. Now came the members of the royal household: the messengers of the chamber, the children of the wood yard, the scullery, the scalding house, and larder, the grooms from the stables, the wheat porters, the men from the counting house. Behind them came the noblemen's servants, the grooms of the chamber, four more trumpeters, the Bluemantle Pursuivant, another Sergeant at Arms, a second royal standard, a second riderless horse. On and on they came: more members of the royal household, the children and gentlemen of the Chapel Royal, the clerks of the Council, the aldermen of London, two standard-bearers carrying the banners of Wales and of Ireland. Then came the office-holders: the Lord Mayor of London, the Chancellor of the Exchequer, the Lord Chief Justice, the Principal Secretary of State. Then the noblemen and clergy in order of precedence. Among them walked Tobie Matthew, Bishop of Durham. Thirty-seven years ago, as a young Oxford student, he had witnessed a royal procession similarly designed to strike awe into the hearts of the spectators in the streets; now he was taking part in one. Behind him came four more Sergeant at Arms, a standard bearer carrying the great embroidered Banner of England, the Norroy King at Arms, the Clarenceaux King at Arms, the Gentlemen-Ushers with their white rods. And behind them came the funeral carriage, drawn by four horses 'trapped in black velvet'.[1]

She had made so many royal progresses in her lifetime, covering almost a third of her kingdom in a deliberate and unprecedented attempt to reach as many of her subjects as possible. This last journey was all too short, from the Palace of Whitehall to Westminster Abbey, just a few hundred metres. On top of the lead coffin was a picture of her, crowned and in her Parliamentary robes; overhanging it was a canopy borne by four noblemen and surrounding them walked the Gentlemen Pensioners, their axes lowered to the ground. And now came the women of the court: the gentlewomen of the Privy Chamber, the noblewomen in order of precedence, the maids of honour of the Privy Chamber. Behind them, bringing up the rear, came the Captain of the Guard. The guardsmen followed five abreast, their halberds also lowered.

Down Whitehall the procession moved in muffled silence, watched by the gathering crowds, and on into Westminster Abbey. Then the great doors swung shut behind it and Elizabeth Tudor, Queen of England, France, and Ireland, Defender of the Faith – the old Plantagenet claims mingled with the new Henrician honour – passed from view to her final resting place. It was the end of a dynasty.

But if this funeral pomp were fitting tribute to the woman who had ruled England for almost half a century, then it was the forty-eight hours or so after her death during which her body was left unattended that more closely matched the mood of the closing years of her reign. Forty-eight hours in which the dead Queen had lain at Richmond virtually forgotten, her corpse covered over with a cloth in 'very ill' fashion, as her ministers and followers scurried to pay court to the new King. This was the reality of a monarch's demise. The Queen is dead! Long live the King! As power shifted from Elizabeth of England to James of Scotland no one could afford to be left behind in the stampede for new favours and new appointments – least of all England's Catholics.[2]

To Sir Francis Walsingham, Elizabeth had been 'the best marriage in her parish, and brought a kingdom with her for a dowry'. To a German diplomat, writing soon after her succession, Elizabeth was 'of an age where she should in reason, and as is woman's way, be eager to marry and be provided for'; for her 'to remain a maid and never marry [was] inconceivable'. But Elizabeth had done the inconceivable and never married.[3]

They had all pressed her to. Parliament had tried to make its granting of payments contingent on her finding a husband. Elizabeth responded by scribbling on the preamble to a finance bill the furious note: 'I know no reason why any of my private answers to the realm should serve for prologue to a subsidies book.' Her ministers and courtiers had pressed her to, even the Earl of Leicester, causing her to round on him, saying, 'she had thought if all the world abandoned her he would not have done so'. To which, gossiped the Spanish ambassador, Leicester 'answered that he would die at her feet, and she said that that had nothing to do with the matter'. Where Elizabeth and marriage were concerned, emotions, on all sides, ran deep. And understandably so: on the issue of Elizabeth's marriage rode the more important issue of the succession; and in the absence of any natural-born heirs of her own, it was by no means clear who should wear the crown after her. This being the case, Elizabeth's death threatened to catapult England into a mad scramble of competing claimants, loyalties and factions – something everyone currently enjoying the privileges of power and the comforts of comparative State stability viewed with alarm. If Elizabeth refused to marry then she must, at least, be forced to name her heir, preferably one on whom all the aforementioned could agree. So as her reign progressed and the prospect of her ever marrying and giving birth faded into the distance, all those who once had pressed her to take a husband now began to press her, even more stridently, to designate her successor. Except that Elizabeth had learned at a very early age the perils of having a clearly designated successor.[4]

While her sister reigned, she, herself, had been talisman to every malcontent in the kingdom: to everyone who hated Mary for her religion, for her marriage. It had almost cost her her life. And only she knew how innocent she had been of provoking their designs. An heir was a magnet for treason. In January 1568 Philip of Spain had his eldest son Don Carlos arrested for plotting against him; Carlos starved himself to death in custody. Closer to home, in Scotland, Elizabeth had watched her cousin Mary Stuart deposed in favour of her infant son. With her claim to her own throne more tentative than that of her sister, her cousin, or Philip to theirs, Elizabeth had had every reason never to discuss the succession. Not that this stopped anyone else from doing so.

Mary Stuart had always been the obvious choice for heir presumptive, as Elizabeth's closest blood relative and granddaughter of Henry VIII's eldest sister, Margaret. Her execution had opened up the field. So much so, in fact, that one commentator would write in 1600, 'this crown is not like to fall to the ground for want of heads that claim to wear it'. The front runner was still Mary's son, James VI of Scotland, but the same legal considerations that had cast a shadow over Mary's hopes also applied to him. As a foreigner he was forbidden from inheriting property in England and this might be said to extend to the Crown. Moreover, Henry VIII had excluded his eldest sister's family from the succession under the terms of his will. This left a number of English heirs for consideration, but though any or all of them might be good for an each-way bet, none looked like staying the distance, chiefly because none was much favoured by anyone in England. 'Either in their worth are they contemptible, or not liked for their sexes,' remarked one nobleman acidly, if accurately.[5]

There was Lady Arbella Stuart, like James a great-grandchild of Margaret Tudor, so barred by Henry's will, but unlike James raised in England. Sadly for Arbella, that meant being raised by her maternal grandmother, the formidable Bess of Hardwick. Bess's rapid advancement through English society had been a triumph of tenacity over tact; she seemed to have inculcated similar talents in Arbella. In 1587 Elizabeth had invited the twelve-year-old girl to Court; watching her elbow aside the other women of the royal household to claim a place in the front row of a procession, she had swiftly sent her home again. Nonetheless, Arbella remained a sufficiently sure candidate for the French ambassador to hint that Elizabeth was about to overturn her father's will and name the girl as her heir.[6]

Next, there were the descendants of Henry VIII's younger sister Mary, through her marriage to the Duke of Suffolk. These were an even sorrier prospect than Arbella. True, there were among them a number of male heirs, notably Edward and Thomas Seymour from the senior branch of the family, but they were both technically bastards, their mother (Mary's eldest daughter) having contracted a secret and (it was widely thought) invalid marriage to their father, the Earl of Hertford. In 1595 the earl had had the temerity to seek legal advice about the possibility of validating the marriage and was promptly arrested. This proved sufficient warning for him never to dabble in the succession again: when in 1602 he received a secret communiqué from Arbella suggesting she marry his eldest grandson and strengthen their joint claims to the throne he immediately told the Council. This really only left the extravagantly named Ferdinando Stanley, Lord Strange, heir to the Earl of Derby, for consideration. Ferdinando was descended from the junior branch of the Suffolk line. Through his father he was a powerful landowner in his own right. In fact the only obvious problem with Ferdinando was that he, himself, showed no interest in the Crown.* Others, though, were interested on his behalf.

There had been whisperings throughout much of the early 1590s, scraps of information filtering back to the Council concerning Ferdinando. In June 1592 a Government agent reported dealings between him and William Allen. That same summer the priest-turned-informer James Younger wrote repeated warnings from prison of an invasion attempt on England to be led by the infamous Sir William Stanley (soon to become even more infamous for the Cullen conspiracy). Stanley's army, reported Younger, would land 'upon that part [of the coast] nearest to Ireland', where Ferdinando and his father would come to its aid.† And Henry Walpole, under interrogation in the Tower, completed the picture. Before leaving for England, said Walpole, he had been asked by Stanley 'to deal with some priest that might get access to [Ferdinando] ... to induce him to the Catholic religion'. The priest that Stanley recommended for the job was John Gerard.[7]

The plan was as follows: Ferdinando was to be persuaded to convert to Catholicism and then pursue his claim to the English throne, with the backing of the Catholic Church. In all this, the least unconvincing aspect was Stanley's selection of Gerard as the man most likely to get access to Ferdinando: the two were distant relatives. Moreover Gerard's elder brother, Thomas, was a Lancashire neighbour of the Derbys at their family seat of Lathom

* Ferdinando's chief passion was the theatre. He kept a professional acting company, Lord Strange's Men, who performed privately and publicly, at the Theatre in Shoreditch and the Rose Theatre on Bankside. At the latter they are believed to have presented Marlowe's *The Jew of Malta*, written *c.*1590, and one of Shakespeare's first plays, *The Comedy of Errors.*

† It seems the Government took none of this seriously: when, the following year, Elizabeth wrote to Derby, as Lord Lieutenant of Lancashire, informing him she was sending troops north against a rumoured invasion, there was little sign in her letter that she regarded him as a traitor.

House.* But Stanley's interest was no less familiar: he, too, was related to Ferdinando – and it appears to have clouded his judgement. It must have been a beautiful daydream for an unhappy Catholic exile, imagining one's cousin occupying the English throne, but like most daydreams it failed to stand up to scrutiny.[8]

In the autumn of 1593, just days after the death of his father, Ferdinando was approached by a Lancashire man named Richard Hesketh, newly returned from the Continent and bearing a message from Stanley. Stanley's instructions to Hesketh were to proceed tentatively at first: 'declare unto him in general that your message concerneth the common good of all Christendom, specially our own country, and in particular himself'. Only if Ferdinando seemed ready to listen was Hesketh to come to the meat of the matter: that if the new earl agreed to convert, his 'friends' overseas were prepared 'to offer him all their endeavour, services and helps . . . to advance him'. There was one proviso, though. Ferdinando had to press his claim now, before Elizabeth's death, and with an army behind him, so as to forestall his competitors. If he hesitated, warned Stanley, he was lost, for 'he hath many enemies that daily seek his overthrow'. Scenting treason, Ferdinando informed the authorities. Hesketh was hanged and the matter closed.[9]

Except that early in April the following year Ferdinando fell ill. He died on 16 April 1594, 'tormented with cruel pains by frequent vomitings of a dark colour, like rusty iron . . . [that] . . . stained the silver Basins in such sort, that by no act they could possibly be brought again to their former brightness'. His dead body, it was reported, 'ran with such corrupt and most stinking humours, that no man could in a long time come near the place of his burial'. The symptoms pointed to poisoning and suspicion fell on Ferdinando's stable manager, who had absconded on the earl's

* Gerard's great-great-grandmother was Margaret Stanley. The connection between Thomas Gerard and Ferdinando was hinted at in a report, claiming Gerard was 'brother to one of his [Ferdinando's] familiars'. Gerard, himself, makes no mention of ever having had contact with Ferdinando.

best horse the moment he took to his bed. But no one could believe the man was acting alone: Catholics were known to have approached Ferdinando, therefore, according to the logic of the time, Catholics must have killed him. Privy Counsellor Sir Ralph Sadler wrote darkly of Jesuit connivance, recording in his journal that whether Ferdinando's death was 'by their practice or no, God knoweth, and time will discover. But that so it was . . . there is nothing more likely.'[10]

In fact Ferdinando's death would have benefited the Catholic cause little. He had already told the Government all he knew of the plot, moreover he had never been an enemy of Catholicism – rather, like his father before him, he was regarded as a religious neutral.* If his murder was an act of Catholic revenge then it did little more than add to the general opprobrium in which the faith was already held. Ask *cui bono?* and the answer surely comes back, Ferdinando's rival claimants.† And in certain quarters rumours were soon being spread to this effect. First to be blamed was Sir William Cecil, who, two months after Ferdinando's death, married off his granddaughter to Ferdinando's brother, the new Earl of Derby. Bess of Hardwick, still pushing Arbella's claim, had a fine time fielding accusations, notably against Ferdinando's brother and against Francis Hastings, a descendant of the Plantagenets, but in her opinion Ferdinando's demise was the result of his 'making himself so popular and bearing himself so against my lord of Essex'. And, from Brussels, Hugh Owen, William Stanley's alleged co-conspirator in the Cullen plot, wrote a surprising letter to Thomas Phelippes, the Government's chief code-breaker, attempting to clear the Earl of Essex of any culpability in the affair.

* In a Spanish report of December 1587, detailing the religious disposition of England's aristocracy prior to the Armada, Henry, fourth Earl of Derby was listed as 'neutral' along with the Earl of Shrewsbury. Shortly before Ferdinando's death, leading Catholics assessed his beliefs thus: his 'religion is held to be . . . doubtful, so as some do think him to be of all three religions [Protestant, Puritan, and Catholic], and others of none'.

† Curiously, John Gerard's cousin, Thomas, son of the former Master of the Rolls, Sir Gilbert Gerard, also benefited from Ferdinando's death, becoming governor of the Isle of Man while Ferdinando's heirs quarrelled among themselves as to who should hold the office. Government of the island had been given to Derby's family by Henry IV.

Perhaps it is significant then that the State papers reveal little effort on the Council's part to track down Ferdinando's killers. Tudor England had already borne witness to a wholesale massacre of pretenders to the throne under Henry VII and Henry VIII; against that, the death of Ferdinando Stanley, Earl of Derby passed almost unnoticed. Indeed, within the year it had been entirely eclipsed by the publication of a small book, revealing why the English succession had become the burning international issue of the day.[11]

A Conference about the Next Succession to the Crown of England was composed in Flanders by committee, some time in late 1593, early 1594. It is an extraordinary piece of work, part Renaissance political theory, part propaganda. The first half treats with the notion of succession in the abstract, concluding that propinquity, though desirable in a claimant, is not founded upon any law of nature or divinity, but only upon human law, 'and consequently may upon just causes be altered by the same'. '[I]t is not enough', wrote the authors, 'for a man to be next only in blood, thereby to pretend a Crown, but that other circumstances also must occur.' These 'other circumstances' included the claimant's suitability, both morally and physically, for the job.[12]

With this in mind, the second half of the book treats with all the individual pretenders in turn. And, tracing the bloodlines of successive royal houses back to William the Conqueror and picking apart the more recent House of Lancaster, the writers were able to add a number of new Spanish and Portuguese claimants to the mix, most notably Isabel, Infanta of Spain, descended from the Conqueror's eldest daughter and from John of Gaunt, Duke of Lancaster.* What had been a British affair had now become wholly pan-European. Not that this was news to anyone on either side of the Channel: after all, the royal houses of Christendom had always

* John of Gaunt's eldest daughter by his first marriage, Philippa, had married into the Portuguese royal family, from which the Infanta was also descended. John's second marriage had been to Constance of Castille; his daughter from this marriage, Catherine, had married Henry, later King of Castille.

intermarried. Rather, it was the forensic way in which the committee examined the issues arising from government by 'Strangers' that left readers in no doubt that, in the race for the English Crown, the authors were backing the foreign competition over the home side: 'it is the common opinion of Learned men', wrote these particular learned men, 'that the World was never more happily governed than under the Romans, and yet were they Strangers to most of their Subjects'. The book might insist upon its impartiality, listing the pros and cons of each claimant, but its bias seeps between the lines. The English claimants were individually insignificant, tainted by bastardy and unlikely to unite the country behind them; James was liable to fill the country with his Scottish followers and divide the spoils between them, causing resentment. The whole was a ringing, if tacit, endorsement of the Spanish Infanta.

It was a sound argument and a percipient one: the English claimants were insignificant, James would cause a degree of resentment by favouring his Scottish followers, England had been and would again be ruled by foreigners, to no greater or worse effect than when it had been ruled by monarchs regarded as English. In making 'good and profitable Government' their priority, the authors were only pleading England's interests. In dismissing English jingoism as the 'common vulgar prejudice of passionate men', they revealed that they, themselves, were experienced expatriates who long ago had learned to ignore nationality in favour of shared ideals and beliefs. In laying out the international and sectarian context of the English succession, they were merely offering a timely wake-up call to the world that at stake here were the balance of European power and the future of the Church of England. But what stuck in English throats – and by late 1594 copies of the book had begun to cross the Channel into England – was that the authors were backing a Spanish claimant, the daughter of that same King with whom the country had been at war for almost a decade. And what was to prove altogether unswallowable was the fact that prominent among its authors – reportedly its author-in-chief – was Father Robert Persons, de facto leader of the English Jesuits. Not only were the Jesuits seeking the destruction of England's Church, but they were also, it seemed, seeking to turn England into a Spanish colony: it was a belief that would unite Catholics and non-Catholics in joint opposition to the Jesuits and split the cause of English Catholicism to its core.

The remaining years of Elizabeth's reign would be played out against a backdrop of vitriol, bitter infighting, and a quite bewildering tangle of political manoeuvring and back-stabbing among the missionaries. But to put this in its rightful context, they would also be played out against a backdrop of vitriol, infighting, and back-stabbing among the power-brokers of Europe, as each tried to make capital from the arguments put forward in *A Conference*. The sixteenth century was going out, not with a whimper, but with myriad voices raised in complaint, or plotting in the corridors of government. And, as so often in this century, they were all talking about religion.

Some time in the summer of 1598 John Gerard headed north out of London in secret. Since his escape from the Tower the autumn before, he had remained closeted in the capital while Henry Garnet decided his immediate fate. For a time the Superior considered sending him overseas in case the search for him intensified, a piece of news Gerard received badly: in March Garnet reported to Robert Persons that 'John Gerard is much dismayed this day when I wrote to him to prepare himself to go.' Garnet, too, was reluctant to lose him: 'he is very profitable to me', he told Persons. In the ten years since Gerard's return to England the raw and undisciplined new recruit had become, after Garnet and next to his fellow traveller Edward Oldcorne, the most senior man on the mission. He had shown initiative and daring; he had withstood torture; providence favoured him. At just thirty-three years of age there seemed no limit to what he might achieve for the mission. It would have been hard to let him go. So Gerard stayed. Safety, though, dictated he could now best serve the mission in some part of the country where his face was unknown, so, on an unspecified date, he left London for a new base in Northamptonshire.[13]

He rode through a landscape blighted by bad harvests and by the relentless pinch of war.* Five years earlier, in the summer of 1593, the peace-mongers within the Government, led by the Cecils, had taken the unlikely step of contacting William Allen in secret (and through a third party) to try to bring the conflict to an end. Allen's brief was to enquire of the Pope whether an offer of religious toleration to England's Catholics might open the proceedings for a settlement with Spain. Allen's response had been jubilant. 'I could think no otherwise', he wrote, 'but that God himself hath stirred up in their hearts this motion . . . I am not so estranged from the place of my birth most sweet, nor so affected to foreigners that I prefer not the weal of that people above all mortal things.' It only needed the Government to let him know 'how far and in what sort they of themselves would condescend in matters of religion' for him to approach the Pope and, through him, Spain, to set the talks rolling. Sadly, Allen's jubilation came to nothing. It is unclear at what point negotiations stalled, but stall they did and the war continued. But set against this continuity were the many details that had altered during Gerard's time in prison.[14]

For a start the old order was changing. Catholics had always held that it was Elizabeth's ministers who were responsible for the worst of the laws against them, rather than Elizabeth herself; now those same ministers were elderly men advancing haltingly towards their graves. Leicester had died in September 1588, Sir Francis Walsingham in 1590; 1596 saw the death of one of the most prominent of the hardline Protestants in Elizabeth's Government, Sir Francis Knollys; and, even as Gerard rode north, the chief architect of Elizabeth's reign, William Cecil, Lord Burghley, was

* From 1594 to 1597 England suffered a run of four bad harvests in a row. In 1596 wheat was in such short supply that prices rocketed to 80 per cent above average; in 1597 they stood at 64 per cent above average.

fading. He died on 4 August 1598, aged seventy-eight. Elizabeth responded 'grievously, shedding of tears'.[15]

There had been another death too. Richard Young, chief magistrate of Middlesex, had been called out 'one rainy night [in November 1594], at two or three o'clock . . . to make search of some Catholic houses'. John Gerard described the event with little regret: 'The effort left him exhausted: he became ill, contracted consumption and died.'* And not dead, but fallen from grace, was Richard Topcliffe, Elizabeth's priest-hunter in chief.[16]

Topcliffe's career had always been marred by incident. In the early 1580s he had quarrelled sufficiently violently with the Lord Chief Justice, Sir Christopher Wray, for it to be recorded in the State papers. In January 1585 there were reports of a scuffle between Topcliffe and one of the Earl of Shrewsbury's men, taking place near Temple Bar. But by 1594 the priest-hunter had begun to over-reach himself. In November that year he sued a young man named Thomas Fitzherbert in the Court of Chancery for money he claimed Fitzherbert refused to pay him. Henry Garnet recounted the proceedings to Persons: 'Topcliffe and Tom Fitzherbert pleaded hard in the chancery this last week. For whereas Fitzherbert had promised, and entered into bonds, to give 5000 [pounds] unto Topcliffe if he would prosecute his father and uncle to death . . . Fitzherbert pleaded that the conditions were not fulfilled.' His father had 'died naturally', albeit in the Tower and under suspicion of treason, while his uncle was still alive and well and in prison. A witness came forward to tell the court of the 'devices' set to entrap the pair and Attorney General Coke testified that 'Topcliffe had sought to inform against them contrary to all equity and conscience.' The matter was put over for 'secret hearing', but the

* On 30 November 1594 Young wrote to Elizabeth, thanking her for the messenger she had sent to visit him on his sickbed. 'I think no subject in the world more infinitely beholden unto his Sovereign, in that in these my aged and extreme or last days it pleaseth you so favourably to respect the weak estate of so poor a vassal, weakened in body with infirmities, but so much revived in heart with your gracious remembrance of me.' He enclosed with this letter the key to a small chest containing all the papers concerning Gerard's arrest, 'as the last fruits of all my endeavours'.

outcome went unrecorded. Whatever it was, it did little to improve Topcliffe's temper. Around Easter the following year he made a series of unspecified allegations against Lord Keeper Puckering, accusing several Councillors of bribery into the bargain, and was led off to the Marshalsea for contempt of court.[17]

From prison he wrote bitterly to Elizabeth, on 'Good or Evil Friday, 1595', outlining the ignominy of his current position.

> 'I have helped more traitors [to Tyburn] than all the noblemen and gentlemen of the court, your councillors excepted. And now by disgrace I am in fair way and made apt to adventure my life every night to murderers, for since I was committed, wine in Westminster hath been given for joy of that news . . . and it is like that the fresh dead bones of Father Southwell at Tyburn and Father Walpole at York, executed both since Shrovetide, will dance for joy; and now at Easter, instead of a Communion, many an Alleluia will be sung of priests and traitors in prisons, and in lady's cloisters for Topcliffe's fall'.[18]

Though he would be released after only a few weeks, the incident marked the beginning of the end for him. In 1597 he was one of the panel appointed to investigate Ben Jonson and Thomas Nashe's satirical play *The Isle of Dogs* (their findings closed all the London theatres for a time and saw both authors arrested on charges of sedition). In 1598 he hunted down his last victim, the Franciscan, Father John Jones; Jones was executed at Southwark on 12 June that year, the first priest to be killed in London since Robert Southwell three years before. Here the butchery ends. Soon afterwards Topcliffe retired to the country, to Somerby and to Padley Hall, the 'delightful solitary place' he had succeeded in winning from Thomas Fitzherbert. His last years were spent complaining of lameness and he died in December 1604, aged seventy-three. He left the tenancy of his farm at Heapham to his bailiff and the rest of his possessions to his son Charles, apart from a doublet, a cloak, a load of wood, and half a deer, shared between the two witnesses to his will and three local acquaintances. His legacy to England's Catholics was considerably greater.[19]

There remains one twist to the priest-hunter's tale. In September 1591 the seminarian Thomas Pormont had been caught reconciling a townsman of London near St Paul's Cathedral. Pormont was taken to Topcliffe's Westminster house for questioning. He was brought to trial in February 1592 and executed on the twenty-first of that month on the site of his arrest. Before he died he smuggled out of prison a record of his interrogation, which soon found its way into Council hands. According to this transcript, Topcliffe had allegedly offered Pormont his liberty if he would confess to being the Archbishop of Canterbury's bastard son. But it is the references to Queen Elizabeth that make the most curious reading:

> Item Topcliffe told (unto the said priest) that he was so [manuscript torn] familiar with her majesty that he many times putteth [ms torn] between her breasts and paps [nipples] and in her neck. That he hath not only seen her legs and knees [. . .] with his hands above her knees. That he hath felt her belly, and said unto her majesty that she h[ath] the softest belly of any woman kind. That she said unto him: 'Be not these the arms, legs and body of King Henry?' To which he answered: 'Yea'. That she gave him (for a favour) a white linen hose wrought with white silk etc. That he is so familiar with her that when he pleaseth to speak with her he may take her away from any company, and that she [is] as pleasant with every one that she doth love. That he did not care for the Council, for that he had his authority from her Majesty.[20]

Witnesses to Pormont's execution reported that the priest 'was enforced to stand in his shirt almost two hours upon the ladder in Lent time upon a very cold day', while Topcliffe urged him to deny his allegations. Pormont refused. The Council seems to have maintained an embarrassed silence about the claims. It seems highly unlikely anyone told the Queen.[21]

North of the River Nene, over the county border from Bedfordshire into Northamptonshire, stood Irthlingborough Manor, belonging to the Vaux family. The house was in a ruinous state, 'old and tumble-down', a victim of old Lady Vaux's liberality, it was said, and of recusancy fines. Four miles to the west, near the town of Wellingborough, was the Vaux family's principal seat of Harrowden Hall, in the village of Great Harrowden. It, too, was 'dilapidated' and, in parts, 'almost . . . a ruin'. Devotion to their faith had cost the Vauxs dear.*[22]

Devotion to her dead husband was costing Elizabeth Vaux, Gerard's newest hostess, even dearer. George Vaux (younger brother of Anne and Henry) had died in 1594, leaving six children under the age of seven and a wife so distraught 'that she hardly moved out of her room for a whole year', recorded Gerard. Even then, she was unable 'to bring herself to enter the wing of the house in which her husband had died'. The Jesuit's first task as the family's resident priest was to ease this distress. 'By degrees', he wrote, 'I healed my hostess's excessive sorrow. I told her our grief for the dead should be tempered; we were not to mourn like men who had no hope. Her husband, I pointed out, had become a Catholic before he died, and a single prayer would do him more good than many tears . . . So I gradually brought her round to turn her old sorrow into a sorrow of another and nobler kind.' This new sorrow, for the godless state of the nation, translated itself into practical action: swearing vows of chastity and poverty, Elizabeth Vaux now dedicated her life to the Catholic cause. 'She was ready', reported Gerard, 'to set up house wherever and in whatever way I judged best for our needs.' With Gerard advising and Nicholas Owen assisting, Elizabeth Vaux moved her young family the short distance from Irthlingborough back to Great Harrowden, scraping together sufficient funds to extend the existing hall by a new three-storey block with custom-built hiding places, to serve as Gerard's

* In a letter to William Cecil, Lord Vaux described himself as 'the unfortunatest peer of Parliament for poverty that ever was, for even my Parliamentary robes are at pawn to a citizen'.

headquarters. It was to be his base for his remaining years on the mission.[23]

Harrowden Hall stood at one corner of a pronounced Northamptonshire Catholic triangle, with Rushton Hall, belonging to the Treshams, 10 miles to its north, and Ashby St Ledgers, belonging to the Catesbys, 28 miles to its west. All three families were related; all three had undertaken, back in 1585, to support the mission financially; more specifically, all three had played host to the first English Jesuits, Persons and Campion. In November 1581 Lord Vaux, Sir Thomas Tresham, and Sir William Catesby appeared before the Star Chamber for refusing to state under oath whether Campion had stayed with them; they were all committed to the Fleet prison. This was the world of entrenched Catholic loyalism that John Gerard had entered and now he set off on a tour of these neighbouring households and the counties surrounding. One of his most prominent converts from the period was Sir Everard Digby of Gayhurst in north Buckinghamshire. Digby was a courtier and a Gentleman Pensioner, one of Elizabeth's official escorts on State occasions. Away from Court, wrote Gerard, 'he had no interest apart from his hounds and hawks', so it was small wonder the pair should have become friends: Gerard describes him as his brother 'in blood'. With Gerard's encouragement, Digby and his wife now took in the Jesuit Father John Percy as their family confessor.*[24]

It is likely that Gerard also used this time to set up a secret press in the area, though he, himself, leaves no record of it. The clue is an English translation of an Italian book, *The Spiritual Conflict*, by Lorenzo Scupoli, dated 1598: the translator was John Gerard, the printer, Henry Owen, brother of Nicholas Owen. Henry was the fourth and final Owen boy to join the Catholic underground. His early years were spent as an apprentice to Joseph Barnes, founder

* Percy was a relative newcomer to the Jesuit mission. He sailed for England in 1596, but was captured en route and handed over to the English authorities. He was imprisoned in the Bridewell, but escaped along with Richard Fulwood. In 1599 he was sent to assist Gerard at Harrowden, before moving to Gayhurst.

of Oxford University Press.* Then the trail goes cold until, in 1595, he turns up in the Clink prison, alongside Gerard; while there, he ran 'a press and printed divers popish books'. He was transferred to the nearby White Lion prison – a former inn on the east side of Borough High Street – from which he promptly escaped. Again his trail goes cold until, in 1600, he appears in Northamptonshire, printing Catholic texts; such was the information Sir Robert Cecil received in May that year. That John Gerard was supporting him seems highly probable.[25]

Over the years, news of Gerard's supposed activities continued to reach the Government. William Atkinson, a renegade priest, wrote to Cecil *c.*1602, saying, 'It is credibly reported that Mr John Gerard, Fisher [Fr John Percy] and Litstar [Fr Thomas Lister] are to be hunting in Beskwood Park . . . and they were determined to go to the Lady Markham, Sir Griffin Markham's wife, and likewise to Francis Tresham; young Vaux . . . was to accompany them.' Beskwood was a royal hunting lodge in Sherwood Forest, of which Sir Griffin Markham was the keeper. Other reports had Gerard disguised, 'with an artificial beard and periwig of a brown colour', back in London and en route to Ireland. 'His beard', added this informant, with an eye for detail, 'is very long cut after the spade fashion.' This fragment joined the growing body of paperwork on the elusive Jesuit, in particular two descriptions of him, dating from soon after his escape from the Tower and from August 1601 respectively. The first was Topcliffe's:

> 'John Gerard the Jesuit is about 30 years old Of a good stature somewhat higher then Sir Tho Layton and upright in his pace and countenance, somewhat staring in his look or Eyes, Curled hair by Nature & blackish & apt not to have much hair of his beard. I think his nose somewhat wide and turning Up,

* In 1584 Barnes was lent a hundred pounds by the university convocation to set up a press in Oxford; the Earl of Leicester was instrumental in securing a licence for him. Barnes's first publication was John Case's *Speculum moralium quaestionum* dedicated to Leicester and singing the praises of the new press, 'which by your means our university has lately obtained'.

Blubbered Lips turning outward, Especially the over Lips most Upwards towards the Nose[.] Curious in speech If he do now continue his custom And in his speech he flowereth & smiles much and a faltering or Lisping, or doubling of his Tongue in his speech'.

The second, from the spy William Byrd, portrayed Gerard as 'high shouldered, especially when his cope is on his back, black haired, and of complexion swarthy, hawk nosed [and] high templed'. And Richard Bancroft, Bishop of London, completed the picture, notifying Cecil in April 1603 that 'Mr Gerard is a tall black man, very gallant in apparel, and being attended with two men and a foot boy is exceedingly well horsed.'[26]

It was a rare luxury for the Government to be in possession of so many descriptions of a single man. A letter from Sir William Cecil to Sir Francis Walsingham, written at the height of the search for the Babington plotters, revealed just how inadequate such information could often be. Coming across members of the local militia patrolling the outskirts of a village, Cecil had asked them why they were watching. They replied, 'To take three young men.' How would they recognize the men? asked Cecil. 'One of the parties hath a hooked nose,' was the unhelpful response. Yet despite fresh intelligence that Gerard was staying with the Countess of Arundel, and an offer from one informant to betray the Jesuit for the sum of a hundred pounds, he remained at large.[27]

One episode Gerard recounts from this period serves as a useful introduction to the messy and confusing conflict that by now was raging between the missionaries. Some time *c.*1602 Gerard was approached by Henry Hastings, grandson of the Earl of Huntingdon and a first cousin of Elizabeth Vaux. Hastings was keen to study the *Spiritual Exercises.* Gerard asked him what had prompted his interest and Hastings answered: 'I read in a book written against the Society that you use this means to persuade people to enter the religious life and then rob them of their property. Among other names mine was mentioned . . . As they have calumniated you so badly I am come to redress their lies.' Hastings was rare in his unwillingness to believe everything he was reading about the Jesuits. Many were not so discriminating.[28]

To set the scene one needs to turn to two rather different locations, the first, Rome, the second, Wisbech Castle near Ely, in England's fenlands. In Rome, the centre of disturbance was the English College. There had always been trouble among the students there; as Robert Persons described it to Henry Garnet, the college seemed 'from the beginning to have had a certain infelicity following it'. The first rector, Welshman Maurice Clenock, had been accused of favouring his fellow countrymen over the English-born students, sparking off an early rebellion. At the students' request, the college was given over to the Jesuits to run, with the Italian Alphonsus Agazzari appointed rector in 1579. But, continued Persons to Garnet, despite this change, Rome seemed to draw men 'more heady . . . and less tractable than others brought up at home'.* Clenock may have gone, but the accusations that followed were all too familiar: now the Jesuits were charged with favouring the more able students, seducing them into the Society and posting them around the world on other, safer missions, while sending the less able, the disposable, home to die at Tyburn. Unfair, was the cry that echoed through the college corridors. It was a cry that was heard back in London. Ever attuned to signs of discord among its opposition the English Government acted quickly, recruiting Solomon Aldred, a former tailor living in Rome, to stir up dissent. His brief, as revealed in a letter to Francis Walsingham dated March 1586, was 'to set a faction' between Robert Persons and William Allen, creating open conflict between the Jesuits and the seminarians. In part he was devastatingly successful. Though Persons and Allen remained undivided, there were among the seminary students many willing to listen to stories

* This was not just an English complaint. In the same letter Persons explains that 'Spaniards, Frenchmen and Flemings and other nations' were all concerned about the pernicious effects time spent in Rome had on their countrymen.

of Jesuit cruelty, favouritism, and deceit. And not just among the students either: Dr Humphrey Ely, one of the foremost of the Catholic clergy in exile, was soon complaining about rumours of Jesuit espionage in Rome. 'Nothing is so contrary to an Englishman's nature,' he thundered, 'as to be betrayed by him whom he trusted. If such spies were in Oxford . . . they would be plucked to pieces.' By February 1595 Robert Cecil was receiving reports that Jesuit rule at the college was so precarious 'as a little help would dissolve it'.[29]

Back in Wisbech, conditions were no better. The castle's birth as a Catholic internment camp dated from the late 1570s.* By the mid-1590s it was packed with priests, including the Jesuit Father William Weston, Garnet's predecessor as Superior, and in this atmosphere of enforced idleness differences quickly arose between inmates: the same month that Jesuit rule in Rome was reported as being so precarious, Weston was charging some of his fellow prisoners with 'whoring, drunkenness, and dicing'. Soon a group of eighteen priests had banded together in pursuit of stricter discipline and had written to Henry Garnet, begging that Weston be appointed their Superior. Garnet, fully aware that a Jesuit had no right of authority over non-Jesuits, wrote back suggesting Weston become their spiritual adviser rather than their Superior and accept the title of 'Agent' from them. This, though, was sufficient to enrage some among the seminarians who complained the Society was getting above itself, and now the Wisbech prisoners split into factions, each demanding separate quarters, separate tables in the refectory, an entirely separate existence in fact.[30]

That such tensions should ever have arisen was not surprising: there existed significant differences between the Jesuits and the seminarians that, fissure-like, were always going to be vulnerable

* Wisbech was one of several castles chosen as detention centres in response to the 1577 census of recusants. The others were Banbury (Berkshire), Framlingham (Suffolk), Kimbolton (Huntingdonshire), Portchester (Hampshire), The Vize (Wiltshire), Melborne (Derbyshire), Hatton (Cheshire), Wigmore (Montgomeryshire) and Barney (Yorkshire). It seems that, for a time, the Government contemplated a policy of wide-scale internment, but in the end this policy came to nothing.

to applied pressure. And pressure was one thing in abundant supply on the mission. For a start, the Jesuits were a religious order akin to the Franciscans, Dominicans, and Benedictines, while the seminarians belonged to the secular clergy – and the latter had long held the trenchant view that the religious should mind their own, preferably monastic business and leave ministering to the laity to them.* Next, the Jesuits were a new order, born out of the Counter-Reformation and unburdened by a cumbersome heritage or a long-established hierarchy. The seminarians were the heirs and successors to the traditions of the English Catholic Church and were bound by its legacy. Both these factors pointed to areas of potential ecclesiastical conflict between the two sides. But there was one further difference that, in the context of the English mission, chased this conflict out into a more worldly arena. The Jesuits were an international order with specific links to Italy, through their allegiance to the Pope, and to Spain, through their founder. The seminarians were exclusively English.

One need only look to the writing of the period to see what the average sixteenth-century Englishman thought of the average Italian or Spaniard. The theatres were full of Italian-set scenes of stabbings and poisonings. Travellers reported that the Italians were 'hypocritical, close, malicious, encroaching and deadly'. Others warned that if the Spanish landed in England, Englishmen would see 'the rape of your daughter, the buggery of your son, or the sodomizing of your sow'. But the worst type of all, or so everyone in Europe seemed to agree, was the Englishman affiliated to Italy: the Italianate Englishman. He, according to an Italian proverb gaining popular currency in England, was 'the devil incarnate'. Most representative of this type were, of course, the English Jesuits – it was a useful belief to foster.[31]

For the Government, genuinely alarmed by the Jesuits' success in England, the more tales it could tell of slippery Italianate English

* The terms secular and religious are, in this instance, false friends. The secular clergy were no less 'religious' than their religious counterparts. Rather the term defines them as being ordinary members of the clergy, not belonging to a specific religious order.

papistry, as peddled by the Jesuits and in marked contrast to its own bluff honest Anglicanism, the more reason it had to press for the eradication of the Society. For those seminarians already nursing a grievance towards the Society (and aware that its links with the enemy fuelled the Government's case that Catholics were traitors), the contrast between their own Englishness and the Jesuits' internationalism served as a possible bargaining counter with the Government. And the coinciding of two events towards the end of 1594 made the playing of this counter almost inevitable. The first was the explosion onto an already troubled scene of the Robert Persons-led *A Conference about the Next Succession to the Crown of England.* The second was the death of William Allen.

Allen had held the mission together. The seminaries were his creation, the mission, itself, was his creation, the impetus to persuade the Jesuits to enter the fray against their general's will was his too. He had made the return of England to Catholicism his personal crusade. His views on how this might be achieved had dovetailed neatly with those of Persons, and since Persons' flight from England in 1581 the two men had worked together in pursuit of this single goal. Allen and Persons, Persons and Allen: despite all the Government's attempts to set a faction between them, both men remained jointly committed to the venture. Jointly *and* controversially committed to it: in the early 1580s they were drawn into a French scheme to invade England and overthrow Elizabeth; for Persons this had placed him in direct contravention of the Society's founding guidelines, which stressed the order's non-political nature. By 1587 the two men had thrown their support behind the Spanish Armada, convinced only Spain had the military might necessary to effect their desired regime change. In March that year they compiled a joint dossier validating Philip II's claim to the English throne through the House of Lancaster. They also drew up guidelines for a document legitimizing the invasion. This document would elaborate on 'the multiple bastardy of this Queen Elizabeth, her wicked mode of life, [and] the injuries she has done to all Christendom'. It would also point out the advantages to England of being placed under Spanish protection. All this 'is evident to us,' wrote Allen and Persons, 'but the matter must be made more convincing by means of this book'. Finally, Allen, like Persons, had been one of the committee behind *A Conference*.[32]

So it was altogether erroneous to hold Persons, alone, responsible for the pro-Spanish sentiments contained in this text. Indeed, it was altogether erroneous to hold Allen and Persons solely responsible for them: eight years after the Armada's defeat Sir Francis Englefield, England's leading Catholic layman abroad, was still writing 'Without the support and troops of Spain it is scarcely probable that the Catholic religion will ever be restored [in England]'. But theirs were exiles' opinions, opinions that failed to take into account how war with Spain had hardened both the English Government's attitude towards Catholicism and English Catholics' attitude towards Spain. In such a context Sir Thomas Tresham's 'unspeakable joy' at the Armada's defeat was more representative of the views of his co-religionists than Allen and Persons' belief that they would 'with the greatest unanimity' embrace a Spanish ruler. Allen's death – on 16 October 1594, aged sixty-two – would leave Persons out in the open, holding the smoking gun of both men's policies.[33]

The year 1595 saw the increase in unrest in Rome and the Wisbech feud; 1596 saw the effects of this poison seeping into the lifeblood of the mission. That August Robert Persons received word from Rector Agazzari in Rome that the rebel English students there nursed 'such a hatred against the Society that I fear that they would be ready to join hands with the heretics in order to be delivered from them'. And ignoring the fact that William Allen had contributed to *A Conference*, they blamed Persons squarely for its unpopular opinions. 'I know not whether they hate the Society more on account of the Spaniards,' wrote Agazzari, 'or the Spaniards on account of the Society.' Meanwhile, a letter of Henry Garnet's to General Claudio Aquaviva that year testified to the growth in anti-Jesuit criticism back in England.[34]

For Garnet these attacks had now become personal. Among a

few of the seminarians, he wrote, he was being traduced as 'a little wretch of a man, marked out to die, who day and night thinks of nothing save the rack and gibbet'. Worse than that, everything he had laboured for since taking over the Jesuit mission was being painted as self-promoting and hostile to the needs of the seminarians. It was only a small percentage who complained, he stressed, but their slurs still hurt, a fact borne out by the lengths to which the habitually modest Garnet went to refute them: his letter was a passionate defence of his own actions and those of his fellow Jesuits. One story serves to indicate the nature of these slurs. Shortly before Allen's death, wrote Garnet, a seminarian began spreading rumours in Rome that the Jesuits were refusing to assist incomers to England. Many seminarians were now almost too scared to cross over, but when they did so, and found 'there was virtually no one to give them help . . . except our own priests, they were staggered at the man's story and told me so themselves, scoffing at the tales he had told them'. Then the rumour-monger, himself, crossed over, taking lodgings at an inn in London. Garnet takes up the narrative: 'It so happened that on the very same day he entered the city, I did too. Before I visited any Catholic, I made for the inn and sought him out in courtesy and friendliness. I could not ask him to come with me, for I was uncertain myself whether any Catholic family would take me in that night. But I had hardly gone twelve yards, when I met a Catholic whom I knew and asked him to have a care of this priest.' And now, wrote Garnet, 'we have made this priest our friend'. So much so, he noted, that unbidden the priest had just written a passionate defence of the Society's rule in Rome.[35]

The catalyst for all-out conflict was the appointment, in March 1598, of George Blackwell (ex-Trinity College, Oxford and ex-English College, Douai) to the post of Archpriest, in charge of the still leaderless seminarians in England. This appointment was Rome's idea, but the handling of the affair caused more problems than it cured. First, the man responsible for the appointment, Cardinal Cajetan, sought Garnet's advice, enraging those priests already critical of the Jesuits. Second, when the brief announcing the appointment was published, it bore Cajetan's name, rather than the Pope's, as though the Pope, himself, had not officially authorized it. Third, the brief paid tribute to the Society's work in England, particularly Garnet's, stipulating that, in view of the latter's 'experience of English affairs', 'the Archpriest will be careful in matters of greater moment to ask his [Garnet's] opinion and advice' – further enraging the anti-Jesuits. Last, and most insulting, was the fact that the office of Archpriest, itself, was an entirely new creation, a stunted version of a bishop's office, with powers to discipline priests, but no concomitant authority to ordain. Dr Humphrey Ely would write angrily of the position, 'I term it a new dignity . . . the lowest and basest dignity in Christ's church.' Rome's attempts to spread 'peace and union of minds' between Jesuits and seculars had only succeeded in uniting those seminarians who nursed a grudge against the Society with those traditionalists concerned to preserve the hierarchy and heritage of the old English Catholic Church, in passionate opposition to the hapless Blackwell. In June 1598 Garnet would write to Robert Persons, saying that Blackwell 'doth very well and is generally liked', but plans were already afoot among some of the seminarians to depose him.[36]

The next few years saw move followed by counter-move, blast followed by counter-blast, most of them characterized by an almost hysterical amount of name-calling, as seminarians and Jesuits slugged it out. It was as though the decades of pent-up fear and frustration, of living in death's shadow, had finally found release in an act of violent self-mutilation. And the English Government, looking on, had only to keep the combatants locked together for this self-mutilation to continue. One Council agent, suspicious of the ease with which the mission seemed to be imploding of its own accord, warned that the whole thing was an elaborate plot devised by both parties to 'gain liberty'. Meanwhile, the Council gathered information on the dispute and waited to act.*[37]

* There exists a document, drawn up in the winter of 1600–1, probably by Richard Bancroft, Bishop of London, which lays out all the known facts of the dispute so far and suggests areas of further enquiry for the Government.

From the moment of Blackwell's appointment, a small body of seminarians refused to accept his authority, appealing to Rome (and earning themselves the title Appellants) for him to be dismissed and replaced with a bishop of their own choosing. Subsidiary to this they asked that the English College in Rome be removed from Jesuit control. Rome refused, reportedly under pressure from Persons, and ordered the rebel priests to make their peace with Blackwell.

Now the Jesuit Thomas Lister, long a thorn in Garnet's side as a result of his erratic behaviour, put pen to paper to accuse the Appellants of the sin of schism, declaring that ipso facto they were excommunicate. It was a view Garnet, himself, shared – the Appellants were defying the Pope in defying the Archpriest – but Lister's public polemic fuelled anti-Jesuit feeling among the seminarians further.[39]

Then Blackwell, himself, joined the fray, agreeing with Lister and demanding the Appellants acknowledge their sin and make reparation for it – fuelling the Appellants' belief that Blackwell was in the Jesuits' pocket.[40]

Meanwhile, from Wisbech prison, one of the leading Appellants, Dr Christopher Bagshaw, was summoned to London's Gatehouse for questioning. Bagshaw was another ex-Oxford man, a former fellow of Balliol College, where he had studied alongside Robert Persons; anecdotally, the two men had never got on. Now Garnet wrote to Persons, saying he had received word from several sources that Bagshaw was behind William Weston's recent and hurried transfer to the Tower of London and that he was busy holding secret talks with the Government. There exists a document, endorsed by Bagshaw, that appears to support Garnet's claim, showing him to have been agitating for the Jesuits' expulsion from England.[41]

At the same time the polemic war intensified. Blackwell was variously described as 'a Jesuitical idol' and 'a puppy to dance after the Jesuit's pipe'. Robert Persons, for whom the greatest contempt was reserved, was described as 'the principal author . . . of all our garboils at home'; he and Campion, the text went on, had 'so acted as to provoke the queen and magistrates to enact most cruel laws, before unheard of, against the seminarists'. The author here was the seminarian John Mush, formerly Margaret Clitherow's confessor, who had applied to join – and been rejected by – the Jesuits. Anthony Copley, Robert Southwell's layman cousin, also weighed into the attack, calling Persons 'the misbegotten [son] of a ploughman', who had sired 'two bastards, male and female, upon the body of his own sister'.*[42]

And now Persons retaliated, attacking Copley as 'a little wanton idle-headed boy' and describing the seminarian William Watson (who had also questioned Persons' parentage) as 'so wrong shapen and of so bad and blinking aspect that he looketh nine ways at once'. Persons' book, *A Brief Apology, or defence of the Catholic ecclesiastical hierarchy*, was an attempt to lay out the case for the Jesuits (Garnet helped correct its factual errors), but its bellicose edge did little to calm the war of words. Neither did the fact that Blackwell appeared to have delayed publishing a papal brief, which condemned the polemicists and prohibited all further publications, until Persons had completed the book. As the historian William Camden recorded: 'With sharp-pointed pens, venomous tongues, and slanderous books, did Jesuits and seculars fight one another.'[43]

Then, in the summer of 1601, Thomas Bluet, a Wisbech prisoner since 1580 and a leader of both the prison's stirs and the Appellant movement, was given permission to come to London to collect alms for his fellow detainees. In London he was placed in the custody of the Bishop of London, Richard Bancroft. Bluet explained what happened next: Bancroft suddenly brought out 'many letters and books of Persons . . . and other English Jesuits, inviting the King of Spain to invade England, as due of right to him, and urging

* Copley, a born rebel, disgraced himself in Rome by appearing at the English College to preach a sermon with a rose between his teeth. He returned to England on a State pardon and, for a time, settled down quietly to write poetry; Southwell's death seems to have removed the chief restraining influence on him.

private men to kill the Queen, by poison or sword'. Bluet was shocked and declared the seculars innocent of such treasons. Bancroft then explained that the sole reason for the Government's severity towards Catholicism was that it believed all Catholics to be 'guilty of these devices, and all [to be] disciples of the Jesuits'.[44]

The precise events of the next weeks are unclear, but it appears that by the end of June Bluet had met with the Council, then the Queen, and had devised a petition for Elizabeth, in which he protested secular loyalty and begged for limited freedom of conscience for England's Catholics. Elizabeth's reported response to this petition should have left Bluet in little doubt his request would come to nothing: 'if I grant this liberty to Catholics, by this very fact I lay at their feet myself, my honour, my crown, and my life'. So it seems certain the Council did not pass her words onto him, for on 1 July 1601 Bluet wrote to John Mush with an extraordinary proposal. He had arranged with the Council, he told Mush, for a party of priests to leave England, under guise of banishment, and travel to Rome for an audience with the Pope. They were to offer the Pope, on behalf of the Government, an easing of the penal laws and the chance to negotiate religious toleration for English Catholics. In return, they were to demand from him a total ban on all Catholic treasons against England. 'It hath cost me many a sweat and many bitter tears 'ere I could effect it,' Bluet told Mush. 'I have in some sort pacified the wrath of our prince against us . . . and have laid the fault where it ought to be': at the Jesuits' doorstep. Where once the Government had used spies to sow dissent between the seminarians and the Jesuits, now it was using seminarians to defeat the Jesuits.[45]

In the autumn of 1601, happy their enemy's enemy was a loyal friend to them, and certain the Jesuits and their purported treacherous dealings with Spain were the sole obstacle to religious tolerance in England, the Appellant priests Bluet, Bagshaw and Mush, accompanied by two others, Barneby and Champney, crossed the Channel to Paris. Anthony Copley would write of 'the departure of the three B-ees [Bluet, Bagshaw and Barneby] onward into their exile and defence against these [Jesuit] hornets'. From Paris the party travelled to Rome, arriving there on 14 February 1602. On 5 March they had an audience with Pope Clement VIII and presented their demands to him. It did not go well. John Mush recorded in his diary Clement's uncompromising view that religious toleration bred heresy and 'that persecution was profitable to the Church'; he also recorded Clement's anger at the Appellants for referring to Elizabeth as Queen, given that earlier Popes had deposed her. In fact, wrote Mush, all 'we proposed seemed to dislike him'.[46]

Now followed months of painstaking negotiations in which both sides, Appellants and Jesuits, attempted to put their case to Clement. At this point a strange selection of allies came to the Appellants' aid. First was the French ambassador to Rome, Philippe de Béthune. France had followed the arguments put forward in *A Conference* carefully – and had no wish to see a Spaniard on the throne of England. If the Jesuits were being identified as pro-Spanish, then it suited French interests to become pro-Appellant, albeit discreetly so as not to provoke Spain, and Mush confidently reported: 'We are safe [from the Society and Spain] under the protection of the King of France.' Next was the seminary priest Dr John Cecil, who had joined the Appellants in Paris, probably as their translator for the trip. Dr Cecil's interest in their cause was as ambiguous as his career to date. Since leaving Oxford in the company of Nicholas Owen's brothers, John and Walter, he had graduated from the English College in Rome and now divided his time between serving the mission on the Continent and serving the English Government at home. It was Dr Cecil who had provided his namesakes on the Council with their description of Southwell and their details of the Jesuits' landing sites in England; he had also implicated the Jesuits in the plans to convert Ferdinando, Lord Strange. His motivation in all this, he wrote, was to show how a good Catholic could also be a good Englishman. More likely, his actions were informed by a strong, if unspecified, grudge against the Jesuits. But with him now pressing the secular cause and Béthune the French, Clement seemed more disposed to listen to the Appellants. In June Mush could write in his diary of a 'favourable audience' with the pontiff.*[47]

The more Clement listened, though, the clearer it became that the Appellants' case was built of straw. They could reiterate their demand that Blackwell's ruling – that they were guilty of schism – be rescinded. They could petition for a bishop of their own. They could proffer the three demands that bore the Council's stamp: that priests be banned from meddling in politics; that known plotters be removed from the mission; and that Catholics be obliged to reveal all plots against the Queen and the State. But they could give, in return, no details of the Government's offer of religious tolerance. In June Bluet wrote to Bishop Bancroft in London, begging just 'three lines of her Majesty's hand' to confirm her intentions. Bancroft duly forwarded the letter to the Council, but Elizabeth's response was not forthcoming.[48]

With no room to bargain, the Appellants could do little else but continue their attack on the Jesuits in the hope Clement might recall the Society from England just to restore peace. Their methods were underhand. Back in March they had written from Rome, asking that a copy of Robert Southwell's *An Humble Supplication* be sent to them: this copy, as they now presented it to Clement, was an expurgated version of the Jesuit's text, specifically printed for their purpose. Gone from it were the tirades against Sir William Cecil and the criticisms of the English Government; what remained were the tributes to Elizabeth's 'goodness' and 'Princely virtues' and the pleas for tolerance. With the balance of the book now distorted, the Appellants showed Clement that Southwell, like they, was loyal to Elizabeth, that he, like they, advocated negotiating with her Government, that he, like they, viewed his fellow Jesuits' uncompromising stance as the only thing standing between

* Notes taken by Francis Bacon during Henry Walpole's interrogation reveal John Cecil was known by Catholics to be a Government spy; Bacon warns the Council to be chary in its employment of him in future. Yet knowledge of Cecil's spying still did not prevent the Pope from recommending his services to the Queen of France in 1616. Few facts illustrate the ambiguity of Cecil's activities more clearly.

England's Catholics and freedom. The Jesuits were outraged. Henry Garnet attempted to prevent publication of the abridged text; Robert Persons petitioned the Pope, begging him 'not to allow an undeserved aspersion such as this to be branded on [Southwell's] reputation and memory'.[49]

In the end, after eight months of tortuous negotiations, Clement produced his final ruling on the dispute in the form of a papal brief, which can best be described as anti-climactic. So much hatred, so much invective had achieved so little. For the Jesuits, they had won a ban on Catholics having any further dealings with the English Government to the detriment of their co-religionists. For the Appellants, they had won nothing more than a ruling forbidding Archpriest Blackwell from communicating seminarian business to the Jesuits. Further, a Royal Proclamation, published the month after Clement's ruling, indicated just how far they had ever been from achieving their hoped-for religious toleration. This Proclamation, issued at Westminster on 5 November 1602, reiterated Elizabeth's stance against Catholicism, calling on all priests to leave England immediately or face the death penalty. As for those priests, it read, who insinuated 'that we have some purpose to grant a toleration of two religions within our realm', God, Himself, was witness to 'our own innocency from such imagination'.[50]

Four hundred years after the event, what remains so intriguing about the Jesuit–secular dispute is not its outcome, so much as the motivation of its participants and the possibilities it raised. It was an unlikely, albeit predictable, coupling that saw Elizabeth and Pope Clement united in opposition to the principle of religious tolerance, but then this was just one of a number of unlikely couplings forged during the conflict. Bishop Bancroft, London's leading prelate and a future Archbishop of Canterbury, was so active in his support of the Appellants' publications that he would later be accused (unsuccessfully) of treason by the Puritans, for promoting Catholic literature. It was on a Bancroft-backed press, run by the Catholic William Wrench, that the abridged version of

Southwell's *Supplication* was apparently printed; Bancroft would later save Wrench from execution for 'Traitorous' book-running.* It was Bancroft who produced the papers allegedly revealing Jesuit treason, Bancroft who served as go-between in the Appellants' negotiations with the Council, Bancroft who (most likely) led the investigation into the origins of the dispute. The bishop was no Catholic, but he was stridently anti-Puritan: the chronicler Lord Clarendon would later write that only an early death had prevented him from extinguishing 'all that fire in England which had been kindled at Geneva'. His support of the Appellants would seem to have been as much an attempt to consolidate the forces of conservative Christianity against the Puritans, as it was a means of defeating the Jesuits. Whether or not the documents he showed Bluet, setting out the Jesuits' treasons, were fake, genuine, or the product of inaccurate and alarmist information remains unclear.[51]

Also unclear is the extent to which the Council, now dominated by Sir Robert Cecil, Elizabeth's Principal Secretary of State, was acting independently of Elizabeth in pursuing negotiations with the Pope. Cecil's had tended to be a voice for peace on the Council. He was not averse to keeping information from the Queen in order to achieve this: he once instructed Lord Mountjoy, commander of the English forces in Ireland, to 'write that which is fit to be showed to her Majesty, and that which is fit for me to know (*a parte*)'. If he hoped to oversee a smooth transfer of power from Elizabeth to whoever was destined to succeed her, then it was imperative he forestall Catholic resistance to that successor, resistance anticipated by the publication of *A Conference*. Indeed, for a time, it seemed, he even considered the Spanish Infanta's claim to the throne: in 1599 he went as far as to commission secret

* Henry Owen was associated with Wrench's press for a time and in March 1605 he and Wrench were both indicted for recusancy in the London parish of St Bartholomew the Great. Curiously, the same Puritan printer who accused Bancroft of treason also testified that the authorities had allowed Owen to run his press in the Clink, while a Government informer alleged that he had been asked to turn a blind eye to Catholic books in Owen's possession. Any direct connection between Bancroft and Owen is unknown.

portraits of her and her husband.* More realistic, though, was the possibility of his securing a guarantee of Catholic loyalty to the Crown in advance of James of Scotland inheriting, a solution that would unite all factions behind the strongest contender in the race. That he continued negotiating with the Appellants even after the publication of Elizabeth's Proclamation (devising an oath of allegiance for them that repudiated the Pope's power of deposition) suggests a determination on his part to bring Catholicism back into the fold, albeit in a limited way, for the sake of State security.[52]

And this remains the most tantalizing aspect of all about the dispute. For the Appellants had alerted Cecil to the fact that there *might* exist a solution to England's Catholic problem, a solution that allied the twin forces of nascent English nationalism and inherent religious conservatism, to forge a new reduced English Catholic Church.† It was a solution that had its echoes in Catholic Europe: in the French Gallican tradition, which had always sought to limit the extent of the Pope's temporal powers in France; in the reluctance of Catholic rulers to see the Council of Trent's decrees, emphasizing Rome's supremacy over them, published in their territories. If the Catholic nations of Europe could chafe at their Roman bit, aware of a conflict between their sovereign right to self-determination and the Vatican's *ultramontanism*, then why should not the Protestant Government of a divided country hope to exploit this conflict? The Cecil-influenced *Protestation of Allegiance*, when it was finally delivered to Elizabeth at the end of January 1603, suggested how this might have worked. In it the

* Cecil was later accused by his chief rival the Earl of Essex of being pro-Spanish and was forced to deny to James, in the first of a series of letters written to the Scottish King in advance of his succession, that he 'was sold over to the Spanish practice'.

† It must be emphasized that this was never the Appellants' aim: for all its political undercurrents, the Jesuit–secular dispute remained an ecclesiastical conflict and the Appellants never lost sight of their core demand, that they be rid of an Archpriest they believed wholly to be in the Jesuits' pocket. Even during their last negotiations with Robert Cecil over the oath of allegiance, they refused to cede any of their priestly powers in return for greater liberty of conscience. To simplify the dispute into a contest of identity between the nationalist seminarians and the internationalist Jesuits is to misrepresent it.

Appellants swore to acknowledge Elizabeth as their true and lawful Queen and to defend her life and realm against all plots or invasions, even those carried out 'under colour of the restitution of the Romish religion'. Here was the answer to the Bloody Question that the Government had longed for, unequivocal, enforceable in a court of law. Here was a Catholic-led movement to limit the bounds of papal supremacy, to place national security above awkward theological uncertainties, to place new England before old Christendom. That this movement was in a minority was indicated by the thirteen signatures, out of some four hundred seminarians then in England, that the *Protestation* garnered.* That it attempted to codify the uncodifiable, a man's loyalty to his God and his country, made its appeal limited. That it would have been stopped in its tracks by Elizabeth's fear of religious tolerance, no less than by the Pope's opposition to it, is unquestionable; its timing was not propitious. But its mere existence pointed to a possible way out of the cycle of paranoia and persecution into which England was locked; and, significantly, the Appellants had identified their Jesuit rivals as the main obstacle to that possibility. It remained to be seen what a Government looking for a way out would make of this.[53]

In the autumn of 1602 a foreign visitor to court watched the Queen taking the air, 'walking as freely as if she had been only eighteen years old'. In her seventieth year there were few signs Elizabeth was slackening her pace. On her annual summer progress that year she went hunting 'every second or third day, for the most part on horseback, and showeth little defect in ability', wrote the Catholic

* Author-in-chief of the *Protestation* was William Bishop, the seminarian who, back in December 1581, had landed at Rye harbour and so failed to convince the port authorities of his false identity. In a report of 1600 he was described as 'tall of stature, flaxen hair, broad-faced, somewhat pale, somewhat gross, and enemy of Jesuits'. In February 1623 the Appellants' demands were finally met when he was created Bishop of Chalcedon, Vicar Apostolic of England, in charge of the seminary mission. He died on 13 April 1624.

Anthony Rivers. But witnesses close-up commented that she 'looked very old and ill'. In October that year she dined with Robert Cecil. 'At her departure,' wrote Rivers, 'she refused help to enter her barge, whereby stumbling she fell and a little bruised her shins.'[54]

That same autumn Henry Garnet sent a circular letter to the Jesuits in England, announcing Pope Clement's ruling and celebrating the 'sweet end of all the controversies which have so long molested not us only, but all other Catholics'. He instructed his men to lay the dispute behind them and urged them to be patient with their detractors. 'And if with such patience we cannot obtain the quiet which we desire, the fault will easily be laid where it shall in deed be found': 'we have a particular obligation to give good example un[to] others'. The letter was a model of calm good humour and tact.[55]

Garnet had every reason to feel optimistic. In May that year he was able to write to Claudio Aquaviva that 'The Catholics are increasing very greatly'. There had been no let up in the penal laws. 'A few days ago', Garnet wrote, 'the Queen rebuked Canterbury sharply and ordered him to carry out real persecution. This, however, was not necessary since they were already proceeding with the utmost severity in all parts of the country.' But even as the Government was tightening the screw, so the mission was winning more converts. Doubtless this had much to do with a certain theological hedging of bets. Three sets of religious settlements had come and gone now, and each had survived only so long as the ruler who had instituted it. The future of Elizabeth's Church was no less uncertain. Garnet carried with him papal breves from Pope Clement, addressed to the Catholic laity and the missionaries, commanding 'that none should consent to any successor upon Elizabeth's death, however near in blood, who would not . . . with all his might set forward the Catholic religion'. These breves Garnet had been instructed to circulate on Elizabeth's death, with the Pope's authority. Robert Persons was now in communication with James VI of Scotland, backing his claim to the English throne (the Spanish Infanta had absented herself from the race). James, himself, was in communication – through various third parties – with the Pope, expressing an interest in converting to the Catholic faith. There was every reason for optimism.[56]

But for Garnet a new anxiety threatened. On 4 August that year he had written to Robert Persons, 'I purpose about Bartholomy-tide [24 August] to travel to St Winifred's Well for to increase my strength.' St Winifred's Well, in Flintshire on the North Wales coast, was 'a standing miracle', according to John Gerard. Legend had it that Winifred was beheaded defending her chastity; where her severed head fell, 'a powerful spring instantly burst forth'; her head was then reattached and she had lived on as abbess to a religious house in Shrewsbury. The site of the spring – in what had once been an 'arid valley' – had become an important place of pilgrimage: Richard I prayed there before the crusades, Henry V before Agincourt. And because the well was seen as medicinal as much as religious, the Reformation had not halted this tradition. Gerard wrote of the healing water: it was 'extremely cold, but no one ever came to any harm by drinking it . . . I took several gulps of it myself on an empty stomach and nothing happened to me'. Many miraculous cures had been recorded there and for Henry Garnet, in need of such a cure now, the well had become a place of hope. Earlier that year he had fallen ill with signs of shaking in his limbs: he feared the onset of the palsy and paralysis. At the age of just forty-seven his body was failing him, even as English Catholics dared to look to the future.[57]

Elizabeth spent Christmas 1602 at Whitehall, amid great festivities. She had once remarked to James of Scotland that she knew the preparations for her funeral were complete; still she seemed in no particular hurry to take part in the event herself. In January she moved to Richmond and the weather turned colder. On 19 January Henry Garnet wrote to Claudio Aquaviva, asking him to prevent Catholics abroad – and Robert Persons in particular – from responding to Elizabeth's November Proclamation against Catholicism. 'If any answer be made,' he wrote, 'we wish the Secretary [Robert Cecil] to be spared as much as may be.' Cecil, meanwhile, corresponded secretly with James and negotiated with the Appellants. By the end of the month the *Protestation of Allegiance* was ready to be presented to the Queen.[58]

At the beginning of February Elizabeth gave an audience to the Venetian ambassador, Giovanni Carlo Scaramelli, who wrote home suitably dazzled by the display she put on. She made no mention of the *Protestation*, then or subsequently, and soon other events had intervened to displace it from Government business altogether. At the end of February the death of the Countess of Nottingham, Elizabeth's cousin, turned the Queen to a deep melancholy; 'she has suddenly withdrawn into herself', reported Scaramelli. She kept to her private apartments, refusing to leave them and refusing to take any of the medicines prescribed by her doctors. She lost her appetite and was unable to sleep. For days she sat motionless in her chamber, staring at the floor, her finger in her mouth. It was, wrote Scaramelli, as if her 'mind was overwhelmed by a grief greater than she could bear'.[59]

On 9 March an English correspondent wrote to Robert Persons in Venice, telling him, 'The Queen's sickness continues, and every man's head is full of proclamations as to what will become of us afterwards.' That same day Henry Garnet noted, 'The Queen is said to be very sick. Arbella [Stuart] is diversely reported of, and is like to be sent up for shortly, to be guarded.' In London rumours were spread that the Council was stockpiling wheat in case of rebellion; Anthony Rivers wrote of a protective trench to be dug 'from the Tower to Westminster for defence of the suburbs'. In Scotland James, receiving regular bulletins on Elizabeth's health, cancelled a scheduled trip to the Highlands. Robert Carey, nephew to the dead Countess of Nottingham and a favourite of the Queen, arranged for fresh horses to be made ready for him the length of the journey from London to Scotland, and sent word to Edinburgh to expect him any hour. From The Hague, Sir Francis Vere, who had just received news that the Queen's condition was serious, wrote to Sir Robert Cecil, saying, 'I never thought to live to see so dismal a day.'[60]

On 21 March William Weston, in the Tower of London, noticed that 'a strange silence [had] descended on the whole city . . . Not a bell rang out. Not a bugle sounded – though ordinarily they were often heard'. On 22 March Elizabeth was finally coaxed to take to her bed. By the following day she had lost the power of speech, though she listened attentively to the prayers read out to her by the Archbishop of Canterbury, John Whitgift. That night she slipped into a coma. She died between two and three o'clock on the morning of 24 March 1603. The diarist John Manningham recorded 'her Majesty departed this life, mildly like a lamb, easily like a ripe apple from the tree'. By nine Robert Carey was clear of London and on the road north to Scotland and to James.[61]

She left no will. She never officially acknowledged that James was to inherit – even to the last she was gripped by the fear of naming her successor – but by now few doubted this to be her intention. Accordingly, at ten o'clock that morning James Stuart was proclaimed King of England, Scotland, France and Ireland, first at Whitehall Gate, then by the High Cross in Cheapside an hour later, then at the Tower of London.[62]

When the reports of Elizabeth's death finally reached Rome they announced that she had died 'with great reluctance'; some added hopefully that she had died 'a Catholic'.* 'God grant the last be true,' wrote one commentator piously; 'and the first also.' Back in London, it took the remainder of that day, 24 March 1603, for any lingering sadness at her death to change to excitement. That evening bonfires were lit and the church bells were rung and their

* Scaramelli, the Venetian ambassador, also reported Elizabeth's supposed deathbed conversion, in a report home to the Doge. '[F]rom her remarks and from her prayers that God would not reckon against her in the next life the blood of priests shed by her, there are some Catholics about court who think that in her inner sentiments her Majesty was not far from reconciliation with the true Catholic faith. This view is confirmed, because it was observed that in her private chapel she preserved the altar with images, the organs, the vestments which belong to the Latin rite, and certain ceremonies which are loathed by other heretics . . . All this stabs the heretics to the heart, and they would fain silence the report, though all agree that the Queen died as she had lived.' Scaramelli concluded: 'Be that as it may, she died a Queen who had lived for long, both gloriously and happily in this world. With her dies the family of Tudor.'

call taken up from parish to parish. Fire answered fire, peal answered peal as the royal messengers rode out across the country. Manningham recalled 'every man went about his business, as readily, as peaceably, as securely, as though there had been no change, nor any news of competitors'. Of James, he wrote, 'the people is full of expectation, and great hope of his worthiness'. It was a sentiment widely echoed. All 'men are well satisfied', wrote Simon Thelwal to his friend Mr Dunn in Bremen, Germany, and on everyone's lips was talk of the 'great hope of a flourishing time'. The Queen was dead. Long live the King.[63]

Chapter Eleven

'they sucked in the sweet Hopes of Alteration in Religion, and drunk so deep thereof, that they were almost intoxicated.'

Arthur Wilson (The Life and Reign of James, the First, King of Great Britain)

ON 23 JULY 1603, his first summer in England, King James I went hunting with the French ambassador. Afterwards, the two men sat down to a discussion about religion and James raised the subject that had preoccupied him for some time now, that of brokering a permanent peace between the Protestant and Catholic Churches. His plan, as the ambassador described it, was to convene 'a general Council . . . in a neutral place and with free access', made up of persons of virtue and learning. He would undertake to bring the Protestant party to the table in a spirit of amity. His question was, could the Pope be persuaded to agree to such a council and to attend it in a similar spirit? The ambassador duly reported the conversation to Henri IV of France.

Henri, once a Protestant, now a Catholic, walking embodiment of the possibilities of spiritual rapprochement (or a protean hypocrite, depending on your point of view) and, like James, a man who had suffered the worst extremes of his country's changing religious fortunes at first hand, sent back a positive if muted response. 'You can say to him', he wrote to his ambassador, 'that

I will always be very glad and happy to support and assist [him].' He shared the Scotsman's belief that the peace of Christendom would 'never be firm and assured so long as the discord in religion is such as we see at present', but in terms of practical suggestions he had none (a result of his own failure to organize a similar forum in France in the 1580s and 1590s).[1]

Four months later James contacted Pope Clement VIII directly, saying 'we do especially desire a general Council'. A year later a leading member of the Spanish Council would reiterate this plea on James's behalf, noting the sincerity of the demand and that it had 'the approval of a large part of the [English] Catholics'. To all these requests the Vatican remained deaf. Cardinal Aldobrandino, Clement's nephew, produced a long list of procedural difficulties – who would preside over a council? where would it meet? when would it meet? – as evidence for why James's request was impractical. Clement, himself, was uncompromising and not a little peevish: Henri IV had converted without need of a council, so why was it so necessary for James to talk?

James, who liked nothing more than to talk, was furious: two years later he was still complaining about the insult to the Venetian ambassador. But the little flurry of letters back and forth had been revealing. First, it suggested that the governments of England, France, and Spain, whose complicated *ménage à trois* had all too often been played out on the battlefields of Europe and who were never short of matter to quarrel over, were tired of using faith as a reason to fight. Political and economic pragmatism was replacing raging religious militarism as the cause closest to their ministerial hearts. Second, it suggested that the Vatican was crucially, painfully unaware of this. It remained to be seen who else had failed to spot the shift.[2]

But in the absence of an official forum for talk, talk, itself, ran riot. Rumour, exaggeration, and misrepresentation all flourished unchecked in James's England, fuelled by James's own willingness to use them as political tools. They contributed to a dangerous, potentially fatal gulf between perception and reality. And into such a gulf a man might all too easily fall and come to grief.

'The King of Scotland has succeeded quietly,' wrote the Venetian ambassador in Rome, back to the Doge and the Senate some fortnight after Elizabeth I's death. The Venetian ambassador to London, who earlier (as Elizabeth lay dying) had noted the arrest of sixty of the city's most prominent Catholics in anticipation of rebellion, now told of their prompt release from prison. The French ambassador to London contacted his counterpart in Spain to marvel at the ease with which James had taken possession of the Crown, 'though for years', he wrote, 'all Christendom held for certain that it must be attended with trouble and confusion'. Of James's reception as ruler, he added, the 'satisfaction is universal among the English, and so miraculous is the unanimity of the King's own nation that one sees his hand or his luck to be great, and his prudence even greater'. John Manningham, writing in his diary on 27 March 1603, recorded his first sighting of the King's fellow countrymen in London: 'I saw this afternoon a Scottish Lady at Mr Fleet's in Lothbury; she was sister to Earl Gowrie.' And on 16 April Henry Garnet wrote to Robert Persons, giving his impression of recent events:

> My very loving sir, Since my last to you of the sixteenth of March, there hath happened a great alteration, by the death of the queen. Great fears were: but all are turned into greatest security; and a golden time we have of unexpected freedom abroad . . .
>
> Yesternight came letters from [James]; but were not to be opened until this day. Great hope is of toleration; and so general a consent of Catholics in his proclaiming as it seemeth God will work much. All sorts of religions live in hope and suspense; yet the Catholics have great cause to hope for great respect, and that the nobility all almost labour for it, and have good promise

> thereof from his majesty: so that, if no foreign competitors hinder, the Catholics think themselves well, and would be loath any Catholic princes or his holiness should stir against the peaceable possession of the kingdom.[3]

'If no foreign competitors hinder': it was astute of Garnet to convey such a warning to Persons. Though James had taken the English Crown with scant protest at home, an adverse response from those still pressing the claims of Spain would throw all into hazard; and if any priest could be said to know Spanish intentions with regard to the throne it was Robert Persons. Garnet urged the case for non-intervention. James was ready for peace with Spain, according to Garnet's (unnamed) sources. He might even be ready to receive a papal envoy into the realm (something Elizabeth had always refused to do). Garnet could report that his co-religionists were content to believe Spanish interest in England had only been 'to set up a Catholic king' on the throne, not to seize it for themselves. Spain – and Persons – would be advised to do nothing to disabuse them of this notion. Indeed, wrote Garnet, his sole remaining worry was the recent talk he had heard of 'some threats against Jesuits, as unwilling to [acknowledge] his majesty's title, ready to promote the Spaniard, meddling in matters of state, and authors, especially, of the Book of Succession'. For this reason, he added, he had taken it upon himself to write to James on behalf of the Society, guaranteeing Jesuit loyalty to the new King. If Persons was chastened by this reference to his work on *A Conference about the Next Succession* he did not say. Instead, he contacted the Papal Secretary of State, then the Pope himself, urging that an envoy be sent to England forthwith. To Garnet, meanwhile, he wrote of the 'great contentment' with which news of James's succession had been received in Rome.[4]

Garnet's communiqué to James, dated 28 March 1603, just four days after Elizabeth's death, was forwarded to a nobleman at court. Sadly, the nobleman's identity is unknown, but evidently he was a man powerful enough to pick his moment with James: Garnet's instructions were to present the letter only when the time was favourable. In it, Garnet assured James of the Jesuits' 'love, fidelity, duty and obedience'. The new King, he wrote, would never have cause to distrust Garnet's fellowship 'of poor, weak and disgraced religious men'.[5]

The new King, himself, was still in transit from Scotland to London. Passage south was slow. Royal protocol required that Elizabeth be buried some four weeks after her death and it would have made poor showing for James's triumphal arrival to have clashed with his predecessor's solemn departure. So James drew out his progress, taking time to go hunting and to test the hospitality of his English subjects. This was not found wanting: on 14 April he was 'bountifully' entertained by Mrs Jennison of Walworth, near Durham, to his 'very high contentment'; Mrs Jennison was mother-in-law to John Gerard's two sisters Mary and Martha.* But true to the spirit of tangled reporting at large, other reasons were soon being given for the protracted nature of his journey. Giovanni Carlo Scaramelli, the Venetian ambassador to England, wrote home that James had no desire to be present at the obsequies because he could never 'expel from his memory the fact that his mother was put to death [by Elizabeth]'.[6]

At York on Tuesday, 19 April James took the opportunity to reward Thomas Gerard, John's brother, for his family's loyalty to Mary Stuart. Witnesses reported that he dubbed Thomas knight with the words: 'I am particularly bound to love your blood on account of the persecution you have borne for me.' To John Gerard, the action showed 'in His Majesty a good consideration of his servants', though the new knighthood, itself, he thought of little worth: the Gerards, he wrote, 'had been [knights] for sixteen or seventeen descents'. Still, for Catholics it was a promising sign: the son of their most conspicuous martyr was even now honouring those who had suffered on her behalf. On 25 April Lord Thomas

* Mary Gerard married John Jennison; Martha, the younger sister, married Michael Jennison. In her will, dated 1604, Mrs Jennison denounced John as being 'contrary in religion', i.e. Catholic.

Howard, heir to the disgraced Duke of Norfolk, and Henry Percy, ninth Earl of Northumberland were sworn onto James's Privy Council at Whitehall. Both men had lost fathers in Mary Stuart's service, Howard's to the headsman, Percy's (as it was rumoured) to an assassin in the Tower. When Howard's uncle was also admitted to the Council and promoted Earl of Northampton, many took this as a clear indication that James was determined to right his predecessor's wrongs. Scaramelli wrote back to the Doge in Venice, 'The King continues to support those houses and persons who were oppressed by the late Queen.' Most promising of all was the news of the imminent release from the Tower and banishment of the former Jesuit Superior, William Weston, as part of a mass clear-out of captured priests from the various internment camps around the country. It was small wonder that Henry Garnet could write of his great hopes of toleration, that John Gerard could note a host of 'hopeful signs of future favour' and that a Mr John Chamberlain, writing to his friend on 12 April, could observe: 'These bountiful beginnings raise all men's spirits, and put them in great hopes.' Hope. To Gerard it was 'a human help of no small force' and in those first few heady weeks of James's reign it bubbled up freely from a great wellspring of Catholic optimism.[7]

It was an optimism not entirely misplaced. Back in Scotland James had, for decades, surrounded himself with a small coterie of Catholic courtiers.* He had made use of Catholic Privy Councillors, a Catholic captain of the guard, and Catholic diplomats in his dealings with foreign powers. All this, despite Scotland's continued drift towards a radical Puritanism and James's own Protestant upbringing. There was talk that James's toleration of Catholicism extended closer to home too. In 1602 a report appeared, claiming that Anne, James's Danish-born and Lutheran-raised Queen, had converted to the Catholic faith some years before. The author of this report, the Scottish Jesuit Robert

* When asked by English Protestants why he kept Catholics around him at Court, James's response, according to the French ambassador, was that 'with one tame duck, he hoped to catch many more wild ones'.

Abercromby, testified that James had received his wife's desertion to Rome with equanimity, commenting, 'Well, wife, if you cannot live without this sort of thing, do your best to keep things as quiet as possible.' Anne would, indeed, keep her religious beliefs as quiet as possible: for the remainder of her life – even after her death – they remained obfuscated. When she declined Communion at James's Westminster coronation in the first summer of his reign English Catholics celebrated, but three years later, in 1606, the Venetian envoy Nicolo Molin would reluctantly conclude that Anne was still a Lutheran, despite her evident sympathy for Catholicism.[8]

More tantalizing, though, than the rumours of Anne's conversion were those of James's own. In February 1601 the Spanish Council of State reported the arrival in Rome of a Catholic 'confidant of the king of Scotland', who, with the latter's apparent blessing, had informed Pope Clement that James was contemplating a return to the Catholic Church, into which he had been baptized. This was a false hare – James had no intention of converting – but over the years it would run far and fast, coursed through the European courts by a diligent pack of Scottish agents, all giving cry in their master's voice. The Spanish council recognized the trick for what it was – an attempt by James to win votes in his campaign for the English throne – and concluded that his actions showed 'a false and shifty inclination'. Less knowing observers, among whom even Robert Persons and the Pope were numbered, were more willing to believe the stories. Indeed, as late as 1605 – two years into the new reign – Catholics could still speak hopefully of James's imminent return to Rome. Meanwhile, Henri IV of France promised to do everything in his power to assist this reconciliation.[9]

Elsewhere, James, too, was making promises. 'As for the Catholics I will neither persecute any that will be quiet, and give but outward obedience to the law, neither will I spare to advance any of them that will by good service worthily deserve it': so he wrote to the Earl of Northumberland in mid-March 1603. It was a statement of tolerance noticeably akin to those made by Elizabeth during the first few years of her reign, an indication of selective blindness towards the private practice of the Catholic faith. But whereas Elizabeth's words had been received by a Catholic majority reeling at the speed with which it had been ousted from power and expectant of few favours from her, James's were seized upon greedily by a minority desperate for signs of hope and full of expectation. Worse, the connection between that expectation and reality was soon being stretched to breaking point as Northumberland's go-between with the King, an impoverished second cousin of the earl's named Thomas Percy, began reporting more than just reasoned assurances from James. According to Percy, James had received him at court warmly and had spoken to him at length. Crucially, said Percy, James had sworn to free England's Catholics from 'bondage and persecution', to 'take them under his complete protection' and to admit them 'to every kind of honour and office in the state without making any difference between them and the Protestants', all this on his word as a prince. It was heady stuff and Percy was quick to broadcast it upon his return home.[10]

What Percy understood from his conversation with James and what James actually meant to convey are matters of dispute. The one was an unreliable witness: a comparatively recent convert to a hardline Catholicism on an embassy above his rank and with a track record of dishonesty (in his dealings with Northumberland's estates). The other was a politician. In all likelihood James gave a series of vague verbal guarantees that could be interpreted favourably and Percy leapt, willingly, to just such an interpretation. But Percy was not the only one to whom James was speaking off the record that month. John Gerard would later write: 'I am well assured that immediately upon Queen Elizabeth's sickness and death, divers Catholics of note and fame, Priests also, did ride post into Scotland . . . At that time, and to those persons, it is certain [James] did promise that Catholics should not only be quiet from any molestations, but should also enjoy such liberty in their houses privately as themselves would desire.' Here, again, was the talk of tolerance England's Catholics longed for. Characteristically, Gerard fails to give the names of those to whom James spoke, a habit born out of the missionaries' need for discretion, but his statement is backed up by an unlikely source: the Appellant priest and self-confessed enemy of the Jesuits, William Watson.[11]

Watson had had a mixed mission so far. He had been arrested five times; he had seen the inside of five different prisons and had escaped from three of them (the first attempt occasioning the execution of the woman who had helped him, Margaret Ward). More recently his had been among the many voices raised in anger in the factional strife between seminarians and Society men, most noticeably in the attacks on Robert Persons. In 1602 he had published his *Decacordon of Ten Quodlibetical Questions*, a bitter denunciation of the Jesuit order.* As an Appellant Watson had agitated for an oath of allegiance to Elizabeth. As an anti-Jesuit he had thrown his weight behind James's candidacy for the English throne (in protest at Robert Persons' involvement in the succession debate). Which was why, on Elizabeth's death, he had sought out James in Scotland, to plead the seminarians' cause.

James's answer to Watson had been 'gracious and comfortable'. It was an answer Watson duly passed on. '[H]is Majesty did bid me tell my friends', he would later explain, 'that himself was neither heretic, as Persons and other Jesuits had blazed him to be, neither would he afflict them as they had been; and therefore wished them by me not to be afraid.' Despite the self-important tone and anti-Society bent, Watson's belief in the 'advancements' coming to England's Catholics was palpable and, like Percy before him, his reporting of his embassy filled his hearers with anticipation.[12]

It was an anticipation that helped palliate the arrest, as James journeyed south, of the seminary priest Father Hill, allegedly for threatening the King. (Hill, acting without authority, had petitioned James to repeal the penal laws against Catholics,

* Curiously, Watson regarded John Gerard as being 'of a good disposition in nature' and wished that he could be persuaded 'to live like a secular Priest'. This did not stop him from naming all those whom Gerard had introduced into the Society.

referring in this petition to the Israelites' disobedience to King Jeroboam when he failed to grant them similar relief; he was quickly removed to London and the Gatehouse prison.) It was an anticipation that helped sweeten James's refusal to empty the gaols of Catholic recusants. (At Newcastle, York, Durham, and Newark he had, in a gesture of political amnesty, 'commanded all the prisoners to be set at liberty, except Papists and wilful murderers'.) It was an anticipation that flourished in ignorance of James's recent correspondence with the man who would become his Principal Secretary of State, Robert Cecil. (The pair had exchanged letters in the build-up to Elizabeth's death, in which James, though stressing his hatred of persecution, had expressed a far stronger hatred of the 'Jesuits, seminary priests, and that rabble where with England is already too much infected' and an overwhelming fear of Catholic numbers increasing. Bluntly, he wrote, 'I [would not] wish to have their heads divided from their bodies, but . . . I would be glad to have both their heads and bodies separate from this whole land, and safely transported beyond seas, where they may freely glut themselves upon their imagined gods.') It was an anticipation that grew daily, as James neared London, fed by a belief among Catholics that it simply made no sense for James to continue persecuting them.[13]

As John Gerard explained, Elizabeth had had reason to outlaw the old faith: Catholic refusal to recognize her claim to the throne and the concomitant danger of foreign intervention that that refusal posed. James, on the other hand, had received messages of congratulation from the Pope and his main Catholic rivals on his succession. Indeed, his foreign policy promised to be singularly less fraught than that of Elizabeth: the new alliance between Scotland and England had closed the door on invasions from the north of the country and, as King of Scotland, James was at peace with Spain, even if England was not. So, wrote Gerard, 'many Catholics . . . could not persuade themselves how it could be possible' for James to turn against them. For so long now they had been arrested and punished not for their faith, they had been told, but for the safety of the realm. With England's safety more assured than it had been for almost fifty years, surely there was little point in continuing this persecution?[14]

On Saturday, 7 May 1603 James reached Stamford Hill, overlooking the English capital. Here the Lord Mayor and Aldermen of London were waiting for him, to present him with the keys to the city. Witnesses reported 'multitudes of people swarming in Fields, Houses, Trees, and highways, to behold the King'. By five o'clock that afternoon he was at the Charterhouse, the former Carthusian monastery just outside the city walls.

His coronation had been fixed for Monday, 25 July, St James's Day. In the end a summer downpour and an outbreak of the plague in the capital combined to do James out of the full pomp and circumstance of the anointing ceremony and the event was carried out in abbreviated style. Those same people who earlier had swarmed to meet him were now commanded to stay away in case they spread disease. Guards were posted at the city gates to keep back the crowds and anyone landing at Westminster from the river was threatened with the death penalty. Outside the Abbey, rainwater streamed through empty stands as the royal party was hurried in and out by an anxious escort. Historian William Camden recorded, with a succinctness that matched his subject, 'The King and Queen are crowned, it being then very bad weather, and the pestilence mightily raging.'[15]

Nothing, though, could dampen the reality of James's triumph. As an infant he had survived the murder of his father and the overthrow of his mother. As a child-king he had survived successive attempts to kidnap him, still more attempts to intimidate him.* His palaces had been stormed, his servants murdered, his person threatened – and all by his own subjects. These traumas may

* On 22 August 1582 the sixteen-year-old James was taken hostage by the Earl of Gowrie and his faction. He was held captive until June the following year, when he was finally able to make his escape.

have left their mark upon him – all his life he would be scared of assassination and he was widely considered a physical coward – but they had also bred in him the instincts of a seasoned campaigner. Now, aged thirty-six, he had beaten off every rival to the English throne and disarmed a powerful European Catholic opposition, charming the Pope, pacifying France and Spain. He had played a cool and calculating game for England and he had won it. And, delighted at the way in which so many Englishmen 'had so generally received and proclaimed him king', he had turned, it was said, to a nobleman next to him, giving voice to his feelings. This nobleman had passed the comment on and, ripple-like, his words had spread out through the Catholic community. Whether the whispers corresponded to what James actually said made little difference in the end: it was what the hearers heard that mattered and the hearers, in this case, heard a sentiment that stunned them. 'No, no, good faith,' James was reported to have exclaimed, as the extent of his victory sank in, 'we'll not need the papists now!'[16]

Queen Elizabeth, an astute judge of human nature, had summed the attitude up with painful accuracy. 'As children dream in their sleep after apples and in the morning when they awake and find not the apples they weep, so every man that bore me good will when I was [princess] . . . imagineth himself that immediately after my coming to the crown every man should be rewarded according to his own fantasy and now finding the event answer not their expectation it may be that some could be content of new change.' Change: within months of James's succession this word had taken on fresh momentum.[17]

On about 13 June 1603 two men met in secret on the north bank of the River Thames near the Horse Ferry, which crossed the river between Lambeth and Westminster. There, caught between the Archbishop of Canterbury's palace and Parliament, respective houses of Church and State, the one took an oath of loyalty to the other, swearing to assist him in a scheme to upset both those estates. The oath-taker was Anthony Copley, Robert Southwell's cousin, active in the Appellant movement. The man behind the action was the seminary priest William Watson.

The plot was farcical from the start. At its heart lay Watson's vanity. Having returned from Scotland and his meeting with James promising imminent freedom of conscience for English Catholics (particularly those Catholics who were anti-Jesuit), the priest had been piqued to find his promises not immediately fulfilled. In May, for example, James, in session with his Council, had ordered the collection of recusancy fines. According to his subsequent testimony, Watson was now forced to suffer the indignity of 'divers upbraiding speeches against [him], . . . as, "Lo! there was Watson's king! There was he that was said to be so well affected to Catholics, as that he would grant toleration!"' According to Copley, Watson returned to court to challenge James on the subject and received an unsympathetic hearing. It was at this point, testified Copley, that Watson learned of James's comment about papist dispensability.[18]

Now Watson began drawing in fellow malcontents: William Clark, a seminary priest and an active Appellant, Sir Griffin Markham, keeper of the royal hunting lodge in Sherwood Forest (and familiar to John Gerard), and George Brooke, brother to Lord Cobham. Their aim was to manhandle James into delivering Catholic toleration and it was Markham who gave them their *modus operandi*: he 'had heard it was usual in Scotland for the king to be taken of his subjects, and kept in strong hold until he had granted his subjects' requests'. Consequently, the plotters, to which group Watson said he was adding all the time, agreed to kidnap James at Greenwich, take him to the Tower of London (which they planned to use as their stronghold) and persuade him to accede to their demands. A date was set for the escapade, 24 June, a month before the coronation, and so that all the conspirators – for Watson was now promising a cast of thousands – were better able to recognize one another it was suggested they 'wear stockings either of yellow or blue'. It was Markham who brought back the bad news that James would not, in fact, be at Greenwich on the day appointed for his kidnapping. Neither, it was soon apparent, were the hordes of plotters boasted of by Watson likely to put in an appearance: Anthony Copley spent a dispiriting day sitting in to wait for them, by the end of which his hopes of success had faded. At eleven o'clock the following day Watson called off the attempt, 'affirming that he despaired of the action, and laying blame on all, that they had not complied with their promises'.[19]

But others had been complying with their promises. It had been less than three months since Garnet had assured James of the Jesuits' fidelity; now he and his men had a chance to keep their word, for some time in late May/early June an attempt was made to recruit John Gerard into the conspiracy. According to Gerard's account of events, he had told his recruiter – probably Griffin Markham – that the intrigue was both 'unlawful' and 'hurtful' to the Catholic cause; then, he had written to Henry Garnet and the Archpriest George Blackwell, begging them to forbid 'their acquaintances from entering the cause, and to stay it what they could'.* Watson would later complain that the Jesuits had prevented him from approaching potential plotters in Lancashire and Wales; and one who did join the plot, a Mr Harris of Pembrokeshire, would explain that 'the Jesuit party . . . [had] . . . gotten some inkling of the action, and so laboured to cross it'. Meanwhile, George Blackwell was contacting all those in his charge, instructing them 'to give stay and restraint' to anyone wishing to join the conspiracy.[20]

* It seems Henry Garnet's immediate response was to contact the Vatican, though his letter has not survived. He, however, would later testify to this effect and a response from the Papal Nuncio in Flanders, of late July, appears to corroborate his evidence. The Nuncio commanded Archpriest Blackwell (Watson's Superior) 'to exhort [his priests] to do nothing against the public peace or anything that may make their religion hateful and suspect. They shall count for joy any contumely then may endure for the name of Jesus'. He continued: 'the Pope will be most ready to use his authority amongst Catholics to secure the safety of his Majesty's person and state and will call out of his kingdom all those whom his Majesty may reasonably judge to be noxious to himself and his state'. It was a clear papal prohibition of Catholic unrest.

It was probably inevitable that Watson, with his near pathological hatred of the Jesuits and the Archpriest, should have ignored all these early warnings; and when Gerard and Blackwell saw 24 June approaching with no obvious sign of the action having been called off they both stepped up their response. Gerard told everything he knew to a Scottish courtier close to James, asking him to bring the matter to James's attention. Blackwell passed on the same information, via a courier, to a priest in prison (a convoluted path, concealing his whereabouts), telling him to bring the matter to the Privy Council's attention. It was Blackwell's message that reached the Government first.*[21]

As news of the plot broke, Watson and company took flight and the Council began the slow process of reeling them back in again. Descriptions were issued of the wanted men; Watson was listed as 'a man of the lowest sort . . . and very purblind, so as if he read anything, he puts the paper near to his eyes'. Anthony Copley was picked up some time in early July, George Brooke soon afterwards, and as the outbreak of plague intensified in the hot weather so, too, did the manhunt: failing to find Markham in London, the pursuivants now raided his Nottinghamshire estate. By early August Watson had been run to earth in Wales, Clark in Worcester, and all the plotters were in the Tower and talking freely.[22]

Watson's line of defence was singular: he had only devised his plot, he argued, in order to save James from a much deadlier Jesuit plot. The Jesuits might act as though 'butter would not melt in their mouth', they might be 'extolling the king's majesty', but in reality, he claimed, they were even now massing arms, horses and men to assist in a Spanish invasion of the realm. This was why they had hindered him, this was why they had 'sent down post-haste into the country, for all Catholics to beware of such and such priests, as were about a most dangerous conspiracy'. He, Watson,

* Gerard would write of his Scottish contact: he 'made haste unto the Court to open the matter unto the King himself; but found it already known the day before he came, and so spake nothing of it, being not then needful, nor he willing without cause to be acknowen of acquaintance with Father Gerard'.

had merely been trying to save James from certain death. As justification for high treason it was not enough to save Watson: on 15 November 1603 he, Clark, Brooke, Markham and Copley were convicted of conspiring to raise rebellion, to alter religion and to subvert the state. In the end, only Watson, Clark and Brooke were executed: Watson went to the scaffold – significantly – asking the Jesuits' pardon; Markham and Copley received the King's pardon and were sent into exile.*[23]

Theirs had been a sorry, scrappy affair. That its fallout was no greater was down to the clumsiness of the plotting and the prompt action of Gerard, Garnet and Blackwell to stop it. James, who, before the discovery of the plot, had told a French envoy he was considering remitting some recusancy fines, continued with this remittance as part of his coronation festivities, reassuring a Catholic deputation led by Sir Thomas Tresham that his feelings towards them remained benign. Privately, though, he complained that in spite of his kindness England's Catholics had still sought to harm him.[24]

How much Watson's plot played upon James's terror of assassination (and brought back humiliating memories of his earlier kidnapping) is a matter of conjecture, so, too, the extent to which Watson's talk of Jesuit scheming was taken seriously by the Government. His was not the only purported intelligence concerning the Society reaching Robert Cecil at this time. Sir John Popham, who operated a wide network of informants, told Cecil, 'I do assure myself the Jesuit faction have their practice afoot as well as [Watson], though carried with more secrecy, and so the more dangerous.' The spy reports coming in to him all agreed 'something [would] be done'; an attempt would be made on James's life; an invasion fleet would be launched: it was not a question of if, merely of when. If governments tend to listen best to the kind of

* A witness to the hangings wrote, 'The two priests that led the way to the execution were very bloodily handled; for they were cut down alive; and Clark, to whom more favour was intended, had the worse luck; for he both strove to help himself, and spake after he was cut down. They died boldly, both, and Watson (as he would have it seem) willing.'

intelligence they already hold to be true, then James's Privy Council was listening to a steady stream of information, a constant chatter of anxiety, that satisfied its worst fears: the Jesuits were plotting. Watson's explanation of his exploits was simply another prediction of future conspiracies, of future Spanish enterprises, of future unrest. Further, he, himself, had effectively demonstrated – to anyone wishing proof of it – that England's hidden priests, and the Catholics in their thrall, were no more to be trusted under James's Government than they had been under Elizabeth's. Meanwhile, in a backhanded compliment to the Society, Anthony Copley (conferring with Griffin Markham about Watson and Clark's inefficiency) remarked 'how much more sufficiently the Jesuit party would have carried the like'.[25]

What, then, would they have made of Henry Garnet's private correspondence that summer? On 15 June, as he tried to prevent discontented Catholics from joining Watson, Garnet had written to Robert Persons, condemning the action as a 'piece of impudent folly'; 'it is by peaceful means', he emphasized, 'that his Holiness and other princes are prepared to help us'. His words chimed with those of Persons himself: a few weeks later Persons would write to Anthony Rivers (Garnet's sometime secretary), urging English Catholics not to give way to their 'passion' and 'break out' in rebellion. 'I do not see possibly here what may be counselled in the present case of our country . . .' he wrote, 'but only to have patience and to expect the event of things.' The event that both men were expecting was an end to the war with Spain.[26]

Hostilities between England and Spain had rumbled on for almost twenty years; neither nation had the upper hand, both were facing creeping financial ruin. One of Robert Cecil's first actions of the new reign had been to tot up the cost of Elizabeth's conflicts. The figures said it all: £49,478,054 had been spent on military action during the reign; less than a tenth of that sum had been granted to the Queen in Parliamentary subsidy. James had inherited not only his cousin's crown, but also her searching need for revenue and her debts. In light of this, his decision to rescind recusancy fines (to those willing to sue for pardon) spoke of a plausible desire for tolerance on his part in a way that his earlier vague promises to Catholics had suggested only political contrivance, for by now the notion that it could profit from papistry had become a mainstay of Government economic planning. The Exchequer receipts for the year 1604 reveal that James's single act of generosity towards England's Catholics would cost him almost £5,700, money he could ill afford to lose. Spain – though still Europe's dominant power – was in not much healthier a condition, moving towards long-term imperial meltdown, haemorrhaging gold in pursuit of an aggressive foreign policy, ruled by a young and indecisive king. Philip II had once said, in reference to his heir, 'God has given me so many kingdoms, but not a son fit to govern them.' This was a harsh assessment. Philip III may not have inherited his father's strengths as a statesman, but neither had he inherited his bottomless treasury. Over the years he would find himself increasingly unable to finance the armies necessary to hold his scattered territories together and wage his father's wars. Which was why, in April 1603, just a month after James's succession (and in answer to his immediate declaration of a cease-fire), Philip's Council drew up its first instructions for peace.[27]

It was always likely that Spain, which for so long had given financial aid to English Catholic refugees and which had made the restoration of English Catholicism its official justification for the war, would explore measures designed to help the English Catholics in the peace. And when the Spanish envoy charged with paving the way to a treaty, Don Juan de Tassis, sailed for England, he carried with him the orders of his King, penned in the margin of the Council's policy document, 'to insist strongly on the liberty of conscience for Catholics'. This, wrote Philip, is 'what I desire the most'. It was a simple wish, clearly stated. Sadly, all Philip's Council did not share it. Set against the King and his call for religious toleration was a body of ministers aware that to demand concessions from England towards its troublesome Catholics risked similar concessions being demanded of Spain towards its troublesome Protestants in the Low Countries. To make any such concessions while the Dutch were still in open rebellion against their Spanish rulers was tantamount to granting them independence, a humiliating climb-down after decades spent fighting to keep them in the empire. So even before Tassis left to pursue the cause of religious tolerance, that cause had been weakened.[28]

In Brussels Tassis halted his journey north to talk with a group of English Catholic exiles about conditions at home. His meeting, he reported, took place 'at a late hour of the night to protect ourselves from certain spies' and, in keeping with this air of paranoia, he was warned that once he crossed the Channel 'divers persons [might] attempt to insert themselves into [my] favour under pretence of being a Catholic'. Trust no one – this was the exiles' message to him. Brussels buzzed with news from England. James was said to hate those he regarded as 'Hispaniolated' or 'Jesuitised', words that seemed interchangeable. This information, set beside dispatches reaching the Spanish Council from Rome that Robert Persons and his fellow authors of the Book of Succession were now 'looked upon as lepers to be avoided by necessity', suggested a worrying scenario. It was as though James could never forgive anyone who had challenged his right to the English throne. Spain had appeared to do so, so too the Jesuits: both linked to this challenge by Persons' contribution to the succession debate. Persons might now be *persona non grata* in Rome – but at least he was in Rome and out of James's reach. The Spanish had congratulated James on his succession and were now on their way to making peace with him. This left only the Jesuits to fill the role of James's personal bogeymen.[29]

Warned to trust no one, Tassis was soon able to put this advice to the test. On 26 June he reported the arrival in Brussels – at the Court of the Archduke Albert, co-regent of the Spanish Netherlands – of a mystery informant, who declined to give his name, but who claimed to be carrying secret intelligence from England.

According to this, Tassis told Philip, the English Catholics believed 'themselves capable of mustering a total of 12,000 men and should your Majesty assist them in time with a similar number, they offer to expel [James] from England'. If this were true, added Tassis cautiously, showing a quick appreciation of the world in which he was now moving, it was 'momentous' news, demanding 'very deep reflection'.[30]

Philip's response to this news was as cautious as his envoy's. 'I charge you', he wrote, 'to investigate with great care and secrecy the basis and resources behind these Catholics.' He had good reason to urge care and secrecy. Recently arrived at his Court were two Englishmen who appeared to be of the same confederacy as Tassis' unnamed man, who were also making extravagant claims about the number of Catholics they could muster in the event of a Spanish invasion and who were even now begging him to launch such an invasion. One of the men was called Anthony Dutton, a name that fades from the records every bit as quickly as it appears. The other was a Yorkshire Catholic called Guy Fawkes.*[31]

Three years later Fawkes would be notorious all across Europe, but in the summer of 1603 he was just one more disenfranchised English Catholic pursuing another enterprise against his homeland. His journey would take him from Philip's Court to a gunpowder-packed cellar beneath the Houses of Parliament, the desperate last fling of a desperate age. It had begun a year earlier in the spring of 1602 with the arrival in Spain of a Worcestershire man, Thomas Wintour, begging money from Philip for impoverished Catholics back home. Subsidiary to this Wintour offered Philip the 'devotion' of all those Catholics – some thousand

* The evidence linking Dutton, Fawkes and the unnamed informant to the same conspiracy is circumstantial but compelling. On his arrival in Spain in May 1603 Dutton claimed that another of his circle was about to visit Brussels, seeking 'arms and munitions against James'. Fawkes, two month later, claimed to have been sent to Spain by an unnamed Englishman who had recently crossed to Brussels to 'give a complete account about England' to the Archduke and to ask for military aid. The presence at Albert's Court of two different individuals both independently agitating for a Spanish invasion of England is unlikely, particularly as Dutton and Fawkes said they were part of a wider group of conspirators, therefore it seems sensible to view the three men as co-plotters.

horsemen or so – in the event of Spain launching an attack on England. So far, so predictable: Wintour's was the standard promise of any exile unhappy with the government at home and eager for change. Spain's response to his request was far from predictable, though. After a succession of surprise meetings with ministers Wintour was assured of the sum of 100,000 crowns and told that Philip 'meant to . . . set foot in England' the following year. The scene was set for invasion.[32]

By the spring of 1603, though, all bets were off. Elizabeth was dead, James was amenable to peace and the Spanish, if ever they had seriously considered a fresh assault on England, were now favouring negotiation. But Wintour had been given an assurance; moreover he had told others about the invasion plans. When Dutton and Fawkes arrived at Philip's Court to hold him to the guarantees made in his name, they had about them the frantic air of men trying to alter destiny. Their claims were lavish, their invective coarse: Dutton promised that 'with work, speed, secrecy and good weather we will have won the game in six days'; Fawkes launched into a bitter diatribe against the Scots. The Spanish Council demurred, then said little, committing to less and keeping its eye firmly fixed on Tassis' progress. For good measure it placed Fawkes and Dutton under house arrest to prevent them derailing the peace process. Fawkes and Dutton had gone looking for Wintour's war-mongers among the Spanish Council and had found only doves. Their hopes of war had been raised, only to be dashed again as this new spirit of diplomatic pragmatism began to prevail.[33]

Of course no political transition is quite that abrupt. There were still enough hawks left both in Spain and England to make peace an option, not a certainty. So, while Fawkes and Dutton whiled away their detention writing lurid proclamations to be read out once the invasion had triumphed, Tassis, on Philip's instructions, crossed to England, seeking to explore their claims every bit as much as he sought a solution to the hostilities. It took him just a month to decide that the plotters had overstated their case. England's Catholics, he reported, 'go about in such timid fear of one another, that I would seriously doubt that they would risk taking to arms'. So far as an invasion was concerned he was adamant: 'I would not dare to trust these people in this question.'[34]

Plague kept the English Court on the move that summer and autumn and Tassis travelled with it, keeping his eyes and ears open. At Oxford he met with Thomas Wintour, who told him that his Catholic horsemen were 'ready as requested', but that none of the money promised to him had yet arrived. This information Tassis duly passed back to Spain, joining reports from Wintour, himself, which stated – ominously – that his fellow plotters could not 'be restrained much longer'. As Tassis continued his delicate negotiations with the English Government, so the cracks in the Spanish position with regard to religious tolerance began to widen. In September the Archduke Albert – ruler of the territories in which so much of the war had been played out and a driving force for the peace – asked Philip not to insist on freedom for England's Catholics until after the treaty was signed. If it were 'placed on the agenda in the first instance,' he wrote, 'it [would] damage ... negotiations'. Meanwhile, the Spanish Primate, Cardinal de Rojas y Sandoval, advised that the 'present case of the Catholics of England is one of charity and not of justice': Spain had no moral duty to assist them. Finding himself more isolated by the moment, Philip now authorized Tassis secretly to offer money to anyone in England prepared to push forward the cause of tolerance.*[35]

James spent the close of 1603 at Hampton Court. Throughout the autumn he and his Government had been reading reports that, with the relaxation of the recusancy laws, Catholic numbers were growing again. In October Robert Cecil's brother wrote from

* It was the widely held belief throughout mainland Europe that the only way to do business with the English was by bribing their Government ministers. 'There is no one', wrote one Venetian envoy to his Doge, 'who sooner or later is not forced to apply to the Council and ... seek the protection of some member and that can only be gained in England by presents and gifts.'

Lincolnshire: 'The plague spreads here in divers places . . . So likewise does . . . popery.' The following month James received a stern warning from Lord Sheffield in York: 'As long as by the laws of this land [Catholics] were kept under, that affection of theirs bred no infection. But since of late the penalty of those laws has not so absolutely as before been inflicted . . . they begin to grow very insolent and to show themselves.' Perception mattered more than fact in this instance, as two opposing viewpoints demonstrated. Tassis, investigating invasion, perceived only a demoralized English Catholic body of insignificant size; the English Government, still possessed of its siege mentality, perceived a worrying proliferation of a threat it had believed to be under control. Most likely England's Catholics had become noticeable to those around them only because now they felt more able to worship openly, instead of behind bolted doors, but the talk surrounding the subject was dangerously heady: a few months later Thomas Wintour would be overheard declaring that there had been some 10,000 converts to Catholicism that year. It was precisely this anxiety that James had voiced to Robert Cecil in the run up to his succession. 'I would be sorry', he had written, 'that Catholics should so multiply as they might be able to practice their old principles upon us.' The liberality that had seen him declare 'I will never allow . . . that the blood of any man shall be shed for diversity of opinions in religion', and that had led him to ease recusancy laws, was now increasingly at odds with any fears he may have had about his safety and that of the realm. Liberality is a luxury for the fearful: on 15 December James instructed the Archbishop of Canterbury to compile a list of all popish recusants countrywide and to enforce the laws against them.[36]

In January 1604 James announced his first Parliament, to meet at Westminster in two months' time. Days later he convened the Hampton Court Conference of Conformity, to discuss the state of the nation's religion. After almost a year of anxious speculation as to James's precise intentions towards them, England's Catholics were about to learn their fate.

And the cruellest aspect of this fate was that they, themselves, were to play no part in the deciding of it. The Hampton Court Conference was to be an entirely Protestant affair, its purpose to address the issue of Puritanism within the official Church; the Catholics were not invited to attend: 'we were put to silence, our mouth was shut', reported John Gerard. In fact, what the conference revealed was that, with its numbers increasing, its powerbase widening and its muscles flexed against anything it saw as man-made hierarchy, Puritanism was replacing Catholicism as the greatest challenge to James's authority. To his son, James had described Puritans as 'very pests'. To the conference-delegates, particularly to the speaker who, rashly, used the term *presbytery*, he was even more forceful. A 'Presbytery', he exploded, 'as well agrees with a monarchy as God and the Devil'. If the Puritans were agitating to replace the episcopacy, argued James, then what might they attack next? 'It is my aphorism "No bishop, no King"', was how he put it. But it was not just the Puritans who came under attack as James laid out his vision of the nation's religious future. Catholics were soon reporting that James had spoken 'emphatically and virulently' against their faith, insults that only added to the injury of their non-attendance. The insults kept coming. On 19 February James talked publicly about 'his utter detestation' of the papist religion. On 22 February he issued a proclamation ordering every Catholic priest from the country. Back in September Robert Persons had urged the Vatican, on the advice of James's own envoy to Rome, to send someone 'to confer with the king . . . before his Majesty [was] definitely committed to' a course of action. The Vatican had not responded and James, it seemed, was now committed to his predecessor's anti-Catholicism despite all his promises to the contrary. The focus turned on Westminster.[37]

On 19 March James entered Parliament House. His address to the assembled MPs took an hour, time in which to set out his concerns for the nation. Foremost in his mind was peace: there was the predicted peace with Spain, which he had made possible 'by

[his] arrival' in England, and there was the peaceful union between England and Scotland, a union 'made in [his] blood'. Only then did he move on to the subject of religion, its place in his speech a prophetic illustration of its destined place in his and Parliament's energies this first and fractious session. That New Year the Catholic poet Henry Constable had contacted the Vatican independently to reiterate James's plea for a council for religious reconciliation, advising Rome to act before Parliament sat. But no one had acted and now the State's priorities had become clear: sandwiched in between debates about the union (and amid unprecedented rows over the respective rights of monarch and Commons) were the customary anti-Catholic measures. It was James who had instigated them, complaining in May that Catholic numbers were up again and that laws were needed 'to hem them in'. It was the Lords who had introduced the required bill in June, with the exception of the Catholic Lord Montague whose objection to the legislation put him in the Tower for a time. It was King, Lords and Commons who had rubber-stamped the new *Act for the due execution of the Statutes against Jesuits, Seminary Priests, Recusants, etc.* The act confirmed all the existing Elizabethan penal laws with some additions: anyone sending their child overseas to receive a Catholic education was to be fined, anyone going overseas for such an education was to be deprived of all possessions, anyone transporting such a person overseas was to forfeit job, goods and liberty. This agreed, King, Lords and Commons went back to arguing the subject of their respective rights and privileges and only the MP who wished to see all Catholics classed as outlaws can have been disappointed by the smooth and rapid passage of the new act.[38]

It is probable James's willingness to extend the penal laws was as much a measure of his need to pacify a Puritan-dominated Commons refusing to grant him subsidies and fighting him on the issue of union as it was of his fears about Catholic increase. To an envoy of the Duke of Lorraine he would admit that he regretted Parliament's actions; to the French ambassador he would say that he had no intention of putting the new penalties into effect. Then again, this may have been bluff – by now James had proved himself the master of the mixed message. But to England's Catholics the message was clear: here was another reminder of their status as hostages to political expediency.[39]

This was further highlighted by the year's other developments. That spring the Anglo-Spanish talks had ground to a halt, drowned in detail. In April Philip contacted Tassis after some four months' silence, telling him to reassure English Catholics that Spain had not abandoned them. This was how it looked, though. Anthony Rivers, writing the same month, noted, 'The Spanish Ambassador has yet done little for and in behalf of the Catholics.'[40]

In May the Archduke Albert's commissioners arrived in London to inject new life into the proceedings and the final push for peace began. It took place over eighteen meetings from May to July and featured lengthy discussions about the rebel Dutch and about overseas trade agreements. England's Catholics barely rated a mention. Toleration was now – officially – off the agenda. Henry Garnet, who had spent the previous autumn begging that, were Spain serious about buying relief for Catholics, then the money must be made available before Parliament sat, now found himself with one eye on the peace talks, the other on Westminster, where MPs were even now discussing the new anti-Catholic laws. By June he was in a gloomy mood, writing to Robert Persons on the fifth of that month, 'I think . . . that it is highly probable the [penal] laws will be confirmed; and that there is little hope of . . . liberty of conscience.'[41]

In August the man formally charged with closing the deal for Spain, the Constable of Castile, landed at Dover and the negotiations moved towards a conclusion. The constable was a realist, so much so that his presents for James – jewels from Antwerp – had been bought sale or return, in case the talks collapsed even at this late stage. As a realist he had stood out against making religious tolerance a condition for peace, believing, as he told Philip, that the Vatican was 'the true portal through which the affairs of the Catholics should be arranged'. The hallmarks of his realism were stamped all over the final peace treaty. It was a triumph for the diplomats. It had been hard fought. Money had changed hands and key members of the English Council, including Robert Cecil, would receive Spanish pensions for years to come in return for their willingness to negotiate. It had preserved just enough ambiguity in its phrasing to ensure that neither side felt it had compromised unduly. It had utterly failed England's Catholics.[42]

The blame for this was swiftly apportioned. The Constable of Castile, writing to Philip, noted, 'I see that the Pope himself, whose principal concern should be this very matter, is . . . silent.' Philip, in response to a Vatican complaint that he had done nothing for the Catholic cause, fired off the testy response 'there is a great difference in not attempting something and not succeeding in it'. Meanwhile, Pope Clement, who had refused publicly to endorse Spain's efforts on behalf of English Catholics and who had dismissed as 'scandalous' all attempts to buy tolerance, reiterated his belief that God had His own time-scale for such matters and that everyone must simply be prepared to 'await the Divine Will'.[43]

The post-peace celebrations were lavish. The negotiating party, giddy with relief, danced, dined, exchanged their presents and watched from the windows of Whitehall as, below them, the King's bears fought to the death with a pack of greyhounds. Then they were escorted back to their residence at Somerset House through the half-dark of an August evening by a troop of fifty torchmen. Days later the constable was on his way back to Dover and a boat to the Continent, leaving Tassis with the unenviable task of continuing the secret attempts to buy relief for English Catholics. Three months later, in November, the constable would write up a report of his brief time in England. It was a realist's summation. 'The temporal resources of the Catholics of this Kingdom', he told Philip, 'are very weak, and they could not and would not dare to attempt anything.' Moreover, he added, 'they do not want foreigners, especially the Spanish, to come in here'. The best way, he wrote, for Spain to assist England's Catholics was by maintaining the peace. For only with lasting peace could the suspicion that Catholics were fifth columnists and traitors be removed from English minds.[44]

The constable would get his peace. The treaty would hold and the Spanish invasion forces that Guy Fawkes and Thomas Wintour had hung their hopes upon would never again sail up the Channel to threaten England's shores. But by now Fawkes, Wintour and their associates had moved on to another plot altogether and the elements of this particular tragedy were almost all in place.

There is little information about the activities of John Gerard and Henry Garnet at this time. Gerard remained in Northamptonshire, travelling the county from family to family as before, from Vauxs to Treshams to Catesbys. It seems likely that alongside his regular pastoral work he was involved in an attempt to ease relations between the Jesuits and the Appellants. He, himself, makes no mention of this, but William Watson, in his evidence to the Council, refers to a series of talks between Gerard and the Appellants in the late spring of 1603 and two letters exist from Henry Garnet, both written that April and both alluding to the possibility of a 'union' between the two factions. One of them, dated 9 April, specifically mentions a hoped-for meeting between Watson and Gerard. Whether or not this last meeting took place is unclear, but Watson was dismayed at the outcome of the others: they 'ended only in breach', he told the Council. It was, perhaps, this 'breach' that led Watson to delate Gerard as one of those Jesuits agitating for a Spanish invasion. According to Watson, Gerard had 'bought up all the great horses he could throughout the country' in preparation for war. This intelligence the Council carefully filed away for future use.[45]

Henry Garnet, meanwhile, was backing Tassis' attempt to buy

relief for England's Catholics. In September 1603 he had written to Persons in Rome, asking him to use his influence at Philip's Court to 'solicit that they [the Spanish] will give money for any little ease'. In July the following year he contacted the papal nuncio in Flanders directly. The 'principal [English] councillors', he wrote, 'give their oath that, on the prompt payment of 200,000 escudos, all the monetary fines will be relaxed against Catholics'; he urged Rome to act upon this information. Three months later, in confirmation of Garnet's claim, Tassis reported home what he believed to be a serious offer from the Council of limited religious tolerance in return for further pensions from Spain.* The offer would come to nothing, but soon Garnet was writing again, this time to Claudio Aquaviva. Now he suggested that English Catholics contribute to the payments for their relief as a further incentive to the Council to help them. 'For otherwise,' he added, 'in my opinion, nothing will be done.' Once Garnet visited Tassis, in secret, at Walsingham House close by the Tower; twice more he entered Somerset House to talk with him there; and in August 1604, at the height of the peace negotiations, he was back at Somerset House to meet with the Constable of Castile. No records exist of these conversations, but doubtless the subject was the buying of tolerance; and, clearly, for Garnet the hope of achieving some relief for his co-religionists outweighed the danger of his being caught. His actions should be viewed in conjunction with the instructions he now carried with him and with the information to which he was now privy.[46]

In July 1603 Garnet had received word from Claudio Aquaviva, exhorting him to extreme caution. The English Jesuits, wrote their general, must in no way meddle 'in matters that did not concern their apostolate'; neither must they allow others to 'attempt

* The Constable of Castile, realistic to the last, instructed him to get the offer in writing immediately. 'I see no assurance for our expense but their word,' he wrote cautiously, 'and although that of a king and such important people could be trusted in very great matters, in the spending of money more resoluteness is necessary for the accidents that can happen.'

anything that ... might bring considerable distress not only on you and me, but on all Catholics in general'. 'I implore you to be prudent', Aquaviva continued with uncharacteristic emphasis. 'Shun every species of activity that might make priests of our Order hated by the world and branded the instigators of tragedy.' These instructions, wrote Aquaviva, came by order of the Pope. On 22 September 1603 Garnet wrote back to Aquaviva, warning him, cryptically, of 'restless men who, even in our name, entice our most intimate friends to rebellion'. He explained that, on learning of the trouble from 'one of our brethren', he had 'sent a message and all was [now] quiet where previously the greatest danger had threatened'.[47]

About 29 September, just days after this letter, one such 'intimate friend' sought Garnet out. His information, as Garnet later recollected it, seems almost colourless now: he said, wrote Garnet, 'that there would be some stirring, seeing the King kept not [his] promise'. But distance is altering: it bleeds the colour out of the significant and, by contrast, saturates the insignificant in lurid hues. Garnet's 'intimate friend' was Robert Catesby, long-time supporter of the Jesuits, fast on his way to becoming (in the words of the Attorney General) 'a pestilent traitor'. The 'stirring' to which he referred would transform itself into 'that most horrid and hellish conspiracy', the Gunpowder Plot. And Garnet's response to Catesby's announcement would prove prophetic. 'I greatly misliked it,' Garnet remembered afterwards as he argued for his life; 'it was against the Pope's express commandment ... [and] I earnestly desired him that he ... would not join with any such tumults. For in respect of [his] often conversations with us, we should be thought accessory.'[48]

Chapter Twelve

'I never yet knew a treason
without a Romish priest.'
Sir Edward Coke, January 1606

ROBERT CATESBY'S 'STIRRING' was of a scope greater than anything that had gone before. The plots – real and imagined – that measured out Elizabeth's reign had shared the same fingerprints, notably the invasion of England by a foreign army and the killing of the Queen. What Catesby planned for James was far more than a marksman's bullet or a dagger in the back. What Catesby planned for England was far more than the loss of its King and a new regime. He planned to kill as many of its royal family as were assembled in one place at one time. He planned to kill its MPs, its peers, its bishops, its law lords; to blow the heart out of its Government, its Church and its judiciary; to leave it staring aghast at the smoking, twisted ruins of its Parliament building, at the windowless, pockmarked façade of its great abbey, at the debris-strewn streets of its capital city. He planned to dismantle it from the top down. As Thomas Wintour, Catesby's cousin and his brother in political violence, later explained, Catesby 'had bethought him of a way at one instant to deliver us from all our bonds and without any foreign help'. '[I]n that place', Catesby told Wintour, referring to Parliament, 'they have done us all the mischief, and perchance God has designed that place for their punishment.'[1]

Catesby and Wintour: it was a connection that encompassed kinship, an early attempt at rebellion (under the Earl of Essex's leadership), and the abortive invasion talks with Spain: Wintour later testified that his journey to Spain had been at Catesby's request.* Its bloodlines reached out through the wider Catholic community, tying both men to the Treshams and the Vauxs, and caught fast in the web of these inter-connections were the Jesuits. Robert Catesby's father had supported the Jesuit mission since its inception, alongside Sir Thomas Tresham and Lord Vaux. Robert himself had loaned Henry Garnet his house at Uxbridge, to which John Gerard had ridden after escaping from the Tower. Robert's cousin, Anne Vaux, had cared for Garnet since his arrival in England, renting for him his latest safe house, White Webbs, near Enfield Chase, where Catesby was a frequent visitor. Catesby's cousin by marriage was Elizabeth Vaux, John Gerard's provider, and Catesby's spiritual confessor was the Jesuit Father Oswald Tesimond, who joined the mission in the spring of 1598 and was stationed with Garnet.[2]

On Wintour's side the connections continued. Wintour's elder brother, Robert, owned Huddington Court, near Worcester, equipped with a hiding place bearing all the trademarks of Garnet's servant, Nicholas Owen. Soon to be based at Huddington was the Jesuit Father Nicholas Hart, who arrived from Rome in 1604. Wintour himself had gone to Rome in 1601 carrying a letter of recommendation from Father Edward Oldcorne. A 'son of mine called Timothy Browne' was how Oldcorne described Wintour, the alias suggesting Wintour had no State licence to travel. It was as Timothy Browne that Wintour would later arrive at the Spanish Court and in Spain he would carry a letter of recommendation from Henry Garnet.[3]

Garnet was in the habit of issuing *laissez passers* to Catholics

* On 8 February 1601 Essex attempted to lead a band of about two hundred swordsmen through the City of London to overthrow his enemies on Elizabeth's Council, chiefly Sir Robert Cecil. Among his followers were several young Catholics, hoping to secure religious freedom for themselves. The coup was a failure.

heading abroad, entrées to a world of international charity and religious aid. Over the years he had written numerous letters to his fellow Jesuits overseas, in order, as he put it, 'to commend friends'. Most of these friends were men and women hoping to join the priesthood or to enter a convent. Some, like Wintour, had other ambitions in mind.[4]

'As I remember', testified Garnet afterwards, 'the first motion of the matter of Spain was between Christmas and Candlemas [2 February] the year before the Queen died [1602]'; he continued, 'Catesby, and Wintour dealt with me about a sending into Spain, and I wrote of their business . . . to Father Creswell.' This information came in a series of sworn declarations by Garnet to the Government, but though the facts contained therein may have been correct they were also noncommittal; in his private letters to Father Oswald Tesimond and to Anne Vaux he was more revealing. To Tesimond he wrote, 'I was moved in it, but would not consent to any invasion, but only to commend them for to receive pensions of the King of Spain.' To Anne Vaux he described Wintour and Catesby as 'conspirators in the Spanish action', adding, 'I utterly dissuaded that intent, and they promised to desist, and . . . they told me they would only sue for pensions in Spain.'[5]

Of the four men capable of confirming or denying Garnet's testimony, the first, Catesby, was dead by the time the so-called 'Spanish Treason' came to light. The second, Wintour, made no mention of Garnet being in any way party to his and Catesby's plans. The third, Catesby's cousin by marriage, Lord Mounteagle, had, by now, turned King's evidence and his own part in the conspiracy was carefully being erased from the relevant documents; if ever he was interviewed about the negotiations with Spain such an interview no longer exists. The fourth, Catesby's cousin Francis Tresham, did point to Garnet's involvement: 'he confesses', his examination read, 'that Father Garnet . . . [was] by them drawn to be acquainted with Wintour's employment into Spain, to give the more credit unto it'. Crucially, he would retract this testimony on his deathbed, saying he had only named Garnet 'to avoid ill usage', or torture.[6]

In his letter to Anne Vaux, Garnet had referred to Catesby and Wintour's promise only to sue for pensions from Spain; to Tesimond he described the outcome of this promise: 'After Mr Wintour's return I perceived he had negotiated other matters and that they intended to get horses.' So, that day in September 1603 when Catesby came to Garnet, telling him of a possible 'stirring', the exchange that followed may, in hindsight, have been cause for alarm. 'I earnestly desired him', wrote Garnet, 'that he and Mr Thomas Wintour would not join with any such tumults . . . He assured me he would not.'[7]

It took Catesby five months to break this promise. In February 1604, as England's Catholics absorbed the several shocks of the Hampton Court Conference and of James's recent proclamation banning papist priests, Catesby sent for Wintour to join him in London. There, in the company of Jack Wright (known to them for his part in Essex's uprising), Catesby told Wintour of his proposed scheme to blow up Parliament with gunpowder. At first Wintour hesitated, pointing to 'the scandal . . . which the Catholic religion might [thereby] sustain', but Catesby was persuasive: 'the nature of the disease', he told his cousin, 'required so sharp a remedy'. Wintour was won over. As yet it seemed neither man had lost hope in foreign help, for at this point Wintour was dispatched to the Low Countries to talk to the Constable of Castile and find out what pressure was to be brought to bear on James during the Spanish peace talks. The constable was characteristically circumspect: he spoke 'good words, but I feared the deeds would not answer', said Wintour afterwards. While there Wintour also met with Guy Fawkes, newly released from his house arrest: 'I told him', said Wintour, 'that we were upon a resolution to do somewhat in England, if the peace with Spain helped us not.' Fawkes returned with Wintour to London.[8]

The fifth member of the core group of plotters joined a few weeks later in early May, his first words a sharp reminder of the part he had already played in souring relations between James and England's Catholics. Thomas Percy, whose reiteration of James's talk of tolerance had raised Catholic hopes the previous year, entered Catesby's Lambeth lodgings with the cry 'shall we always, Gentlemen, talk, and never do anything?' In Catesby, Wintour, Wright, and Fawkes he had found men similarly tired of reacting, and ready to act.[9]

On Sunday, 20 May 1604 the five met again in a private room, one of several above the Duck and Drake, an inn on the east side of St Clement's Lane, just off London's Strand. There, prayer book in hand, they took an 'oath of secrecy'. Then they went 'into the next room and heard Mass, and received the Blessed Sacrament'. The repercussions of this single event – the conjunction of a regular act of Sunday worship with the planned destruction of the entire English Government – would be felt long after their plot had been discovered. In his opening speech to his first Parliament James, perhaps unconsciously alluding to his own darkest fears, had said of the Catholic clergy in hiding, the 'point which they observe in continual practice is the . . . murders of Kings'. In this conflation of paranoia and rumour he had fed the belief that England's secret priests were all killers-in-waiting. Now, by walking from the oath-taking chamber to the room next door where mass was being said, the plotters had succeeded in fleshing it out further and giving it features: and the features it bore were those of the man in whose rooms they had met, John Gerard.* It mattered little that the plotters, when questioned later, denied that Gerard had any knowledge of their activities, or that Gerard, himself, could swear under oath that they were mistaken in identifying him as the officiating priest that day.[10]

With the taking of this oath of secrecy, the Gunpowder Plot

* This St Clement's Lane house was one of a series of London houses used by Gerard since his time in prison. 'It was a convenient and very suitable place,' he wrote, 'with private entrances front and back, and I had some very good hiding-places constructed in it.' Of London's fashionable Strand he wrote, 'Most of my friends lived in that street, and I could visit them more easily, and they me.'

proper was set in motion. From that day until Guy Fawkes's arrest in the early hours of 5 November 1605, it would be the quietly ticking accompaniment to the ongoing business of James's Council, to the Jesuit mission, to the thousand little ordinary actions of everyday existence up and down the country.

In September 1603 Henry Garnet was confident he had prevented his friends from stirring against the Government – so he would testify. The following midsummer he was given indication that this was not the case. Some time in late June 1604, about a month after the secret oath, 'Catesby and Wintour, or Mr Catesby alone, came to [Garnet] at White Webbs and told [him] that there was a plot in hand for the Catholic cause against the King and the State'; 'they entered into no particulars', Garnet confessed afterwards.[11]

Catesby appeared to have absorbed Garnet's earlier warning that unrest was 'against the Pope's express commandment'. He now referred Garnet to two breves, one to the clergy, one to the laity, issued by Pope Clement in the run up to Elizabeth's death and sent secretly to Garnet. The purpose of these breves was to stop England's Catholics supporting a non-Catholic claimant to Elizabeth's throne; James's wooing of the Vatican had ensured they were never put into effect: 'when I saw the Queen dead,' wrote Garnet, 'I burned them.' But not before he had shown them to Catesby and, perhaps, Wintour too (he would later admit that he 'had no commission to divulge' the contents of the breves). It was an instance of carelessness from a tired man, sick with palsy and distracted by the ongoing Appellant controversy, or the sinister act of a politicized priest meddling in State affairs: Garnet's defenders and detractors were quick to pick their favoured motive. But knowledge of the breves had given Catesby 'an invincible argument . . . for his purposes', wrote Garnet. Catesby reasoned that, 'it being lawful by force of the said Breves of the Pope to have kept the King out, it was as lawful now to put him out'. 'I still reproved [him]', testified Garnet, reminding Catesby of the Pope's prohibition against violence, 'and he promised to surcease.'[12]

There are no surviving letters from Garnet of this period. His actions and attitudes must be pieced together from scattered comments and from various later attempts to exculpate him. That Easter it had been business as usual as he travelled through Lincolnshire; reports filtered back to the Council that he had said mass at Twigmore (at Jack Wright's house), at Thornham and at Glandford Brigg, near Lincoln. But that summer he was in and out of Somerset House in an effort to secure relief for English Catholics. It seems he was uncharacteristically incautious about discussing these efforts; at the year's end Juan de Tassis wrote to Philip of Spain that his secret negotiations for tolerance were being hampered: 'the Jesuits', he complained, 'are a little imprudent in not knowing how to keep silent'. From a letter quoted subsequently by John Gerard, but not extant, Garnet appeared increasingly desperate to win concessions for his co-religionists. 'If the affair of toleration go not well,' he noted, 'Catholics will no more be quiet. What shall we do? Jesuits cannot hinder it. Let Pope forbid all Catholics to stir.' According to Gerard, Garnet wrote this on 29 August, two months after his meeting with Catesby.[13]

From Father Oswald Tesimond, whose own involvement in the plot was about to become controversial, came evidence that Catesby was sidelining Garnet. Catesby was reported to find Garnet's 'lukewarmness displeasing' and his sermons on long-suffering 'unpalatable'.* He 'began to say openly . . . that the Jesuits were getting in the way of the good that Catholics could do themselves'. Leaving a supper at which Garnet had repeated his injunction to patience, Catesby was heard to say 'that some . . . had grown tired of putting up with ill fortune . . . They were asking if there was any authority on earth that could take away from them

* Anne Vaux would later give an example of one of these sermons. 'She remembereth he would use these words, "Good gentlemen, be quiet, God will do all for the best, we must get it by prayer at God's hands, in whose hands are the hearts of Princes."'

the right by nature to defend their own lives'. From now on, said Tesimond, Catesby began to hold aloof from Garnet. He was not the only one doing so. In March 1605 Tassis reported that he had been approached by two anonymous Englishmen offering money for the relief of Catholic recusants, but insistent that Garnet should not 'share in the merit of a work which they are doing before the Lord'. Though Tassis would call upon Garnet to substantiate the men's claims to funds, he also admitted it was far easier to do business with them than with the Jesuit, 'for he is always in hiding, or in flight'. His comments suggested growing tensions between Garnet and certain of the Catholic laity.[14]

While the English Catholic community fragmented, amid renewed frustrations that the Spanish peace had failed it, James's Government was contemplating its next steps. In September 1604 Tassis sent back to Spain the leaked minutes of a private meeting between James and his Councillors to discuss 'the mitigation of the recusancy laws'. The Earl of Northampton, probably the source of the leak, opened the meeting, repeating James's position that he did not 'desire the blood of Catholics'. From this starting point the other Councillors gave their opinions. Sir Robert Cecil advised: 'I would not counsel your Majesty to use such rigour against Catholics in that they govern themselves moderately and well for they intend nothing against the state of our country.'* Thomas Egerton, a lapsed Catholic, put the opposing view: 'the Papists are a dangerous people'. Baron Kinloss followed suit, speaking to James's anxieties: 'should your Majesty not wish the laws to be

* Cecil's views on Catholicism could vary according to whom he was speaking: he was a consummate statesman with many political enemies. But if his words are viewed in conjunction with his actions, then, just as with his father Sir William Cecil, a pattern is discernible. Broadly, both men recognized English Catholics had split loyalty; both saw this as a potential threat to State security; both sought to neutralize this threat, devising oaths outlining where Catholic loyalty lay; both realized needless persecution did more harm than good; both shared a dislike of bloodshed. Within these parameters, the Cecils behaved with consistency, and with a surprising moderation. That posterity has judged them harshly seems less a case of just deserts and more a reflection of the distaste in which they were held by their less successful, largely aristocratic, often crypto-Catholic contemporaries. Robert Cecil was created Earl of Salisbury in May 1605.

enforced against Papists undoubtedly their increase will be . . . so great that they will rise up against your Majesty and expel you from this kingdom'. Cecil responded: 'I have no fear of the rebellion of Catholics for sake of their religion as I have never understood that any people has rebelled for sake of religion but more for politics and matters of state under pretext of religion.' The Earl of Dorset, James's treasurer, put the financier's point of view: 'It is necessary that there be an increase of either Papists or Puritans. I prefer the increase of the Catholics and not the Puritans for the Papists are a peaceful people and in their increase your Majesty will derive much money.' With this display of Government at work James wrapped up the meeting. Two months later, on 28 November, he ordered the collection of recusancy fines again.[15]

Away from the Hampton Court session, Henry Garnet was back at White Webbs. About this same time the Frenchman Charles de Ligny visited him there. He 'found Garnet in company with several Jesuits and gentlemen, who were playing music: among them Mr William Byrd, who played the organ'. But it was not all music and masses. In November Garnet had written to Aquaviva: 'The Catholics are havering as before between hope and fear. Some of them seem over impatient, though all the better and graver of them persevere in patience. It would be well if Clement wrote to console them and to restrain the unquiet minds, so that their impatience be not harmful to all.'[16]

The year 1605, like that of the Armada before it, was hedged with ill omen. The almanac-makers wrote cheerful warnings of dire events to come. The Government, quick to stamp out such prophecies, hauled in the gloomier of doom-mongers for questioning. One, William Morton, talked of 'great troubles to happen within the kingdom this year' and of 'fire and sword in divers parts'. These prognostications, he said, he had had from a man, who had had them from a man, who had 'had the judgements of 26 ancient writers therein'. That October there would be an eclipse of the sun, the month preceding an eclipse of the moon; William Shakespeare, then working on *King Lear*, gave the Earl of Gloucester the ominous line:

> These late eclipses in the sun and moon portend no
> Good to us . . .[17]

In March that year Elizabeth Vaux wrote to Agnes, Lady Wenman of Thame Park, telling her of the proposed marriage of her young son to the Earl of Suffolk's daughter. This marriage had been beset by problems, not least because Vaux and her son were known to be Catholic, but now, she told her friend, it looked sure to happen: 'for ere it were long there should be a remedy or a toleration for religion'. 'Fast and pray that that may come to pass which we propose,' wrote Elizabeth, 'which if it do, we shall see Tottenham turned French.'* The letter was opened by Lady Tasborough, Agnes's mother-in-law, who promptly showed it to Agnes's husband (who blamed Elizabeth for corrupting his wife in religion). The pair subsequently lost the letter, but remembered its contents – and passed them on. On 5 November, even as Guy Fawkes was being questioned for the first time, Sir John Popham was informing Robert Cecil that Elizabeth Vaux was privy to the plot: she 'expected something was about to take place', he wrote. He also took care to remind Cecil that 'Gerard and Garnet, the Jesuits, make her house their chief resort'. It was an odd little episode of uncertain meaning (and Elizabeth Vaux pleaded forgetfulness when questioned about it), but its significance was far-reaching.[18]

By April, in spite of the almanac-writers' worst predictions and the resultant jittery start to the year, life had settled down to

* This was popular slang for a miraculous or unlikely event. The Duke of Norfolk used the phrase in 1536 after the execution of his niece, Anne Boleyn, writing to Thomas Cromwell: 'A bruit doth run that I should be in the Tower of London. When I shall deserve to be there Tottenham shall turn French.'

normal. That month one correspondent reported to his friend, with a certain ennui: 'For news here is none, but only of matches, marriages, christenings, creations, knightings and suchlike, as if this world would last for ever.'[19]

In June Garnet was in London, in a rented room on Thames Street, a claustrophobic thoroughfare stretching westwards from the Tower, in line with the river and overrun with tradesmen. There, on 9 June, Robert Catesby visited him, asking him the following question: whether, 'in case it were lawful to kill a person or persons, it were necessary to regard the innocents which were present lest they also should perish'? For months Catesby had been telling friends he was raising a regiment for Flanders – thereby explaining his new interest in horses and armaments. Garnet's reply to him was couched in military terms. 'I answered', wrote Garnet, 'that in all just wars it is practised and held lawful to beat down houses and walls and castles, notwithstanding innocents were in danger', so long as 'the gain . . . of the victory' outweighed the number of innocents killed.* Afterwards, Garnet testified that he 'never imagined' it more than 'an idle question' on Catesby's part. Until, that is, Catesby made 'solemn protestation that he would never be known to have asked me any such question so long as he lived'.[20]

If Garnet is to be believed, then up to this point he had had no reason to suppose Catesby was pursuing his 'stirring'. The previous autumn Thomas Wintour had come to him, promising that he and Catesby had stopped 'intermeddling in . . . tumults'. In May 1605 Robert Persons informed Spain that there had been 'difficulties' in England, but the crisis had been 'dampened'. Catesby's question suggested this was not so. Days later Garnet sought Catesby out to challenge him. He found him in the company of his cousin Francis Tresham and his cousin by marriage, Lord Mounteagle, and once more he reiterated Vatican instructions that Catholics 'be quiet'. He also quizzed the men. Did they think they were able to muster

* This was the standard theological argument to excuse collateral damage in warfare, based on Aquinas' doctrine of double effect.

sufficient forces to rise up against James? Mounteagle answered 'if ever they were, they were able now' because James was 'so odious to all'. This was a conditional answer, not a definitive one, replied Garnet: did they have sufficient forces? Their answer was no. Then why, asked Garnet, did they blame the Jesuits for preventing Catholics helping themselves, when they were obviously incapable of helping themselves? 'So', Garnet testified later, 'I concluded that I would write to the Pope that neither by strength nor stratagems we could be relieved, but with patience and intercessions of Princes.'[21]

The result of this meeting, and of a subsequent one between Garnet and Catesby alone, was that the latter agreed to inform the Vatican 'how things stood here'. This was a hard-won compromise. Garnet had tried to get Catesby to tell the Pope of his plans; Catesby had refused 'for fear of discovery'. Both men had trodden delicately around the details: each time they met, wrote Garnet, 'Catesby offered to tell me his plot'; each time, he added, 'I refused to know, considering the prohibition I had [from Aquaviva, to keep the Jesuits clear of political unrest]'. They did agree on a messenger to deliver Catesby's news and Catesby 'promised . . . he would do nothing before the Pope was informed'. Now Garnet issued another *laissez passer*, this time to Sir Edward Baynham, introducing him to the papal nuncio in Flanders.[22]

By midsummer Garnet was homeless: 'betrayed in both our places of abode', as he described it in a letter written on 24 June, and 'forced to wander up and down until we get a fit place'. Spies were closing in on him: White Webbs was under suspicion, so was a house newly leased by him at Erith on the banks of the Thames near Dartford. His enforced wanderings took him to Fremland in Essex, home of the Catholic Sir John Tyrrel, where he spent the Feast of Corpus Christi with 'great solemnity and music'. Here, too, the spies were watching him. A month later he was back at Fremland, where, 'a little before St James' tide [25 July]', Father Oswald Tesimond 'revealed to him . . . [the] conspiracy of blowing up of the Parliament House with powder'.[23]

The baldness of the statement gave no taste of the agonized discussion that, according to Garnet, had preceded this revelation. Tesimond, wrote Garnet, had come to him perturbed: 'it was', he said, 'about some device of Mr Catesby', but Catesby had 'bound [him] to silence'. Garnet admitted he knew Catesby was up to something. The two men 'walked long together', deciding whether Tesimond should tell and Garnet should listen. Garnet concluded that if Tesimond had 'heard the matter out of confession', then he might safely break his silence since Catesby himself was happy Garnet knew of the device. Tesimond concluded that he would do so, but only in confession. Then, 'because it was too tedious to relate so long a discourse in confession kneeling', Tesimond asked if he might make his confession 'walking'. Garnet agreed.[24]

English Catholics had become accustomed to a conflict of loyalty. But Henry Garnet's conflict had just been made untenable. As an Englishman, subject to common law, he was bound to disclose the plot to the Government. As a Catholic priest, subject to canon law, he was bound to inviolable secrecy; he had learned of the plot *sub sigillo confessionis* (under the seal of confession): to reveal it would be sin and sacrilege both.

On 24 July, hours after this meeting, Garnet wrote to Aquaviva. Two versions of this letter exist. The first is in the Public Record Office, in a hand not Garnet's own and with no explanation of how it came to be there.* In this version Garnet warned of the dangers of a Catholic uprising. 'There were some', he wrote, 'who dared to ask . . . whether the Pope could prohibit their defending their lives.' He hinted at the tensions between himself and Catesby: 'some friends complain that we put an obstacle in the way of their plans'. He explained he had persuaded these friends 'to send

* It is unclear who made this copy. It is written out on the same piece of paper as a letter, in the same hand, to Garnet from Aquaviva; the one is listed as an 'Example' of Aquaviva's correspondence, the other as Garnet's 'Response'. The copy of Aquaviva's letter was originally (wrongly) dated '1606'; this has been scored out and 1605, the correct date, inserted in its place. This would suggest that both copies were made *after* 1605. It should be noted that Garnet was scrupulous about destroying his correspondence.

someone [Baynham] to the Holy Father . . . at least to gain time, that by delay some fitting remedy may be applied'. And here the letter ends, with an '&c', indicating something missing. Version two exists only in Jesuit accounts of the period. It continues where version one leaves off, with Garnet adding a second warning of an even 'worse' threat: 'the danger is lest secretly some treason or violence is shown to the King'. He offered his judgement: that the new Pope, Paul V (Clement had died in March), must indicate what was 'to be done' and, publicly, must 'forbid any force of arms . . . under censures'. He ended with a call for speed: 'as all things are daily becoming worse, we should beseech His Holiness soon to give a necessary remedy for these great dangers'.[25]

Garnet's original letter is no longer extant. Of the two existing versions – State Paper transcript of uncertain provenance and Jesuit apologia – one has deliberately been tampered with. 'There is one thing that makes us very anxious,' reads the first version; 'two things make us very anxious', reads the second. Both letters warn of trouble, both contain a degree of ambiguity, room for that warning to be misinterpreted; but one, in its very generality, suggests tacit ambivalence and one, in its call for urgent action, conveys alarm. Which message had Garnet sent?

High summer 1605: King James left London for the country. In early August there were flash floods in the capital, 'such as the like had not been seen in the memory of man'. The 'channels and water courses rose so high,' recorded John Stow, 'that many cellars by them were over-flowed'. In late August Thomas Wintour and Guy Fawkes found that the gunpowder they had hidden in a cellar beneath the Lords' debating chamber had 'decayed'.* More

* The plotters had accessed this space by the simple expedient of leasing the house next door to Parliament, the cellar of which lay directly below the House of Lords.

gunpowder was brought in, ready for the first sitting of the new session of Parliament on Tuesday, 5 November.[26]

While Londoners mopped out their houses, James continued his summer tour, spending a night at Harrowden Hall as a guest of Elizabeth Vaux. On 27 August he entered Oxford on a State visit. Thirty-nine years ago almost to the day his predecessor had ridden down these same streets, heard similar speeches of welcome, smelt the fresh paint on the casements, posts, and pumps as he did now, and greeted the cheering ranks of students. Queen Elizabeth's visit had been a calculated charm offensive, James's showed how times had changed: under the steady, almost thirty-year helm of Oxford's Professor of Divinity the university had become the Protestant seminary the Government had hoped for, grooming mildly Calvinistic students for the national Church. There were pockets of covert papistry, pockets of a more defiant Puritanism, but for the time being Church and State combined happily in the quadrangles of Oxford – which was fortunate, for this was a messy, charmless visit. The nightly plays alternately shocked and bored the royal party; James fell asleep during one and had to be persuaded to stay through a second; he was late to the disputations, then interrupted them with argument. On Friday, 30 August, as he rode out of town, he 'seemed not to see' the verses 'set upon the [college] walls', celebrating his stay.[27]

That same day another party took to the road: Henry Garnet, Anne Vaux and a handful of friends and servants, including Nicholas Owen, set out for Wales and the shrine at St Winifred's Well. 'I hope in this journey (which I undertake . . . both for health and want of a house) I shall have occasion of much good,' Garnet wrote to Robert Persons two days before. With his letter to Aquaviva on its way, with the expectation that any day now the Pope would respond, and with the promise exacted from Catesby not to attempt anything meanwhile, had Garnet relaxed his guard? Or was he being less than honest with his old Jesuit colleague? '[F]or anything we can see,' he told Persons, 'Catholics are quiet, and likely to continue their old patience.'[28]

The journey west took Garnet from safe house to safe house. In Warwickshire he stayed at Norbrook, near Stratford-upon-Avon, belonging to John Grant, husband of Dorothy Wintour; from there he moved on to Huddington as a guest of Robert Wintour. Both Grant and Robert Wintour were now in the thick of the plot. The party grew in number; Elizabeth Vaux, John Gerard, Edward Oldcorne, and Oswald Tesimond, and three friends of Gerard's, Elizabeth, wife of Ambrose Rookwood, and Sir Everard and Lady Digby: all seized the chance to make this pilgrimage. The return journey took them to the Treshams at Rushton Hall and from there Garnet, still homeless, rode on to Gayhurst, Digby's house in Buckinghamshire, arriving towards the end of September. Rookwood, Tresham, and Digby: each new name bound the Jesuits closer to the plot; on 29 September Catesby approached Rookwood to join him, on 14 October he approached his cousin Francis Tresham and in late October he approached Everard Digby.* The strength of the Jesuit mission lay in its ability to build a network of secret enclaves across the country protected by ties of consanguinity and by a shared vulnerability to exposure. This strength was now going to be its undoing: safe houses these no longer were.[29]

On 4 October Garnet wrote again to Robert Persons. The persecution, he explained, had become 'more severe' and rumour was that James 'had hitherto stroked Papists, but now [would] strike'. He still believed that 'the best sort of Catholics [would] bear all their losses with patience', but he offered Persons a dark warning: 'how these tyrannical proceedings . . . may drive particular

* There exists an undated letter from Digby to Robert Cecil among the State Papers. It is worth quoting as an example of Catholic ill-feeling towards James's Government: 'If your Lordship and the State think it fit to deal severely with the Catholics, within brief there will be massacres, rebellions, and desperate attempts against the King and State. For it is a general received reason amongst Catholics, that there is not that expecting and suffering course now to be run that was in the Queen's time, who was the last of her line, and last in expectance to run violent courses against Catholics; for then it was hoped that the King that now is, would have been at least free from persecuting, as his promise was before his coming into this realm, and as divers his promises have been since his coming. All these promises every man sees broken.'

men to desperate attempts, that I cannot answer for'. Throughout their pilgrimage Anne Vaux had been struck by the quantity of horses stabled with her cousins, telling Garnet she 'feared these wild heads had something in hand'. But in the absence of any public response from the Vatican, Garnet seems to have retreated into numb officialdom: his orders from Aquaviva were to avoid meddling in anything that did not directly concern his apostolate, so this is what he did. He later explained: 'I . . . cut off all occasions (after I knew the project) of any discoursing with [Catesby] of it, thereby to save myself harm both with the State here, and with my Superiors at Rome.' With his pilgrimage over his ambition now was to get back to the capital: 'we are to go within few days nearer London', he told Persons.[30]

Garnet never made it to London. In late October Anne Vaux came to him 'choked with sorrow'. She told him she feared 'disorder', because 'some of the gentlewomen [probably some of the conspirators' wives] had demanded of her where they should bestow themselves until the burst was past in the beginning of Parliament'. 'Whereupon', testified Garnet afterwards, 'I gathered that all was resolved.' If this statement is true, then Garnet now did an extraordinary thing: instead of removing himself from danger he took himself right to the hub of it. He accepted Everard Digby's offer to join him at Coughton Court in Warwickshire; and he did so knowing that Digby was drawn into the plot and suspecting that the plotters wanted him with them 'for their own projects'.[31]

It was a small party that set out to Coughton on Tuesday, 29 October, just Lady Digby, Garnet, and Tesimond, Anne Vaux and her sister, and Nicholas Owen. Sir Everard Digby was to ride over from Gayhurst a few days later, for a 'hunting party' (the purpose of which was to kidnap Princess Elizabeth, James's daughter, from her nearby lodgings, ready to proclaim her Queen the moment her father had been killed).* Catesby too, wrote Garnet,

* The plotters seem to have presumed that James's sons Henry and Charles would go with their father to Parliament, though Thomas Percy was deputed to snatch Charles, in the case of his not going.

had promised to come to Coughton. Had he kept this promise, Garnet intended to enter 'into the matter with [him], and perhaps might have hindered all'. It was a desperate hope. More realistic was the next sentence in Garnet's statement: 'Other means of hindrance I could not devise, as I would have desired.' On Wednesday, 6 November Robert Catesby's servant Thomas Bates rode into Coughton, bringing news of Guy Fawkes' arrest and of the plot's discovery. As Garnet read a letter from Catesby and Digby, which excused their rashness but begged him to help them raise a party against the King, Bates heard him turn to Tesimond and say 'we [are] all utterly undone'.[32]

According to a subsequent Government publication, the news of an imminent attack on Parliament was first revealed to it late on Saturday, 26 October. It so happened that Catesby's cousin by marriage, Lord Mounteagle, was at his Hoxton house that evening, a house in London's suburbs he seldom visited (this was the same Mounteagle who earlier had exclaimed that the time was ripe to rebel against James because he was 'so odious to all'). A stranger approached his servant in the street with an anonymous letter.*

* It is not known who wrote the Mounteagle letter, but everyone involved in the plot, from the Jesuits to Mounteagle to Cecil himself, have been suggested as its author. Nor is it clear why it was written. The two most plausible explanations are 1) that the author wished to sabotage the plot, while giving the plotters the chance to save themselves (Mounteagle's servant was connected to the plotters and promptly revealed the existence of the letter to Catesby) and 2) that the letter was a device, fabricated either by Mounteagle alone, in an effort to further his own ambitions, or by the Government and Mounteagle together, to give substance to otherwise unsubstantiated intelligence about the plot and allow the Government to act. The text was as follows: 'My lord, out of the love I bear to some of your friends, I have a care of your preservation. Therefore I would advise you, as you tender your life, to devise some excuse to shift of your attendance at this Parliament; for God and man have concurred to punish the wickedness of this time. And think not slightly of this advertisement, but retire yourself into your [county] where you may expect the event in safety. For though there be no appearance of any stir, yet I say they shall receive a terrible blow this Parliament; and yet they shall not see who hurts them. This counsel is not to be condemned because it may do you good and can do you no harm; for the danger is passed

Mounteagle had trouble deciphering the letter and asked his servant to read it out to him while he ate supper, so that everyone could hear. The letter's meaning was unclear, so Mounteagle immediately took it to Robert Cecil. Cecil, looking through it, was 'put in mind of divers advertisements . . . from beyond the seas . . . concerning some business the papists were in', but he and his fellow Councillors, with noticeable sang-froid, decided to wait until the King returned from hunting before informing him. James returned from hunting on Thursday, 31 October. The afternoon of the following day, Friday, 1 November, Cecil showed him the letter. James, 'who was always very fortunate in solving of riddles', no sooner read the letter but divined the true meaning of it – that there was a plot to blow up Parliament – and ordered that the buildings be searched. A preliminary search, which did not take place until some three days later, on Monday, 4 November, revealed a suspicious quantity of firewood in one of the cellars. A second search, some time late on Monday/early on Tuesday, 5 November, revealed the gunpowder and Guy Fawkes, hiding in the shadows. This is what happened, said the Government.[33]

The exact details of the Gunpowder Plot have never been established. Any account of them must pick its way carefully between two extremes. On the one hand there is the version, favoured by the conspiracy theorists, that holds that the plot was a deliberate Government invention, designed to destroy English Catholicism. On the other hand there was the official version released subsequently by James's Government, some of the less believable details of which have been outlined above.

This account holds that there was a genuine plot, led by Robert Catesby, to detonate a quantity of gunpowder beneath the House of Lords as Parliament met for its new session, but that the Government-authorized version of events, with all its omissions, elisions, and obfuscations, was a calculated means of making

---

as soon as you have burnt the letter. And I hope God will give you the grace to make good use of it, to whose holy protection I commend you.'

capital from Catesby's crime. In which case, the nine-day gap between Robert Cecil first reading Mounteagle's letter and Guy Fawkes' arrest is crucial to any understanding of the events that followed – because it is inconceivable that Cecil, even assuming he was ignorant of the plot before 26 October, did nothing at all during this period (he, himself, would later admit that the gap had allowed the plot time 'to ripen').[34]

It is the breaks in the pattern of normality over these few days that attract attention. For example, on 31 October the new Spanish ambassador, Don Pedro de Zúñiga, wrote that he had just received an unexpected message from Robert Cecil, saying that if the Pope could guarantee Catholic loyalty to the State under pain of excommunication, then James would remit the penal laws, allowing Catholics to 'live as they please'.* Assuming that Zúñiga passed this important piece of news on promptly (and, since his instructions were to do everything to further the cause of religious tolerance for English Catholics, then this is probable), then Cecil's message to him was written *after* the discovery of the Mounteagle letter. So what did it mean? It could not mean that Cecil hoped the Pope would step in and save the day – there was no time for this and Cecil was more than capable of stopping a plot himself. It could have been statesman's bluff, a way of making England's Government appear reasonable in the face of soon-to-be-uncovered Catholic treachery; but the gains from this last were few. A more tantalizing conjecture is this: if Cecil were still looking to neutralize the Catholic threat by finding a tidy legal answer to the Bloody Question, then the plot had given him a powerful new tool, outrage, with which to force the matter through. In which case, as with his previous dealings with the Appellants over an oath of allegiance, there was one group of Catholics that might

* Zúñiga's predecessor Juan de Tassis was summoned home in June 1605. He died early in 1607 and was buried in the chapel of the Augustinian Convent, Valladolid. He would always be blamed for his failure to achieve religious freedom for English Catholics, but only after he had left England for good did the new Pope, Paul V, instruct that no obstacle was to be put in the way of Philip III's efforts to buy toleration.

usefully serve as a scapegoat to help him achieve this end: the Jesuits.[35]

While Cecil waited to act, the intelligence reports began rolling in. On 3 November, some thirty-six hours before Guy Fawkes' capture, the informant John Bird was identifying a possible hide-out for Garnet and Gerard, adding, 'most like it is that they . . . have been the hatchers and plotters of this damnable stratagem'. Two days later, with Fawkes in custody *but refusing to name his accomplices*, Attorney General Sir Edward Coke was writing that Thomas Wintour's connections had been sent for, to be questioned. That same day came Sir John Popham's comments on Elizabeth Vaux's letter and his observation that Gerard and Garnet were often at her house, and also a note from the informant George Southwicke, explaining that he had been hunting the plotters for eight days now (since 29 October) and he needed a warrant 'for their apprehension'. The following day, Wednesday, 6 November, the King ordered that Fawkes be tortured.[36]

It took only a short while for the nation to wake up to its recent escape, helped in part by the Government placing all London 'under arms' on 5 November. One correspondent, writing on Thursday, 7 November, described how on Tuesday night the church bells had rung 'and as great [a] store of bonfires [had been lit] as ever I think was seen'. His letter mingled shock with rumour: 'some five or six Jesuits', he wrote, had been arrested. Shock and rumour would characterize the public response to the plot, swiftly followed by righteous anger: 'this most devilish treason', wrote a second correspondent, this 'most horrible and detestable treason', wrote a third, this 'abominable conspiracy so inhumanly contrived by the devil', wrote the Scottish Council to James.[37]

On 7 November Sir William Waad, Lieutenant of the Tower of London, informed Robert Cecil that though Fawkes was still not talking, he had now given a reason for his silence: an unbreakable oath of secrecy sealed by a holy ceremony. Little is unbreakable under torture: two days later Fawkes was ready to confess – but to Cecil alone. His testimony was worth the wait: 'Gerard, the Jesuit,' said Fawkes, 'gave them the sacrament, to confirm their oath of secrecy'. Though Fawkes was insistent Gerard was unaware of the plot, the Jesuit's name was now passed on to the informant George Southwicke to add to his newly issued arrest warrant. Fawkes had a second piece of information: the plotters had made use of Garnet's lodgings, White Webbs, as a meeting place. A search of the house was ordered, but the pursuivants were disappointed: they found 'popish books and relics' and 'many trapdoors and passages', but no papers, no munitions and no Garnet.[38]

The search for Gerard, ordered on 10 November, was begun two days later. 'I have used all possible expedition for my repair to Mrs Vaux, her house at Harrowden,' wrote the man charged with it, William Tate, 'whither I came with as much secrecy as could be on Tuesday, the 12th of this instant month, between twelve and one of the clock.' Gerard takes up the story: 'They were to search [Harrowden] scrupulously and if they failed to find me, stay on until they were recalled. Day and night guards were set at a distance of three miles round, with orders to arrest any passing stranger.'[39]

'I was in my hiding-place,' wrote Gerard. 'I could sit down all right but there was hardly room to stand. However, I did not go hungry, for every night food was brought to me secretly. And . . . when the rigour of the search had relaxed slightly, my friends came at night and took me out and warmed me by a fire.' Tate detailed the 'unprofitable endeavours' of the searchers: 'I examined every corner,' he explained to Cecil, 'though there [was] no appearance to give the least suspicion.' A few days into the search Elizabeth Vaux opened one of Harrowden's hides for the pursuivants. 'Her hope', wrote Gerard, 'was that they would think that, if a priest was in the house, he would be hiding there, and that they would then call off the search.' 'I entered and searched the same,' wrote Tate, 'and found it the most secret place that ever I saw, and so contrived that it was without all possibility to be discovered. There I found many Popish books . . . but no man in it.' On Saturday, 16 November Tate left Harrowden for London, taking Elizabeth

Vaux with him but leaving his servants to continue the search. It was not until 20 November that they finally abandoned the house: 'they thought I could not possibly have been there all that time without being discovered', explained Gerard.[40]

It must soon have become apparent to the Government how convenient it would be to make this a Jesuit-inspired conspiracy. The same day the plot was revealed to the nation the Council instructed London's Lord Mayor to quash an 'evil bruit' that Spain was behind the attack. Four days later James went to Parliament, to give MPs his account of the plot's discovery and to clear himself of any responsibility for it: the plotters' actions could not have been a 'work of revenge', said James, because 'I scarcely ever knew any of them'. James would not be blamed: under his 'wise temperament' and his 'indulgent hand', the papists, said the Government, had never had it so good. And Spain could not be blamed: the recent peace treaty and subsequent alliance made this diplomatically impossible, no matter the details now emerging about the Spanish Treason. So the plotters became, in the Government's wording, 'mad zealots' and, just as all the dangerous conspiracies of Elizabeth's reign had been 'incited . . . by the Jesuits', so this action, too, was laid at the Society's door. A supporting cast of exiled – and reviled – Catholics, notably Hugh Owen and Sir William Stanley, was drafted in to flesh out the conspiracy, but the impetus for the plot, implied the Government, was the Jesuits'. This suited James, for by now his loathing of the order was well documented: 'Puritan-papists' was his description of its members.[41]

The problem was that as each surviving suspect was examined – some of their number, including Catesby and Percy, had been killed during capture – not one of them implicated the Jesuits in the action. On 12 November Thomas Wintour was brought to the interrogation room for the first time, only to say that 'they had no priest amongst them'; Francis Tresham, questioned specifically about Gerard and Garnet, declined 'to say what [had] passed between them'; Elizabeth Vaux denied all knowledge of Gerard. So, with no one giving it the information it wanted the Government was forced to go hunting for itself. Thomas Wilson, Cecil's secretary, dug up an out-of-date, inaccurate list 'of the haunts and residence which Jesuits were wont to have . . . whereof haply there may be some use made at this present'. George Southwicke reported rumours from Norfolk that Gerard had said mass for the plotters. Sir Edward Coke fleshed this out hopefully, scribbling on the margin of Elizabeth Vaux's confession, 'Gerard the priest ministered the sacrament to all the traitors, etc., as well for execution as for secrecy'. Meanwhile, Everard Digby's servants provided evidence about the pilgrimage to St Winifred's Well and a mass said by Garnet for the Digbys at Coughton.[42]

On 4 December Cecil wrote to James's Clerk of the Signet, Nicholas Faunt, voicing his frustration. It was, he said, logical to assume the plotters' spiritual confessors knew of the conspiracy, 'seeing all men that doubt resort to [a priest] . . . and all men use confession to obtain absolution'. But 'most of these conspirators have wilfully forsworn that the priests knew anything in particular and obstinately refuse to be accusers of them, yea what torture 'soever they be put to'. However, there had been a breakthrough: 'you may tell his Majesty that if he please to read . . . what this day we have drawn from a voluntary and penitent examination, the point . . . shall be so well cleared . . . as he shall see all fall out to that end whereat his Majesty shoots'. The 'penitent' was Catesby's servant, Thomas Bates.[43]

Bates had been privy to the plot. Through Catesby he had been introduced to Father Oswald Tesimond, alias Greenway (Tesimond, like all his fellow priests in hiding, used a false name), and to Tesimond, he testified, he had revealed his troubled conscience. 'And Greenway, the priest, thereto said that he would take no notice thereof; but that he, the said examinate, should be secret in that which his master had imparted unto him, because it was for a good cause.'[44]

Behind the scenes the activity mounted. On 8 December

William Waad reported back to Cecil that Tesimond had been a contemporary of Guy Fawkes and Jack Wright at the free school known as Le Horse Fayre, on the outskirts of York. Some time towards the end of the year a royal proclamation was drafted for the arrest of Gerard and Tesimond. Christmas and New Year came and went; in late December the Government received the bill for the ironwork, 23s. 6d., on which Catesby's and Percy's severed heads had been mounted; in January the first poems about the plot began to appear. On 13 January Thomas Bates was back in the interrogation room. Now he described bringing Catesby's letter to Garnet on 6 November, its purpose, he said, 'to crave [Garnet's] advice what course they were to take in their proceedings'. He described the panicked conversation between Garnet and Tesimond, Garnet saying they were undone, Tesimond that 'there was no tarrying'. He described Catesby's request that Tesimond come to him and he described Tesimond's answer, 'that he would not forbear to go unto him, though it were to suffer a thousand deaths, but that it would overthrow the state of the whole society of the Jesuits'. He described taking Tesimond to Catesby at Huddington, where the pair talked for half an hour privately, before Tesimond rode away. Then came his final blow. By now the Government knew of Edward Baynham's mission to Rome, though its understanding of his purpose there was rather different from Garnet's; Guy Fawkes, for example, had testified that Baynham was in Rome to inform the Pope of the plot's success, not ask his counsel – this Catesby had told him. Bates now added a further detail: according to him, Thomas Wintour had said that Baynham was in Rome, waiting only for Garnet's letters before approaching the Pope with news of their triumph. Wintour himself would contradict this just a few days later, but it was too little, too late. On 15 January the Government decided it had enough information to proceed.[45]

James's proclamation reflected none of the uncertainties that existed. '[I]t is now made plain and evident by divers examinations of many of those prisoners that have been the principal conspirators in the barbarous practice . . . that these three Jesuits under named, John Gerard alias Brooke, Henry Garnet alias Walley alias Darcy alias Farmer, Tesimond alias Greenway, have all three peculiarly been practicers in the same.' It exhorted his subjects to do their duty and assist in the Jesuits' capture and it threatened any that concealed them. Then it attached a detailed description of each of the wanted men.[46]

Oswald Tesimond learnt of the arrest warrant some forty-eight hours after its pronouncement and straight away headed for London. He travelled 'by day, and through public streets' and in 'almost every parish he found the proclamation posted up in which he was described to the life' (so he later wrote). 'Of a reasonable stature,' read the posters: 'black hair, a brown beard cut close on the cheeks and left broad on the chin, somewhat long-visaged, lean on the face but of a good red complexion, his nose somewhat long and sharp at the end.' In London he was recognized and arrested.[47]

The arresting officer was unwilling to ask for help from anyone about him, so the two men set off unaccompanied. As they walked Tesimond 'argued with [his captor], and urged him to be careful' and so he 'gradually led him away from the more frequented streets'. As soon as they were clear of the crowds Tesimond took to his heels and ran. He left England soon afterwards, hidden among a cargo of dead pigs bound for Calais. From there he made his way to St Omer, near Calais, then on to Rome. He died in 1635 aged seventy-two, leaving behind him an exculpatory, if patchy, account of the Gunpowder Plot and a series of unanswered questions.[48]

The case against Tesimond is confused, not least by the fact that the Government, in combing through old intelligence reports, mistook him for another man called Greenway who was widely regarded as a Catholic agitator.* Catholic witnesses recorded

* Anthony Greenway was born in Buckinghamshire *c.*1575 and educated at Eton and Oxford. He was converted to Catholicism by 'reading', so he said, and served for a time as a soldier in the Catholic regiment in Flanders, travelling back and forth between London and Belgium, conveying refugees and information as he went. In January 1606 he enrolled to train for the priesthood in Rome; later he became a Jesuit.

Wintour's attempt on the scaffold to clear Tesimond of guilt, a detail lacking from the Government-authorized account of his execution. John Gerard recorded a letter from Thomas Bates, hinting at the psychological pressure put on him by his examiners; Bates excused the evidence he had given, explaining, 'I did it not out of malice but in hope to gain my life.' But from Father Edward Oldcorne came a further revelation. According to Oldcorne, Tesimond had ridden straight from his 6 November meeting with Catesby to nearby Hindlip Hall, where Oldcorne was based, saying, '"that he brought them the worst news that ever they heard," and . . . "that they were all undone"'. He told them of Catesby's plot, adding that 'now they were gathered together some forty horse at Mr Wintour's house . . . and told them, "their throats would be cut unless they presently went to join with them"'. When Oldcorne refused, 'Tesimond said in some heat, "Thus we may see a difference between a [phlegmatic] and a choleric person!" and said that he would go to others . . . for the same purpose as he came to Hindlip'. There is no information about his activities from this moment until his appearance in London, just prior to his escape.[49]

While Tesimond fled, John Gerard engaged in clearing his reputation. Almost immediately after the Harrowden search he had written an open letter addressed to a friend in which he had maintained his innocence. 'I had many copies of the letter made', he explained, 'and had them scattered about the London streets in the early hours of the morning.' With the publication of James's proclamation he now took a more direct approach, writing three more letters – to Robert Cecil, to the Duke of Lennox, and to a third unidentified Councillor. Again he insisted upon his innocence: 'I was not privy to that horrible Plot of destroying the King's Majesty,' he wrote, begging 'that full trial . . . be made, whether I be guilty therein or not. And', he added, 'if so it be proved . . . then all shame and pain may light upon me'. He asked that the conspirators be questioned again before they were executed and he enclosed a letter to Sir Everard Digby, exhorting his friend to defend him 'from a most unjust accusation'. 'And if', he repeated, 'this protestation be not sincerely true, without any equivocation, and the words thereof so understood by me, as they sound to others, I neither desire nor expect any favour at God's hand when I shall stand before His tribunal.' It was astute of Gerard to mention equivocation: the subject was about to obsess the Government's interrogation team. But it made little difference to the charges against him – the hunt continued.[50]

In the months that followed Gerard was variously reported to have been arrested in Gloucester (masquerading under the name of Valentine Palmer), to be at liberty somewhere in England, to be at liberty somewhere in Italy, to have narrowly escaped from Lord Montague's house and to have been taken in Warwickshire. All the while, he remained in London, living quietly. The pursuivants were never far away; on 17 April the Lord Mayor led a raid on a house he was using, just as mass was being said. Gerard was not there, but the officiating priest had time only to bundle himself and the altar things into a hiding place before 'an uproarious mob' swept through the house. 'It was so close', wrote Gerard afterwards, 'that the Mayor and men with him smelt the smoke of the snuffed out candles.'[51]

From the moment Thomas Bates clattered into Coughton on 6 November, bringing Catesby's letter, Henry Garnet's trail goes cold. Nothing is known of his movements for the next few weeks, though it seems likely that he continued on at Coughton. Certainly this was where Father Edward Oldcorne wrote to him towards the end of November, hearing that Garnet was there and, as he put it, 'in some distress'. And on 4 December Garnet travelled the 18 miles to Hindlip Hall, just to the northeast of Worcester, accompanied by Anne Vaux and Nicholas Owen.[52]

Hindlip was a new house, begun in the 1570s by the owner Thomas Habington's father. For sixteen years it had been Oldcorne's headquarters; in this time he had made it, in John Gerard's words, 'like one of our houses in some foreign country – so many Catholics flocked there'. It was set on the highest ground in the neighbourhood with sweeping views in all directions and, with multiple hiding places cut deep into its masonry, it was regarded as one of 'the safest [houses] that existed, not merely in the county but in the whole of England' – this from Oswald Tesimond's description. A later description gave a clearer picture still: 'There is scarcely an apartment that has not secret ways of going in or going out; some have back staircases concealed in the walls; others have places of retreat in their chimneys; some have trap-doors, and all present a picture of gloom, insecurity, and suspicion.' The Habingtons, added Tesimond, 'had had good experience of . . . many searches. [But] not once in all of them had [the pursuivants] ever been able to find a priest'. Garnet was now to test this to the limits.[53]

For six weeks he remained quietly at Hindlip, living in a small chamber on the ground floor near to the dining room, attended by Nicholas Owen and joining the Habingtons at meal times. He used this period of calm to write an open letter to the Privy Council, in which, like John Gerard before him, he maintained his innocence. Unlike Gerard, whose letter bristled with indignation, Garnet applied measured reason to his task. He stressed his 'obedience to his General and to the Pope, even in this particular case', reminding the Council of the Vatican's 'express prohibition of all unquietness', issued in response to Watson's plot. The Pope, he said, would stand witness that he himself had secured this prohibition. He emphasized the recent lengths to which he had gone to secure a further 'prohibition under censures of all violence towards his Majesty' – here was the reference to his letter to Aquaviva. He challenged the Council to inform him of the charges being made against him, so that he might defend himself; he acknowledged his part in furthering the Spanish peace; he spoke of the damage to 'our whole Society . . . if we had been faulty'. And then, picking his words with deliberate care, he denied all knowledge of the plot: given the Pope's earlier prohibition and the Jesuits' own vow of 'holy obedience', it was, he wrote, 'in no way probable . . . that the author of this conspiracy durst acquaint me or mine with their purposes'.[54]

In early January, just a few miles to the north of Hindlip, the last remaining suspects in the plot were rounded up by the pursuivants, helped in their task by a drunken poacher and an overly observant cook. Bargaining for his life, one suspect, a minor player in these events, offered to part with the names and hide-outs of 'certain Jesuits and priests which', he claimed, 'had been persuaders of him and others to these actions'. The first name on his list was Oldcorne's.[55]

On Saturday, 18 January a neighbour arrived at Hindlip, warning Mrs Habington – Thomas Habington was away on business – that the house was to be searched. The following day, Sunday, he wrote her a note, saying the search was to be 'one day in that week' and he 'prayed her to be careful'. Neither Garnet nor Oldcorne took this opportunity to leave, nor, it seemed, to prepare themselves for what was coming next. At daybreak on Monday, 20 November one hundred armed men surrounded the house, led by the local magistrate, Sir Henry Bromley.[56]

Bromley's instructions, from Robert Cecil's secretary, were meticulous. First he headed to the dining room. '[I]n the east part of that parlour,' read the instructions, 'it is conceived there is some vault, which to discover you must take care to draw down the wainscot.' The 'lower parts of the house must be tried with a broach [rod], by putting the same into the ground some foot or two, to try whether there may be perceived some timbers, which if there be, there must be some vault underneath it. For the upper rooms, you must observe whether they be more in breadth than the lower rooms, and look in which places the rooms be enlarged . . . If the walls seem to be thick and covered with wainscot, being tried with a gimlet, if it strike not on the wall, but go through, some suspicion is to be had thereof'.[57]

On Monday night Thomas Habington returned home, 'hoping', he wrote, that his presence there 'would dissolve the search'. Bromley showed him a new warrant for Oldcorne's arrest, signed by Cecil, and the January proclamation for Garnet's arrest; Habington denied any knowledge of the priests. Bromley was unimpressed – 'I did never hear so impudent liars as I find here,' he wrote to Cecil; 'all resolved to confess nothing, what danger 'soever they incur.' The search continued. It took until Wednesday for the first of Hindlip's hides to be uncovered, testimony to Nicholas Owen's skill; they were found to be full of 'Popish trash', but no priests. Early on Thursday morning guards stationed in the house spotted two men stealing away from them down the Long Gallery. When challenged the men admitted that they were 'no longer able . . . to conceal themselves: for they confessed that they had but one apple between them', their only food in the last four days. They refused to give up their names and Bromley wrote hopefully to Cecil, 'surely one of them, I trust, will prove Greenway [Tesimond], and I think the other be Hall [Oldcorne]'; in fact they were Nicholas Owen and Oldcorne's servant, Ralph Ashley. Possibly theirs was a bold attempt at escape, possibly they hoped that in giving themselves up they would distract attention from Garnet and Oldcorne – an old ruse that had worked in the past. It did not work now. 'I have yet presumption that there is one or two more in the house,' wrote Bromley to Cecil, 'wherefore I have resolved to continue the guard yet a day or two.'[58]

The search now intensified and gradually, with bare force and wrecking tools, Hindlip was made to relinquish its secrets. In all eleven hides were broken open. Thomas Habington denied knowledge of each of them in turn 'until at length the deeds of his lands being found in one of them' he was forced to concede defeat. '[T]here were found two cunning and very artificial conveyances in the main brick wall,' read the search report, 'so ingeniously framed, and with such art, as it cost much labour ere they could be found . . . Three other secret places contrived by no less skill and industry, were likewise found in and about the chimneys.' These last had 'the entrances into them so curiously covered over with brick, mortared and made fast to planks of wood, and coloured black like the other parts of the chimney, that very diligent inquisition might well have passed by without throwing the least suspicion upon [them]'. Closely examined the chimneys were found to have funnels 'to lend air and light downward' into the hides below. On Monday, 27 January, in the morning, Garnet and Oldcorne were discovered.[59]

They were taken to Worcester, where late on Wednesday night a newly arrested priest was able to give Bromley the information he needed. The January proclamation had been detailed. It described a man 'of middling stature, full-faced, fat of body, of complexion fair, his forehead high on each side, with a little thin hair coming down upon the middest of the fore part of his head; the hair of his head and beard grizzled. Of age between fifty and threescore. His beard on his cheeks cut close, and his gait upright and comely for a feeble man'. Eight days in hiding had altered Garnet beyond recognition, but now Anthony Sherlock – an Appellant priest, a man whom Garnet had cared for upon his return to England, securing him a safe chaplaincy at Stonor – confirmed Garnet's identity. Bromley had got his man. The following day, Thursday, 30 January, he wrote to Cecil with the good news.[60]

Much had happened in London between Monday, 27 and Thursday, 30 January. On the twenty-seventh the eight surviving plotters, Thomas and Robert Wintour, Sir Everard Digby, Ambrose Rookwood, Guy Fawkes, John Grant, Thomas Bates, and Robert Keyes, were brought to Westminster Hall for trial. Robert Cecil, in his instructions to the Attorney General, Sir Edward Coke, spelt out the minefield that was the prosecution's opening speech. Coke must be sure to mention the Spanish Treason, but without upsetting the Spanish. Neither must he upset James: 'some men there are', wrote Cecil, 'that will give out, and do, that only despair of the King's courses on the Catholics and his severity' had driven the plotters on. Coke must make it 'appear' that their treason was set in motion 'before his Majesty's face was ever seen, or that he had done anything in government'. So the plotters' arraignment became, in effect, an arraignment of the absent Jesuits: 'the said Henry Garnet, Oswald Tesimond, John Gerard and other Jesuits did maliciously, falsely and traitorously move and persuade' the plotters 'that it was lawful and meritorious to kill' the King. The Jesuits had conceived the plan to blow up Parliament, the Jesuits had helped the plotters hire a cellar, the Jesuits had provided Guy Fawkes with the very 'touchwood and match, therewith traitorously to give fire' to the gunpowder.[61]

On 30 January, as Henry Bromley wrote to Cecil informing him of Garnet's capture, the first four conspirators were executed outside St Paul's Cathedral; the remaining four were killed the following morning outside Westminster. All eight died as Catholics and Catholic witnesses to their deaths reported that Thomas Wintour, Guy Fawkes, and Everard Digby had each acquitted the Jesuits of any guilt in their conspiracy. A few days later Henry Garnet was called to London for examination.[62]

Conventional wisdom might have dictated that the plotters' trial and execution be delayed until such time as they and the wanted Jesuits could be made to confront each other, conventional wisdom and a desire to see justice done fairly. But this Government was being driven by neither: when Parliament met that month to consider how best to respond to the plot, attempts were made to convict Garnet, Gerard and Tesimond of treason even before they had been arrested. This was a Government driven by its emotions and, in some quarters, by the cold political realization that there was capital to be made from a narrowly averted catastrophe. When James addressed the new session of Parliament his opening speech gave indication of how fairness had given way to fear. Not all Catholics were guilty, he reminded MPs, yet 'no other sect . . . [not] Turk, Jew, nor Pagan, no, not even those of Calcutta, who adore the Devil, did ever maintain . . . that it was lawful, or rather meritorious as the Romish Catholics call it, to murder Princes'. Robert Cecil, writing the day of the January proclamation, saw no irony in detailing his dislike of the Jesuits, then giving his reader the good news that those same Jesuits were 'discovered . . . to be persuaders and actors' in the plot. James's Government was not unduly savage – over the course of this period several Jesuits would fall into its hands: all were questioned, some were tortured, few were killed – but it was capable of political expediency and it did want revenge.[63]

Henry Garnet's progress towards London was a slow, painful affair – Bromley would describe the priest to Cecil as 'a weak and wearisome traveller' – but for all that, it was surprisingly cheerful. In a secret letter to Anne Vaux, written at the beginning of March, Garnet told of the journey and of the events surrounding it. 'We were very merry and content within [our hide],' he wrote, 'and heard the searchers every day most curious above us, which made me indeed think the place would be found.'* His capture, he said, had been by chance – the searchers had not known he was at Hindlip; neither, he wrote, had he known about the proclamation for his arrest: 'if I had . . . I would have come forth, and offered myself to Mr Habington . . . to have been his prisoner'. He described the respect with which Sir Henry Bromley had treated him: 'we . . . were exceedingly well used, and dined and supped with him and his every day'. His health had suffered much during his eight days' hiding, and at Westminster's Gatehouse prison, where first he was lodged, he 'was much distempered . . . and could not eat anything, but went supperless to bed'; the Tower, to which he was moved on 14 February, was 'far better'. 'I am allowed every meal a good draught of excellent claret wine,' he wrote, 'and

* Garnet was frank about conditions in hiding. Had they had 'a close-stool [chamber pot]', he wrote, they 'could have hidden a quarter of a year. For all that my friends will wonder at, especially in me, that neither of us went to the stool all the while, though we had means to do *servitii piccoli* [urinate]'. His captors were equally frank: 'Now in regard the place was so close, those customs of nature which of necessity must be done, and in so long time of continuance, was exceedingly offensive to the men themselves, and did much annoy them that made entrance upon them.' The hide had been provisioned for the search: 'Marmalade and other sweetmeats were found there lying by them, but their better maintenance had been by a quill or reed through a little hole in the chimney that backed another chimney into the gentlewoman's chamber, and by that passage . . . broths and other warm drinks had been conveyed in to them.' In light of this, and of the warning they received, it is surprising that Owen and Ashley's hide was not similarly provisioned and that Garnet's hide was not clear of the 'books and furniture' cluttering it, so that he could stretch out his legs.

I am liberal with myself and neighbours for good respects.' Of his life, he was 'careless'. For Garnet, who was never destined to be Jesuit Superior, who had sought to resign this position believing himself unsuited to it, who, in Rome, was called the 'poor sheep' on account of his shyness, the mission was over. His ordeal was just beginning.*[64]

On 13 February, the day before his transfer to the Tower, Garnet was led the short distance from the Gatehouse to Whitehall to meet the Privy Council. The streets, he noted, were packed with 'a great multitude', all come to stare at him; the Councillors found him no less of a curiosity and throughout this first examination they treated him with an elaborate 'courtesy'. He was kept there for four hours. During that time he acknowledged receipt of Catesby's letter at Coughton, but denied any intention to inform Edward Baynham of the plot's success: this was the sum total of Bates's evidence against him. He further denied any knowledge of the Spanish Treason. The bulk of the questions, he wrote to Anne Vaux, 'were about the authority of the Pope', the issue at the heart of the Bloody Question. '"You see, Mr Garnet,"' Garnet quoted Robert Cecil as saying, '"we deal not with you in matters of religion, or of your priesthood . . . but in this high point in which you must satisfy the King, that he may know what to trust unto."' Subsidiary to this line of enquiry was a long debate about lying. On the table in front of the Council was a manuscript copy of a treatise on equivocation, found in Francis Tresham's rooms and annotated by Garnet. In fact, though the Government was unaware of it, the treatise was Garnet's own work, originally written in 1598. Its handwritten title, *A Treatise against Lying and Fraudulent Dissimulation . . . published in defence of Innocency*, gave some clue as to

* In July 1585 Claudio Aquaviva had written to Robert Persons, expressing concern about Garnet's suitability for the mission: 'Further, about Father Henry Garnet, I shall have to think much [before sending him to England]. For, not to mention the need I have of him here [in Rome] – this is something I am too easily tempted to put before the advantage of England – a number of reasons occur to me why we should think this Father more suited to a quiet routine than the wandering and ever anxious life that the mission of England means.'

Garnet's discomfiture with the subject; not, though, to the extent to which he was already putting his theories into practice. At the end of this first session, ill and exhausted though he was, Garnet felt he had acquitted himself well: 'I am sure I have hurt nobody,' he informed Anne Vaux.[65]

For the following few days his captors' courtesy and his own cheerfulness continued. In his letter to Vaux he recounted 'a pleasant discourse' between himself and Sir Edward Coke; he described the 'kind' usage shown him by Sir William Waad, Lieutenant of the Tower, and the sudden moment when Sir John Popham had recognized him as a fellow trainee printer from their youth. He did not describe, and perhaps should have, the kindness of his gaoler, who had placed him and Oldcorne in adjoining cells, telling them they might communicate with each other (by means of a small hatch) and offering to smuggle out letters for them. But there were moments of anguish: he told of his realization that his friend Catesby 'had fained . . . things for to induce others'; he revealed his confusion at the charges made against Tesimond by Bates, and his certainty that Tesimond was innocent of them; he spoke of his anxieties. 'I thank God I am and have been *intrepidus*,' he wrote, '[but] I often fear torture.' He was right to.

On 19 February the Privy Council instructed Waad to commit 'the inferior sort' of prisoners 'to the manacles'. Among the first to the wall was Nicholas Owen. Owen had been in prison before; he had been tortured before. His preliminary examination, taken on 26 February, showed practised restraint. He denied knowing Garnet or Oldcorne. He had met Ralph Ashley in a pub, he said. He had only arrived at Hindlip two days before the search and he refused to say where he had come from.[66]

On 1 March he was examined again. Now he admitted to being Garnet's servant. He admitted to being with Garnet at Coughton, at Hindlip and at White Webbs. He gave detailed information about life at Hindlip – about the room in which they had stayed, about the fires he had lit each day for Garnet, about their eating arrangements. He said nothing incriminatory. At 5 p.m. on the afternoon of the same day the Tower bell rang, calling an end to visiting hours, and a Somerset gentleman, James Fitzjames, said goodbye to a friend and left the prison, taking with him the breaking news that Owen had just been tortured to death. (Later he passed the news on to another friend. Later still both men would come up before the Star Chamber charged with 'treasonable speech'.)[67]

The official report told a very different story from Fitzjames. On 2 March – somehow twenty-four hours had been lost in the process – Sir William Waad's evening meal was interrupted by Owen's gaoler, running in to say that his prisoner was dying. Waad had gathered up his dinner guests and they had all hurried off to Owen's cell, in time to hear him confess to killing himself rather than face more torture. The gaoler filled in the picture. Earlier Owen had complained of feeling unwell and the gaoler had fetched him a knife to cut his meat; he had also complained that his soup was cold and had asked the gaoler to warm it for him next door. The gaoler had agreed, but as soon as his back was turned Owen had ripped himself open with the knife; Waad and his witnesses could see for themselves the two deep tears in the dead man's abdomen.[68]

It was not too long before this official report was being questioned. On 13 March the Venetian ambassador reported home in careful cipher: 'Public opinion holds that [Owen] died of the tortures inflicted on him, which were so severe that they deprived him not only of his strength, but of the power to move any part of his body.' Owen's friends, meanwhile, were incredulous. How had Owen got hold of a knife? asked John Gerard; 'knives are not allowed . . . [except] those such as are broad at the point, and will only cut towards the midst. And if one be sore tortured . . . he is not able to handle that knife neither for many days'. Moreover, added Gerard, Owen had a hernia, 'taken with excessive pains in his former labours; and a man in that case is so unable to abide torments, that the civil law does forbid to torture any man that is broken'. Every Catholic agreed it was inconceivable that Owen could have committed so mortal a sin as to take his own life.[69]

Both John Gerard and Oswald Tesimond reported Owen's death and their anger burns fresh off the page even at this distance. 'They tormented him with hideous cruelty,' wrote Tesimond, and 'the result of all this brutal, indeed bestial, torture was that, in the course of it, Owen's belly burst open, his bowels gushed out, and in a short while he died.' Gerard noted that his interrogators had 'girded his belly with a plate of iron to keep in his bowels, but the extremity of pain (which is most, in that kind of torment, about the breast and belly) did force out his guts, and so the iron did serve but to cut and wound his body, which, perhaps, did afterwards put them in mind to give it out that he had ripped his belly with a knife'. Tesimond fired off this question: 'Does William Waad seriously expect us to believe that, even after many days' torture, a man like Owen would abandon his hope of salvation by inflicting death on himself – and such a death?'[70]

It seems certain the suicide story was a fiction concocted by a Government deeply embarrassed to find itself with a corpse in its custody. In which case the clue to Owen's treatment lies in the comments reported (by Catholics) at his capture, comments indicating his identity as a hide-builder had been betrayed. 'Is he taken that knows all the secret places?' one Councillor was said to have asked. 'I am very glad of that. We will have a trick for him.' Robert Cecil's rumoured response was stronger still: 'No dealing now with a lenient hand. We will try and get from him by coaxing, if he is willing to contract for his life, an excellent booty of priests. If he will not confess he shall be pressed by exquisite torture, and we will wring the secret from him by the severity of his torments.' For those few grim days of February 1606, as the Government tried to break the Jesuit mission, the fate of almost every English Catholic lay in Owen's hands. In life he had saved them, in death he would too: not a single name escaped him.[71]

On 26 February, as Nicholas Owen was being taken to the torture room for the first time, Henry Garnet was writing a secret letter in orange juice ink to his friends outside, arranging Jesuit affairs until such time as Aquaviva could appoint his successor. He seemed confident, explaining, 'they have nothing against me but presumptions'. On 3 March he wrote again; still he was confident: 'I see no advantage they have against me for the powder action.' That same day Robert Cecil wrote to Sir Henry Brouncker in Ireland, assuring him 'that ere many days he should hear that Father Garnet . . . was laid open for a principal conspirator' in the plot. Five days later, on 8 March, Garnet testified to knowing about the plot in advance.[72]

What had happened in those few days? Garnet's supporters said torture: John Gerard reported rumours that Garnet had been confused and 'so drowsy, as not able to hold up his head' in interrogation, suggesting deliberate sleep deprivation and possible drugging. He also believed that Garnet had been manacled; Garnet, himself, would refer to his 'next torture' in a subsequent letter, a form of phrase suggesting a previous torture. It is also possible that, having learned of Owen's death, Garnet wanted to prevent others – Oldcorne, Ashley, the White Webbs servants – from suffering the same fate. Whatever the reason he now broke silence – and the effect was dramatic.[73]

'You may confidently affirm that [Garnet] is guilty,' wrote Robert Cecil to the English ambassador in the Low Countries that same day. Next day he wrote to the Earl of Mar with the same news and the observation that Garnet's life was of no 'value'. The 'important thing', he noted, 'is to demonstrate the iniquity of Catholics, and to prove to all the world that it is not for religion, but for their treasonable teaching and practices that they should be exterminated'. His private notes revealed how he proposed to do this.[74]

Garnet had testified that he had learned about the plot in confession and was therefore unable to reveal his information; for a Protestant, Cecil showed a fine grasp of the principles of confession. 'In penance', he wrote in a series of aides-mémoires to himself, 'the first act is *confessio, contritio, satisfactio,* which order not being observed it is no penance'. Neither was it true penance if 'the penitent does not assure reformation and desist from the evil act'. By this line of reasoning the process of confession by which Garnet had heard of the plot was technically invalid; and this being the case, it could be argued, as Cecil did now, that Garnet had concealed and 'abetted' the plot. His notes concluded with a fine rhetorical flourish: 'We are now therefore not to arraign Garnet the Jesuit . . . but to unmask and arraign that misnamed presumptuous Society of Our Saviour Jesus.' This was what the Government had been seeking since William Allen's seminarians first began arriving home in the 1570s: the chance to prove that the Catholic Church sanctioned murder and that England's forbidden priests were secret agents of assassination; and as ideal, as image-laden a chance it could not have hoped for. 'We may now defend that the priests' hands are full of bloody sacrifices', wrote Cecil triumphantly, '[and] we prove not their treasons by wit and inference, but by confession.' For there, argued Cecil, in that strange, shadowed, sacred whisper between confessor and confessant, he had found death, the drop of poison in the ear that said killing was no sin.[75]

The trial of Mr Henry Garnet – in true paradoxical style the Government wished to try him as a priest without affording him the courtesy of that title – took place at London's Guildhall on Friday, 28 March 1606. Garnet arrived there by 9 a.m., brought the short distance from the Tower in a covered coach to keep him 'safe'. Already the crowds were gathering, packing the entranceway to catch a glimpse of him. They saw a balding, bespectacled, somewhat overweight fifty-year-old, whose face bore the signs of long-term illness and recent hardship. But no one was very interested in Garnet the man – it was what he represented that mattered. The commissioners arrived soon afterwards and took their place on the bench – the three Howard earls, Nottingham, Suffolk and Northampton, Lord Somerset, Robert Cecil, Sir John

Popham, Sir Thomas Fleming, Sir Christopher Yelverton and London's Lord Mayor, representing the King: England's most powerful.* James, himself, was there '*incognito*', so the Venetian ambassador reported, as were most of his courtiers, squeezed in tightly to catch the proceedings. Two days earlier William Waad had complained to Cecil about the difficulty of getting a seat: 'there is a place provided in the Guildhall for the prisoner, but none for me', he wrote plaintively. Now he took his specially allotted position next to Garnet, before the courtroom. At about nine-thirty the show trial began.[76]

'Henry Garnet, of the profession of the Jesuits, otherwise Walley, otherwise Darcy, otherwise Roberts, otherwise Farmer, otherwise Philips, (for by all those names he called himself) stood indicted of the most barbarous and damnable treasons': the arraignment began in thunderous fashion. And so it continued, throughout that long March day. First the charges against him were read out: that on 9 June, the date Catesby had consulted him about the killing of innocents, he, Tesimond, and Catesby had all three conspired to murder the King and destroy the commonwealth. Garnet pleaded not guilty and Sir Edward Coke began the case for the prosecution.[77]

'[S]ince the Jesuits set foot in this land,' Coke informed the court, 'there never passed four years without a most pestilent and pernicious treason.' Conspiracy after conspiracy was dusted down and hung about Garnet's neck. Then Coke moved on to the Gunpowder Plot: 'because I speak of several treasons, for distinction and separation of this from the other[s], I will name it the Jesuits' Treason', he told the jury helpfully. Of course, with no proofs to support the assertion that Garnet was author of the plot, Coke was forced to rely on judicial interpretation of the evidence to make his case. So Garnet's letter to Aquaviva 'for the staying of all commotions of the Catholics' was written, claimed Coke, solely 'to lull us asleep for security'. His conference with Catesby

* Sir Christopher Yelverton, a Justice of the King's Bench, was half-brother to Edward Yelverton of Norfolk, John Gerard's first host.

and Tresham about 'the strength of the Catholics in England' – as hero of the hour Lord Mounteagle's name was omitted here – was to promote rebellion, not prevent it. And the pilgrimage to St Winifred's Well 'was but a jargon, to have better opportunity, by colour thereof, to confer'. From manipulation he turned to character assassination. Of Garnet's many aliases he observed, 'I have not commonly known . . . a true man, that has so many false appellations'. Garnet's orange juice letters – carefully provided by his solicitous gaoler – were shown to the court as evidence of his cunning. Garnet, said Coke, was 'a doctor of five DD's, as dissimulation, deposing of princes, disposing of kingdoms, daunting and deterring of subjects, and destruction'.

The subject of equivocation – dissimulation, in Coke's terminology – had dominated Robert Southwell's trial. It decimated Henry Garnet's. The Jesuits, explained Coke, 'equivocate, and so cannot that way be tried or judged according to their words'. Over the course of the day the prosecution would shred the credibility of Garnet's words. State witnesses appeared, swearing to having eavesdropped on conversations between Garnet and Oldcorne in the Tower; then evidence was produced revealing Garnet had denied, in interrogation, that such conversations had ever taken place. More damaging still was the dramatic flourishing of Francis Tresham's deathbed statement. It had been Tresham who had accused Garnet of complicity in the Spanish Treason, then retracted his evidence as he lay dying, saying he had only implicated the Jesuit to avoid torture (he died in the Tower on 23 December, of an inflammation of the urinary tract). Now his statement was read out, in all its fatal ambiguity. '[T]o give your Lordship a proof [that Garnet was not involved],' Tresham had sworn, 'I had not seen him in sixteen years before.' The meaning was unclear: had Tresham meant 'before' 1602, the date of the Spanish Treason (the subject to which he was referring), or 'before' 1605, the date of the plot? The first was true, the second false, as independent statements from Anne Vaux and Garnet, himself, confirmed. But for the prosecution there was no lack of clarity:

even as Tresham breathed his last, Coke exclaimed, he had lied – instructed how to do so by pernicious Jesuit teachings. By the time the prosecution moved on to the crucial subject of Garnet's knowledge of the plot, his every utterance had been cast in doubt.[78]

The questions and comments about Garnet's response to Tesimond's confession came in waves. Robert Cecil asked why Garnet had not informed Claudio Aquaviva of the plot. The Earl of Northampton wondered why Garnet was unable to mention the confession before his arrest, but was able to discuss it now. Cecil returned to his notes to enquire about the process of confession, contrition, and absolution, to convince the court that Tesimond's confession was invalid and therefore needed 'no secrecy'. Then he switched tack, asking why Garnet had never thought to disclose the plot using the general knowledge of it he had received from Catesby; he also asked why Garnet had refused to listen every time Catesby had offered to tell him about the plot.

There was no lawyer to speak for Garnet: he stood in the dock alone, without notes and without access to the various examinations read out as evidence against him.* His word had been discredited by the prosecution's repeated attacks on equivocation. His defence – that 'he was bound to keep the secrets of Confession' – was a Catholic defence in a Protestant court, a court thoughtfully reminded all day of the Pope's alleged power to depose princes (a subject guaranteed to provoke waves of anti-Catholic sentiment). Moreover, in the crucible of public opinion he had already been tested and found wanting. His confession, reported John Gerard, had been 'censured by many; and even by some of his friends and well-wishers esteemed a weakness in him'. His

* It was not just all mention of Lord Mounteagle that was missing from the various examinations read out at Garnet's trial; from copies marked up by Sir Edward Coke it is clear many statements were subtly manipulated to boost the prosecution's case. No mention was made, for example, of Guy Fawkes' assertion that Gerard was ignorant of the plot and all references to Garnet's 'mislike' of the plot were omitted. It was also striking that the prosecution made no attempt to prove the actual indictment against Garnet; rather it worked on cracking open Garnet's admission of misprision of treason, implying that his concealment of the plot indicated his approval of it.

'partisans' in the Low Countries were claiming he had only cracked 'after having suffered great torments . . . but he would retract the same at his arraignment'. London's gossips had sharpened their pens on him ('He has been very indulgent to himself,' noted the inveterate letter writer John Chamberlain, '. . . and daily drunk sack so liberally as if he meant to drown sorrow'). And from the heart of Government Robert Cecil had begun a carefully targeted smear campaign (on 19 March he had written to Sir Henry Wotton, ambassador to Venice, announcing that Garnet had confessed the plot 'justifiable by divinity'). Witnesses sympathetic to Garnet reported a bear-pit-like atmosphere to the proceedings: 'he was often of set purpose interrupted'; 'great laughter' greeted his defence of the seal of confession; suggestive comments were made about his friendship with Anne Vaux; he was 'clean wearied out with so long standing at the bar'. Official reporters painted a different picture: '[he] made no great answer'; '[he] faintly answered'; '[he] denied to answer'; '[he] began to use some speeches [not recorded] that he was not consenting to the Powder-Treason', but could give no proofs.[79]

He did his best. He invoked the common law principle that no man should be forced to incriminate himself, to justify his use of equivocation. He insisted that, to him, Tesimond's confession was valid and that therefore he was bound to secrecy. He explained that he had been given permission to break the seal of confession if ever the plot were discovered. He laid out his obligation under canon law to 'labour to divert' the plot, which, he said, he had tried to do. He described again Catesby's many promises to him and, again, how he had been 'loath' to hear more about the plot from him. But these answers were of interest only in human terms, not in the superhuman contest between good and evil being fought about him by his defenders and detractors. And his most human moment of all, when the tension of the last months' conflict of loyalties burst from him 'passionately' (in the wording of the official court reporter), went unrecorded by his Catholic sympathizers: 'I would to God', he exclaimed, 'I had never known of the Powder-Treason.'

It took the jury only fifteen minutes to reach their verdict of guilty as charged. London's correspondents agreed. The letter writer John Chamberlain (though not present at the trial) thought it satisfactorily proven that Garnet 'had had his finger in every treason' since his return to England. The Venetian ambassador (who was present) railed against equivocation, concluding that Garnet had 'caused great outcry against the Roman religion'; he too believed Garnet guilty. Yet the Government hesitated. Indeed, for the next month it was as though Garnet had never been tried at all as his interrogators pushed for a greater admission of his crimes. Trickery was its chosen weapon: at the beginning of April Garnet was told that Tesimond had been captured – and had testified to telling him of the plot *out* of confession. There followed a series of agonized letters from Garnet. His defenders have questioned the authenticity of some of them, but all contain material similarities. Again and again he repeated himself: 'that knowledge I had by [Tesimond] I took it as in confession'. Again and again he wrote of his abhorrence of the plot: '[it was] altogether unlawful and most horrible'; 'I never allowed it [and] I sought to hinder it more than men can imagine, as the pope will tell.' To Anne Vaux (this letter is extant and uncontested) he explained how the conspirators 'used my name freely' to win support for their cause, a comment repeated in the letter to his Jesuit brethren (extant, but contested). Then came the Government's breakthrough. Its beginnings can be traced in his letter to Tesimond (extant, uncontested): 'I wrote yesterday . . . [to testify] that indeed I might have revealed a general knowledge had of Mr Catesby out of confession'. Its progress can be traced through his letter to the Council (not extant, contested): 'I acknowledge that I was bound to reveal all knowledge that I had of this plot . . . out of sacrament of confession.' Finally, in the letters to Tesimond, the Council and the Jesuits, came Garnet's explanation for his silence: it 'proceeded from hope of prevention by the Pope', he wrote, and 'for that I would not betray my friend'.[80]

'Greater love hath no man than this,' Christ had said near

death, 'that [he] lay down his life for his friends.' Catesby had been Garnet's friend, his provider, his companion in that secret, underground, and alienated society that was English Catholicism. For Catesby Henry Garnet had laid down life and reputation – and not just his own reputation, but that of everything he held most dear. In those last anguished weeks as he, himself, neared death, it seemed the doubts about his actions crowded in upon him – worse than any torture he might have suffered at the Government's hands.[81]

On Wednesday, 30 April carpenters began constructing a scaffold in the churchyard outside St Paul's Cathedral. It was higher than usual so that the expected crowds might have a good view, and around it jostled several spectator stands. On Saturday, 3 May these stands were full of people come to see a Jesuit die. Back in February the Venetian ambassador had been confident Garnet would 'not be executed in public': 'he is a man of moving eloquence . . . and they are afraid [of] his constancy and the power of his speech', he explained. Now he took his allotted place among the onlookers, content to see a 'partner in that villainy . . . being extinguished'.[82]

Garnet was drawn on a hurdle the short distance from the Tower, 'his hands together . . . his eyes shut'. At St Paul's he opened his eyes to look around. 'All [the] windows were full,' it was reported, 'yea, the tops of houses full of people.' Eighteen years earlier he, himself, had stood discreetly among a similar crowd at St Paul's watching Queen Elizabeth celebrate the Armada's defeat; now he and his Jesuits had been denounced as 'purveyors and forerunners' of that Armada and he was the one being watched. Watched and harangued, for even at this hour of his death there was to be no let up. The deans of St Paul's and Winchester pressed him to acknowledge the errors of his Catholic faith and recant; the Recorder of London, Sir Henry Montague, representing the King, pressed him to confess his treason and acknowledge himself 'justly condemned'. Witnesses to the scene, both Protestant and partisan, reported 'that his voice was low', his strength gone, but, just as before, both parties disagreed on the content of his replies. The Venetian ambassador described 'the fury of the mob' at his mention of the word 'pope'; sympathizers observed that the crowd was 'much moved . . . by his protestations of innocence'. 'It is looked he will equivocate at the gallows', one correspondent had written gleefully the day before (and sure enough the old charge of lying was levelled against him), 'but he will be hanged without equivocation.'[83]

He had seen so many men die – standing as near to the scaffold as he might to whisper the last rites over them. 'Should it come to pass that we have to suffer for His sake and attain high honour in this way,' he had written to Aquaviva, '. . . we hope God will turn everything to our greater good, making us more like Him whom it must be our aim to resemble.' But to Robert Persons he had mentioned the gnawing doubt that he was 'unfit for the combat'. The official records of his execution described his palpable 'fear of death'; Catholic records described his 'undaunted countenance'. But in the end Henry Garnet died as he had lived, as an English Catholic, praying for his King, country and God, no matter the conflict existing between them. And on that everyone could agree.[84]

He hung for fifteen minutes, the same time it had taken the jury to convict him. Government witnesses reported that he was not cut down until he was dead, on James's orders; Catholics reported a somewhat different scene: the crowd, they said, had surged forward, pulling him by the legs, 'to put him out of his pain, and that he might not be cut down alive'. And the Spanish ambassador recalled, 'When they cut out his heart, which they show to the people with the head, where it is the custom for everyone to shout loudly "God save the King", there was not a sound to be heard.' His 'limbs were divided into four parts, and placed together with the head in a basket, in order that they might be exhibited according to law in some conspicuous place, [and] the crowd began to disperse'. In time his head was set upon London Bridge.

His clothes were taken by the Spanish ambassador as a relic. Later in the year his belongings were disposed of by Thomas Wilson, Cecil's secretary, who rode out to White Webbs to see 'the chief things conveyed away'; 'the remnant of small value', he reported, 'was quickly bought up by the neighbours'. Wilson then continued on to Hertfordshire to holiday with his brother, closing his uncommonly chatty letter to Cecil with the pious observation that White Webbs was 'next neighbour to Theobalds [Cecil's country house], and unfit it should be again a nest for such bad birds as it was before'.[85]

The same day that Garnet was executed John Gerard left England for the Continent and safety. It was, he explained, 'a time for lying quiet, not for working'. He placed his friends in the care of his Jesuit brothers, tidied his affairs – by now the chief burden on his conscience, Elizabeth Vaux, was released from custody – and, with typical aplomb, arranged passage out of England in the company of the Spanish envoy sent to congratulate James on his narrow escape from the Gunpowder Plot. The plan almost failed, as he recalled afterwards: 'When I arrived by arrangement at the port from which I was to pass out of England with certain high officials, they took fright and said they could not stand by their promise.' Then 'suddenly they changed their mind. The ambassador came personally to fetch me and helped me himself to dress in the livery of his attendants so that I could pass for one of them and escape'. Of this change of mind, Gerard wrote, 'I have no doubt that I owed it to Father Garnet's prayers.'[86]

From England Gerard went to the Jesuit school at St Omer, where for a time he remained, too ill to travel. It was not until high summer that he was able to leave for Rome and a long awaited meeting with the man who had sent him to England all those years before, Claudio Aquaviva. He never returned home again. His exile took him from the Italianate splendour of St Peter's Basilica, burial place of the popes (where for three years he served as a Penitentiary), to the Low Countries (to train Jesuit novices for the English mission), to Spain, and then back again to Rome (as Confessor to the English College), worlds far removed from the hunting and hawking fields of his native Lancashire.* Some time in the spring of 1609 he was ordered by his Superiors to write a private account of his time on the mission, probably to inspire the novices in his charge; he was also asked to write a second, public account, dealing specifically with the Gunpowder Plot and his part in those calamitous events. He died in Rome on 27 July 1637, a few months short of his seventy-third birthday.[87]

One question remains. Throughout his *Narrative of the Gunpowder Plot* Gerard insisted, under oath, that he was not the priest who had said mass for the plotters at the house off London's Strand. In his *Autobiography* he expanded upon this, writing of the house: 'I might have stayed there without the slightest risk or suspicion for a very long time, had it not been for some friends who made very indiscreet use of [it] while I was out of London.' Perhaps this was equivocation, but it would have been to little purpose: both Guy Fawkes and Thomas Wintour, the two plotters who named him as the officiating priest, had sworn he knew nothing of their plans, so Gerard was in no danger of incriminating himself. Perhaps Fawkes and Wintour were mistaken in their identification: it is not clear either of them had ever met Gerard (both had spent many years serving as soldiers overseas); also, two other priests are known to have used the house as a base. Was it that Catesby had instructed them to meet at 'Gerard's house' and that therefore they simply assumed the priest in front of them was Gerard? Was Gerard condemned on assumption alone?[88]

If there can be no satisfactory conclusion to Gerard's story,

* On 15 July 1606, two months after his escape, Gerard wrote to Robert Persons, suggesting possible employment for himself. 'I could have care of the garden,' he told Persons, 'for I am excellent at that (if you will permit me to praise myself) for that was much of my recreation in England, and I hope my brother will witness with me that he has seen a good many plants of my setting, and tasted the fruit of some of them.' Curiously John Gerard was also the name of a famous Elizabethan botanist, author of *The Herbal, or General History of Plants*, and supervisor of Sir William Cecil's London gardens.

no definitive proof of his involvement in, or innocence of, the Gunpowder Plot, then there exists one last document in the case to cast doubt upon everything else that might be thought definitive: upon every comment made by every Government eyewitness, upon every signature on every examination, upon every supposed handwritten statement of fact.

Towards the end of 1606 a man called Arthur Gregory wrote to Robert Cecil, telling a sad tale of poverty and desperation. He took the opportunity to remind Cecil of his past employment for 'King and Country'. It was not the only letter of this sort Cecil received that year: the informant George Southwicke, who had spent most of February searching for John Gerard in East Anglia, would also write in, complaining of similar poverty and begging some form of recompense for his efforts.* Such was the habitual lot of the disaffected drifters used by the Government as intelligencers. But Gregory's letter was somewhat different from the norm. Instead of regaling Cecil with all the lurid information he might provide the State at some future date (if only his debts were paid and his release from prison secured), Gregory wrote of the services he had already provided, 'the secret services ... [that] none but myself has done before'. One of these services was 'discovering the secret writing being in blank, to abuse a most cunning villain in his own subtlety, [and] leaving the same in blank again. Wherein, though there be difficulty, their answers show they have no suspicion'. If this 'secret writing' was Garnet's, contained in his many orange juice letters from the Tower (and these are the only 'blank' letters recorded leaving the Tower at this time), then Gregory's next 'service' becomes even more suggestive: it was, he reminded Cecil, 'to write in another man's hand'. And as an example of his skills at forgery he appended a postscript. 'Mr Lieutenant [Waad]', wrote Gregory, 'expects something to be

* Southwicke's efforts deserved some form of recompense. On 9 February Dr Dupont, Vice Chancellor of Cambridge University, contacted Cecil, saying he had arrested Southwicke for lurking 'suspiciously' and for 'being a stranger'. Southwicke was subjected to several days of questioning before Cecil was able to secure his release.

written in the blank leaf of a Latin Bible which is pasted in already for the purpose. I will attend to it and whatsoever else comes.' A Latin Bible was a Catholic Bible. What Waad meant to do with his doctored copy is not known.[89]

Epilogue

Treason doth never prosper: what's the reason?
For if it prosper, none dare call it treason.
attrib. Sir John Harington

THE MOST IMMEDIATE CONSEQUENCE of the Gunpowder Plot was that Robert Cecil achieved his Oath of Allegiance. The text was based on the Appellants' abortive oath of just three years earlier, for which Cecil had campaigned so stealthily, but this new one came wrapped in the rank scent of fear. The oath read:

> I A.B. do truly and sincerely acknowledge . . . that our sovereign lord King James is lawful and rightful king of this realm . . . and that the Pope, neither of himself nor by any authority of the Church or See of Rome, or by any other means with any other, has any power or authority to depose the King, or to dispose any of his Majesty's kingdoms or dominions, or to authorize any foreign prince to invade or annoy him or his countries, or to discharge any of his subjects of their allegiance and obedience to his Majesty, or to give licence or leave to any of them to bear arms, raise tumult, or to offer any violence or hurt to his Majesty's royal person, state or government.[1]

The oath was floated in on a raft of fresh anti-Catholic legislation, as the Government considered its response to Catesby's outrage. Parliament, so troublesome to James in its first incarnation, now showed itself foursquare behind him, raising the subject of popery as soon as it met. In the opening speech of the new session Sir George More, MP for Loseley in Surrey, called upon the House to make the safety of the King its paramount concern. Just ten days later a Commons sub-committee drafted its first bill. The acts that followed were wide-reaching. No known Catholic recusant might enter a royal palace. No known Catholic recusant might come within 10 miles of the City of London. No known Catholic recusant might practise the law or medicine, or hold a commission in the Army or Navy; neither might a known Catholic recusant, nor anyone with a recusant wife, hold public office. Known Catholic recusants might possess neither arms nor armour (except those necessary for their own immediate defence), though, perversely, they were required to provide both at the county musters. It was made lawful for any Crown officer, 'if need be', forcibly to enter any house in the country in pursuit of a known Catholic recusant. Recusancy fines were stiffened and new penalties were introduced targeting those Church papists who attended their local church, but refused to receive communion. It was decided to make 5 November a day of national thanksgiving.* Looming over the entire proceedings – affixed to spikes topping Parliament House – were the severed heads of Robert Catesby and Thomas Percy, grisly sentinels now sightlessly guarding the building they had once sought to destroy.[2]

It was not, stressed the King in his post-plot speechmaking, that all Catholics were dangerous: 'many Catholics were good men and loyal subjects.' But how could you tell a moderate, 'good' Catholic from a 'malignant and devilish' one? Surely it was safer then to make Catholics identify themselves, forcing them to attest to their moderation? Surely it was an act of kindness, too, to give those moderate Catholics an opportunity to demonstrate their loyalty? The oath, wrote James later, 'was only devised as an Act of great

* The verse of the National Anthem, calling on God to confound the King's enemies and 'frustrate their knavish tricks', was reportedly first sung in the hall of the Merchant Taylors' Company, in London's Threadneedle Street, as part of just such an act of thanksgiving.

favour and clemency towards so many of our subjects, who though blinded with the superstition of Popery, yet carried a dutiful heart towards our obedience'. Parliament simply wanted to discover 'whether any more of that mind [i.e. the Gunpowder Plotters], were yet left in the country'. So the Oath of Allegiance enjoined its takers to defend the King against all conspiracies, and to 'abhor, detest and abjure as impious and heretical this damnable doctrine and position, that Princes which be excommunicated or deprived by the Pope may be deposed or murdered by their subjects'. 'Heretical': it was an injudicious choice of word. Were those who accepted the oath thereby condemning to perdition every Pope who had ever argued the case for papal supremacy? This was theologians' battleground. For good measure the new legislation had also enshrined in print – so making it official – the supposed 'doctrine' that the papacy sanctioned murder.*[3]

It seemed at first, though, that many of those Catholics to whom the oath was tendered – those already up before the law for their recusancy – were prepared to swear to it, eager to distance themselves from Catesby's actions. The Government's biggest coup came when the Archpriest George Blackwell, arrested in June 1607, declared himself willing to take the oath, and advised the secular priests under him to do likewise; earlier, Blackwell had expressed his revulsion at the plot, calling it 'intolerable, uncharitable, scandalous, and desperate'. But by now the Vatican had responded, issuing a Brief forbidding all English Catholics from putting their names to the oath. And suddenly Europe's printing presses exploded into action, with Churchmen and non-Churchmen alike rushing to add their comments to the ensuing debate about the authority of Popes, the authority of Kings and the right of an English Parliament to decide upon such matters. George Blackwell was upbraided by the Vatican for his 'fear or imbecility' and

* In a further piece of spin, the laws by which the English Catholic Church had progressively been dismantled since the time of Henry VIII's first quarrel with Rome became the 'sundry necessary and religious laws for [the] *preservation* of Church and State'.

dismissed from office. From a prison cell in the Clink, one Catholic priest, claiming to speak for the Jesuits (but actually writing for the Government), explained how Catholics might in all conscience take the oath. And John Donne, an ex-Catholic but one with impeccable credentials (two of his uncles were Jesuits, while his great-great-uncle was Sir Thomas More), was moved, probably at James's suggestion, to explain to Catholics why martyring themselves for the oath was a wasted cause.[4]

Amid this confusing welter of contradictory opinion (from some of Europe's finest minds), England's Catholics floundered. Because, in attempting to codify their loyalty for them, the English Government had, in reality, forced them to ask of themselves questions to which no one had ever been able to find satisfactory answers. No theologian, ancient or modern, had been able to agree on the extent of the Pope's power to depose. No priest going to his death had been able to agree on *what* had greater call over *which* parts of his loyalty, country or conscience. And every Catholic knew this. And now every Catholic charged with recusancy was expected to do the impossible and find an answer, or be accused of treason and suffer the penalty of praemunire: life imprisonment and the loss of all possessions.

The evidence suggests that the Government was sympathetic to their plight and that the oath was not administered as vigorously as it might have been; in 1608 James instructed his judges to show a 'mild inclination' and tender the oath only to newly converted Catholics or to those considered troublesome. Some Catholics seem to have taken the oath; many (according to John Donne) seem to have refused; a few wrote desperate letters seeking advice. In the summer of 1606 Lady Jane Lovell wrote to Robert Cecil, explaining that she had just been diagnosed with breast cancer and that she had been advised by her doctor to go abroad for treatment; however, she had also just heard rumours that the oath was to be extended to those travelling overseas. A new clause was introduced to this effect in late August that year, followed, a few months later, by a clause extending the oath to all those entering the country. Lovell's anguish was this: she would rather die at home, she wrote, 'than do that thing which a religious and Catholic conscience cannot justify [i.e. take the oath]'; could Cecil help? It seems that Cecil did help: the last in her acutely painful series of letters to the Secretary was written from Brussels. Lovell was profuse in her gratitude.[5]

It seems pertinent to ask what the Oath of Allegiance actually achieved, as distinct from the achievements of its companion anti-Catholic legislation. Certainly, Catholics seem to have fallen away from overt demonstrations of their faith in the months immediately following the plot. From the English ambassador in the Low Countries came a report, dated August 1606, that recusants once happy to support missionary priests had 'retracted their contributions'. From Father Richard Holtby, who succeeded Henry Garnet as head of the Jesuit mission in England, came a letter, dated October 1606, revealing that 'whole countries and shires [were now running] headlong, without scruple, unto the heretics' churches to services and sermons'. And from the informer William Udall came the news that some three hundred Catholic households were about to write to James, begging leave to flee the country for the American colony of Virginia, rather than continue facing persecution at home. Their desperation to escape England can be measured by the fact that an earlier group of would-be Virginian settlers had been found to have disappeared without trace, leaving little more than a 'Gone Away' note behind them. But how much this weakening of Catholic resolve was in consequence of the Oath of Allegiance, and how much it was the result of an unprecedented tightening of all the other laws in place against them in general, is unclear.[6]

What is clear is that the oath did nothing at all to reassure Protestants that their Catholic compatriots could be trusted. Even with legislation in place to separate those Catholics who believed the Pope could depose a ruler from those Catholics who did not (and might therefore be regarded as loyal), Protestant paranoia remained strong. And for this reason alone the oath must be adjudged to have failed – for what is the point of introducing safeguards if nobody feels safe once they are in place? So the anti-Catholic persecution continued. In 1610 new laws were passed by Parliament targeting married women recusants; in 1613 a bill was introduced into the House of Commons to force Catholics to wear a red hat (like Jews in Rome), or multicoloured stockings (like clowns), so that they might better be recognized and 'hooted at' in the street. The Great Fire of London, in 1666, was blamed upon Catholics, just as Robert Southwell had predicted when he wrote in his *An Humble Supplication*: 'If any displeasing accident fall out, whereof the Authors are either unknown or ashamed, Catholics are made common Fathers of such infamous Orphans, [even of] . . . the casual fires that sometimes happen in London.' And in 1678–9 the rantings of a fantasist called Titus Oates led to a spate of Catholic executions and imprisonments, as London once again rang to the rumours of a 'popish plot', this time of an 'intended general massacre of all the Protestants in the world'.[7]

Only time, a changing political scene and a chance occurrence could undo this pattern. By the close of the eighteenth century Britain at last felt sufficiently trusting of its Catholics to permit them, in limited form, to celebrate mass, and so the first Catholic Relief Acts were passed. Then in June 1828 a Catholic Irish barrister called Daniel O'Connell, a leading voice in the emerging Irish anti-Union lobby, stood for, and won, a County Clare by-election: the law preventing Catholics from becoming MPs did not, it transpired, prevent them from standing for election. The Prime Minister, the Duke of Wellington, hitherto an opponent of Catholic emancipation, realized that to pacify Ireland he would need to make concessions. And in April the following year, 1829, a final Catholic Relief Act was passed. Now Catholics could worship freely, could vote again, could sit in Parliament, could follow a profession, could be a part of that society from which they had been excluded for so long. Three hundred years earlier, almost to the month, King Henry VIII had stood before a small papal court in London's Blackfriars, and had failed to win from it his troublesome divorce, thereby setting in motion every event in this story.[8]

Robert Cecil lived on until 24 May 1612. His death, just a few days short of his forty-ninth birthday, was celebrated with public glee and a flood of denigratory ballads. As the letter writer John Chamberlain noted: 'I never knew so great a man so soon and so generally censured'; but then Cecil, like his father before him, had never courted popularity. Among Catholics he was, and is, blamed for many of their woes. Yet Cecil was not alone in regarding England's Catholics as the primary threat to their country's security. Neither was he alone in the methods he chose to defuse that threat: judicious manipulation of evidence, the orchestration of smear campaigns, the singling out of a scapegoat for the nation's hostility. These were, and remain, practised tricks of government. Where Cecil was rare was in his seeming lack of personal animus in so doing. His correspondence is remarkably free of the ugly clatter of religious prejudice so common at this period. Rather it often surprises with its tone of tolerance, a point that should inform any assessment of his role in these proceedings.

James survived his principal minister by thirteen years. He died on Sunday, 27 March 1625, aged 58. True to form, his final moments were hedged with rumours of assassination (by poisoning), and his funeral, just as his coronation had been, 'was marred by foul weather': 'there was nothing to be seen,' wrote John Chamberlain, 'but coaches and torches.' James had never had the knack for ceremony.

He was succeeded as King by his second son, Charles (Prince Henry had died the same year as Robert Cecil). Twenty-four years later, on 30 January 1649, that same Charles was led out onto a scaffold in front of Whitehall Palace and executed for treason by order of his Parliament; during the Civil War preceding this event England's Catholics had been among his staunchest supporters. Two years later James's grandson, Charles II, escaped Parliament's

Army by seeking shelter in Catholic houses, all of them equipped with Catholic priest-holes. It has been suggested, though it cannot be confirmed, that the hides in question were the work of Nicholas Owen.[9]

It is widely held that Charles II converted to Catholicism on his deathbed. His successor, his brother James II, was openly Catholic. In 1688 James's Dutch Protestant son-in-law, William, invaded England and seized it for himself, and James fled; within weeks a compliant Parliament had offered William and his wife, Mary, the Crown jointly, claiming that James had willingly abdicated it. In 1701 Parliament took steps to ensure that the succession would thenceforward bypass James's British-born Catholic son, and fall instead on his German Protestant first cousin once-removed. The king-killer, the king-deposer was not, it seemed, an autocratic, absolutist Pope, but Parliament, seat of democracy, acting with the cold efficiency of English law. And government by 'Strangers', as recommended by Robert Persons in his Book of Succession, and at the time so generally reviled, was now become palatable when advocated by Protestant tongues. William Shakespeare has the phrase for it: 'And thus the whirligig of time brings in his revenges.' In years to come a descendant of Cecil, and one of Sir Francis Walsingham, would join the Jesuits.[10]

Anne and Elizabeth Vaux escaped serious punishment for their involvement in the events surrounding the Gunpowder Plot. Anne was released from the Tower of London in August 1606, allegedly 'much disappointed' that she had not been allowed to die alongside Henry Garnet. Some time later she moved back up to Leicestershire, where she and her sister Eleanor Brooksby continued harbouring priests just as before. She died in 1635 aged 73. Her sister-in-law Elizabeth would also remain loyal to the mission. After her release from house arrest in London in April 1606 she returned to Harrowden, where she quickly installed Father John Percy as resident Jesuit in place of John Gerard. In 1611 an unfounded rumour that Gerard was back in England saw Harrowden raided again and Elizabeth rearrested. The following year she was indicted for refusing to take the Oath of Allegiance and condemned to Newgate prison in perpetuity, but in July 1613 she was freed on the grounds of ill health. She died some twelve years later.[11]

Father Robert Persons died in Rome in April 1610, aged 64, and General Claudio Aquaviva in January 1615, aged 72. Persons left behind him a reputation complicated by his involvement in the Jesuit–Appellant conflict and by his repeated intervention in matters seen as political rather than purely religious. He remains an ambiguous figure in the history of the Jesuit mission, admired for his doggedness and zeal, criticized, even by his defenders, for his willingness to engage in polemic and in his back-room dealings with statesmen and generals. Aquaviva's reputation is more straightforward: he is regarded as one of the chief architects of the Jesuits' long-term survival and success.

On about 21 March 1606 Father Edward Oldcorne was transferred from the Tower of London to the country gaol in Worcester. He was brought to trial at the Lent assizes, charged with inviting Henry Garnet to Hindlip and with approving of the Gunpowder Plot. He was found guilty – he was far too close to the plot's central characters for this verdict to have been in doubt – and on 7 April he was led the mile out of Worcester to Redhill, where, in the company of his servant Ralph Ashley, he was hanged, drawn, and quartered. Thomas Habington, Hindlip's owner, though also condemned, received a State pardon. He lived on to become a well-regarded antiquarian, dying in 1647 at the age of 87.

On 15 December 1924 Edward Oldcorne and Ralph Ashley were beatified by Pope Pius XI, as part of a long-running campaign within the Catholic Church formally to recognize its English martyrs; on 25 October 1970, forty of those one hundred and thirty-six beatified martyrs received canonization. Among their number were Cuthbert Mayne, Edmund Campion, Henry Walpole, Margaret Clitherow, Robert Southwell, and Nicholas Owen. Though Henry Garnet's name was put forward for consideration, the evidence for his case was found to be insufficient to permit official acknowledgement of his martyrdom. Centuries on, the Gunpowder Plot continues to cast a long, almost impenetrable shadow over his reputation.[12]

With no obvious word or gesture from Rome in Garnet's defence – either in the months immediately following his execution or, indeed, subsequently – it was left to various Catholic writers to try to clear the Jesuit's name, and various Protestant ones to try to tarnish it further. It was a dramatist, though, who is generally believed to have given Garnet his most widely spoken epitaph. The precise dating of *Macbeth* is unclear, though it seems certain that Shakespeare was working on the text between the years 1603 and 1606. The likely reference to Garnet comes in the one comic scene in this dark and most dreadful of plays. Macbeth and his wife, just come from murdering King Duncan, are disturbed by a vigorous knocking at the gate; a drunken porter stumbles sleepily to answer its summons and as he goes he imagines he is gatekeeper in hell, welcoming in the newcomers:

> Knock, knock. Who's there . . . [in the] devil's name?
> Faith, here's an equivocator, that could swear in
> both the scales against either scale; who committed
> treason enough for God's sake, yet could not
> equivocate to heaven: O! come in, equivocator.

In the same speech the porter refers to 'a farmer, that hang'd himself on the expectation of plenty'. Farmer was one of Garnet's aliases.[13]

Playwrights and pamphleteers were among the Jesuits' most bilious critics in the years following the plot. 'If about Bloomsbury or Holborn,' warned one writer in 1624, 'you meet a good smug fellow in a gold-laced suit, a cloak lined through with velvet, one that has gold rings on his fingers, a watch in his pocket, which he will value above 20 pounds, a very broad-laced band, a stiletto by his side, [and] a man at his heels . . . then take heed of a Jesuit.' Another advised that Jesuits could transform 'themselves into as many shapes as they meet with objects', to 'beguile' the unwary. Seventeenth-century audiences flocked to the Globe Theatre to watch Thomas Middleton's scurrilous satire *A Game at Chess*, which featured a host of Jesuit villains (and an Induction delivered by 'Ignatius Loyola'). Eighteenth-century audiences laughed their way through Oliver Goldsmith's *The Good-Natured Man*, with its comic references to 'a damned jesuitical, pestilential plot', and through the even less subtle *The Wanton Jesuit*, in which they learned that:

A Jesuit is a clever man
When a maid comes to confession,
He first does absolve, and next trepans
And brings her to oppression.
Then he kisses
And does all he can,
To multiply transgression.

The Jesuits, went the myth, were liars, lechers, and, of course, king-killers. Over the course of its history the Society has variously been accused of plotting the murders – some realized, some not – of one Dutch head of State, three French, four English, and no fewer than five American presidents, including Abraham Lincoln.* It seems only the assassination of the Catholic John F. Kennedy has not been laid at its door. Few other organizations have been so widely despised and in such lurid and hysterical fashion (obscuring both the Jesuits' extraordinary achievements and whatever genuine censures they might deserve). But then, as Robert Persons reported from England back in 1581, just months into his mission, there was already 'tremendous talk here of the Jesuits and more fables perhaps told about them than we were told of old about monsters'.[14]

The English Jesuit mission continued after Garnet's execution,

* These were William of Orange; Henri III, Henri IV, and Louis IV of France; Elizabeth I, James I, Charles I, and Charles II of England; and Presidents Harrison, Taylor, Garfield, McKinley and Lincoln.

riding out the worst storms of the post-plot persecution. The last person to be executed under the 1585 act (making entering England as a Catholic priest an act of treason) was the Welsh Jesuit David Lewis, a victim of Titus Oates's accusations, killed on 27 August 1679. Thereafter King Charles II, independently of Parliament, ordered that no priest should suffer this fate again. The Revolution in France, a century later, sent many Catholic institutions running from its shores for safety, among them the Robert Persons-established English boys' school at St Omer. This took advantage of the first Catholic Relief Acts and moved to Lancashire in 1794, to the Stonyhurst estate of a former pupil, where it has remained ever since. When Catholics were permitted full freedom of worship in 1829, the Society spread freely, building more schools and colleges across the country. In 1896 the Jesuits returned to Oxford, founding a Private Hall there that in time became Campion Hall, a recognized part of Oxford's university. In a note of symmetry, the Hall's first buildings were rented from St John's, Edmund Campion's old college.[15]

Nearly seventy Catholic martyrs from the Tudor and Stuart period came from Oxford University, some fifty of these quitting Oxford during Elizabeth's reign. This last figure makes up just less than half of the one hundred and twenty-four Catholic priests executed under Elizabeth. Thereafter the university was always vulnerable to accusations of popery. It was purged again during the Puritan Interregnum and subjected to heavy scrutiny in Charles II's reign. Still, its students warmly welcomed the Catholic James, Duke of York (soon to be James II) on an official visit in 1683, and when James attempted to emancipate his co-religionists during his short reign, University College was quick to open a Catholic chapel. Curiously, educators looking to reorder Oxford's teaching at this period would pay close attention to the Jesuit schools and colleges on the Continent: 'the Jesuits,' wrote one Oxford graduate admiringly, 'do breed up their youth to oratory the best of any in the world.'[16]

Over the centuries the religious checks and balances keeping

Oxford conformist remained in place: a survey conducted in the mid-1820s found that Oxford and Cambridge were the only two universities in the world so generally to test the faith of their students. As late as 1854 Oxford undergraduates were still required to subscribe to the Thirty-Nine Articles at matriculation. This did not stop one group of early nineteenth-century Oxford scholars from hankering after the 'awe, [and] mystery' of the medieval Church, in the words of their leader, John Henry Newman of Oriel. Soon the group, which would become known as the Oxford Movement, had adopted many of the doctrines and rituals of Catholicism, provoking a flurry of agitated letters accusing its members of 'designing the reintroduction of Popery'. In time several of these scholars would go over to Rome, including Newman. Later, he would be made a cardinal, just like that other Oriel man, William Allen. The novelist Evelyn Waugh, who was also a Catholic convert (embracing 'the Scarlet Woman' in the parlance of the day), would be taught by Oxford's first Catholic don since the Reformation – this in the early 1920s.[17]

If changes to the law releasing Catholics from their pariah status were slow in coming, then the popular attitude towards Catholicism seems still not quite to have caught up. On an evening in March 1998 the Prime Minister, Tony Blair, was seen alone in Westminster Cathedral, England's foremost centre of Catholic worship. The press seized upon the story with extraordinary vigour. Was Blair thinking of converting? Could Britain have a Catholic Prime Minister? 'Could a modern, democratic leader really declare that he orders his spiritual life within such a dogmatic framework [as the Catholic Church]?' This last question was asked by the Anglican Bishop of Rochester, sparking a wider discussion about the conservative, autocratic nature of the Roman Church versus inherent British liberalism. It was as though the voices of long dead Tudor, Stuart and Hanoverian MPs, churchmen, and pamphleteers had suddenly crackled into life again and centuries on were venting their old bias. Their convictions then had been that the Catholic Church was a 'foreign arbitrary power', asking for 'implicit faith and blind obedience' (both of which were incompatible with 'Civil Society'), and that Britain, with its 'mildest laws in the universe', was 'the main bulwark against [such] arbitrary encroachments'. Their modern counterparts appeared to share many of these same sentiments.[18]

In a recent *Times* article about whether a Catholic Secretary of State could head a department responsible for stem-cell research, the author made little attempt to address the subject of conflict of interest, preferring to highlight the 'zealotry and fervour' of the spiritual organization Opus Dei, with which the politician in question is supposed to have a connection. At the foot of the piece, in bold typeface, was the phrase 'so zealous a militant Catholic'. It is unclear whether these comments about Catholics and Catholicism constitute part of a new debate, informed by Britain's increased secularism, or whether they belong to an older one, informed by a pattern of centuries-old fear and prejudice. What does come across, though, is that the language of the debate seems suffused with a hostility disproportionate to the subject being debated. At the time of writing the Downing Street Press Office is still denying rumours that Blair is on the verge of converting. A pre-election pledge made by the Prime Minister in June 2001, to review the 1701 Act of Settlement by which Catholics are debarred from acceding to the British throne, appears to have stalled, amid concerns about how this might affect the relationship between the Church of England and the State.* Meanwhile, the media continues to preserve the myth that England's Catholic missionaries were trained assassins – witness director Shekhar Kapur's 1998 film *Elizabeth* and the BBC's documentary on Elizabeth for its 2002 *100 Great Britons* series, both of which made reference to killer-priests.[19]

* Curiously, John F. Kennedy was called before a meeting of Protestant ministers to reassure them that he would not become a Vatican puppet if elected to the presidency. His response appears eminently Elizabethan in its essential ambiguity and in its recognition of a divided loyalty: 'If my Church attempted to influence me in a way which was improper or which affected adversely my responsibilities as a public servant, then I would reply to them that this was an improper action on their part and that it was one to which I could not subscribe.'

Unsurprisingly, the country's Catholics remain locked in their 'otherness', in that sliding scale of self-definition required of them when England and Rome first split. When the Earl of Denbigh converted to Catholicism in the nineteenth century, he had his coat of arms altered to declare: 'First a Catholic, then an Englishman'. The MP Ann Widdecombe (who converted to Catholicism in 1993) described herself, in conversation with the author Dennis Sewell, as 'a Catholic, British, Conservative, woman from Kent . . . in that order'. Evelyn Waugh summed the position up in his novel *Brideshead Revisited*, in which the Catholic Sebastian Flyte confirms that Catholics are simply not 'like other people'. The Elizabethan propagandists' victory, in turning England's Catholics into a sub-species in their own country, is not diminished by the fact that it is Catholics who now label themselves as 'different', rather than the State.[20]

It is a truism that history books say as much, if not more, about the period in which they are written than they do about the period *about* which they are written. We are creatures of our time, moulded by the multifarious ideas and images thrust our way, and by the preoccupations of the day; we cannot look at the past save through the prism of our own immediate present. I have been struck again and again during the writing of this book by the parallels between this period (and the events contained therein), and our own. Yet I have been wary of making such parallels explicit, believing that to do so would be an act of disrespect towards the men and women about whom I have been writing. Nonetheless, to ignore them also seems to be, at some level, an act of negligence.

On 8 November 2001 the Harvard law professor and prominent civil liberties lawyer Alan Dershowitz wrote an article in the *Los Angeles Times*, arguing in favour of 'torture warrants': mandates, issued on a case-by-case basis by a US high court judge, permitting the use of torture on a detainee (in fact, 'torture warrants' precisely akin to those issued by England's Privy Council in the sixteenth and seventeenth centuries). In December 2001 the hastily compiled new British Anti-Terrorism, Crime and Securities Act made permissible the indefinite detention without trial of certain suspects. On 11 August 2004 the British Court of Appeal ruled that evidence extracted under torture in third countries was admissible as evidence, provided that the UK Government had 'neither procured the torture nor connived at it'. On 20 November 2004 the *New Scientist* magazine featured a wide-ranging discussion about the accuracy of evidence elicited under torture. These four instances fit into a wider pattern of debate about the acceptable treatment of potential terrorists, as the West seeks to respond to the trauma inflicted upon it by the events of 11 September 2001. Once again the country stands trembling at the spectre of young men of a contrary religion, trained in martyrdom, hurling themselves at these shores. There are as many dissimilarities between these two situations as there are similarities. In the former category one can place motive and means: the myth of the assassin-priest has been replaced by the reality of the suicide-bomber. But the terms of engagement with the problem remain much the same – and they revolve around that one loaded word 'potential'. For in this clash of ideologies, in which the battle-lines have been identified, in an act of risky over-simplification, as purely faith-based, how does one distinguish those members of that contrary faith engaged in the conflict from those members with no such designs?[21]

It is the unenviable task of the British Government to find a method of distinguishing combatant from non-combatant, within an acceptable legal and ethical framework. It is the even less enviable task of British Muslims to see any overt denotement of their faith and culture taken as evidence of their disloyalty to the State, and every gesture of dilution rewarded with the sobriquet 'moderate', meaning loyal; to see suspicion in every glance; to see their co-religionists held without trial, no demonstrable evidence brought against them; to see an identity being forced upon them, without their having any say in the matter. 'How can a man truly swear that he does abjure a position which he never held?' asked English Catholics of the Oath of Allegiance. We seem in danger of asking British Muslims to do similarly.[22]

We are closer to our sixteenth-century forebears than we might care to admit, in our willingness to assume that the values by which we order our lives are incontestable. We are no less likely than they were to inflict suffering on any given minority of our population. Indeed, it might be argued that we have now factored the probability of minority suffering into our ethical decision-making: utilitarianism, the perfect philosophy for the politician, depending as it does on majority consensus, now appears to have become our default moral position in every crisis. And our rampant defence of the majority good – it is this that makes the use of torture justifiable, argues its new wave of defenders – permits all too much room for minority pain. More worrying still, it also gives unwarranted scope for majority fear to dominate the process of moral reasoning.

If these thoughts seem to sit uneasily at the tail end of what anyone would, correctly, regard as a popular history book, then I should explain that I offer them only because they have dominated so much of the process of this book's writing.

They had come home as missionaries, most fresh out of seminary college: young men yearning to save their country from the 'heresy' into which it was plunged, Rome's army of arguers, burning with the force of their rhetoric and the certainty of their beliefs. Some were idealists; some unsure what else to do with themselves in a country bent on denying them advantage. Some were hopeful; some disaffected. Some longed to die; some sought only the stability of tradition. Some were regarded as the most able men of their generation; some were plodders, whose quiet labours went entirely unrecorded.

Their Government had termed them spies and assassins, secret agents of the enemy, complete with the trappings of their dubious profession, false papers, aliases, disguises, and ciphers; and as such it had hunted them down. 'Shall no subject that is a spy . . . against his natural prince be taken and punished as a traitor, because he is not found with . . . a weapon, but yet is taken in his disguised apparel with . . . other manifest tokens to prove him a spy for traitors?' This Sir William Cecil had asked back in 1583 – and it was a good question. How could you tell apart the man who behaved like a secret agent *and was a secret agent*, from the man who behaved like a secret agent, but was a man of God (even if you, yourself, had forced that mode of behaviour upon him by your laws)?[23]

No doubt some of the missionaries grew to share the same sense of seething resentment felt by many of the Catholic laity with whom they consorted; no doubt some were privy to information, plans, and plottings, to which they had no right. But, as Oswald Tesimond asked, 'How many things do priests know of which they do not approve?' And Henry Garnet would write to Anne Vaux: 'who can hinder but he must know things sometimes which he would not.'[24]

Indeed, what is surprising, on examination of their story, is how few of the missionaries can be held accountable for anything other than their State-forbidden priesthood, a fact recognized by Catholic commentators of the day. A letter of *c.*1592 makes the point clearly: 'If some priests have fallen, yet can it not be much marvelled at, considering the rigour of the persecution: but, sure, it is a manifest miracle, that, among so many, so few scandals have risen . . . [For] their attire, conversation, and manner of life must here, of force, be still different from their profession; the examples and occasions that move them to sin, infinite: and therefore, no doubt, a wonderful goodness of God that so few have fallen.'[25]

Religion has been used to justify too many acts of inhumanity to enumerate. Robert Catesby, said Garnet, 'was so resolved . . . that it was lawful . . . to take arms for religion, that no man could dissuade [him]'; paradoxically, Catesby needed to claim Jesuit backing for his plot every bit as much as did the Government. But fear has inspired just as many acts of inhumanity. Elizabeth's Government was, with reason, supremely fearful of the Catholic Church; James's inherited that fear. Both preached regular sermons on their own essential decency and reasonableness (in marked contrast to what they perceived as the king-killing doctrines of Rome), both endorsed State-sanctioned acts of inhumanity: forced internments, show trials, revenge punishments, and the erosion of the common law. The argument goes that it is reductive to judge the past by the standards of today. Still, that does not mean we cannot examine the choices made in a fearful and uncertain past, better to evaluate those available to us in a fearful and uncertain present.

Appendix

Of those houses mentioned in this book several are open to the public. Oxburgh Hall is near King's Lynn in Norfolk; Baddesley Clinton is near Knowle in Warwickshire: both are managed by the National Trust. Coughton Court, near Alcester in Warwickshire, also managed by the Trust, has a permanent exhibition detailing the house's links with the Gunpowder Plot. Stonor Park, near Henley-on-Thames, has a permanent exhibition illustrating the life of Edmund Campion, and visitors there can see the rooms believed to have housed the press on which Campion's *Decem Rationes* was printed. Sadly there have been casualties among the houses no less than among their owners. Hindlip Hall is now the site of the headquarters of West Mercia Constabulary; Harrowden Hall is home to Wellingborough Golf Club.

Lastly, I'd like to draw the reader's attention to a house not featured in this book, but worth visiting. This is Harvington Hall, near Kidderminster in Worcestershire. Harvington belonged to Humphrey Pakington, a recusant and a close friend of Thomas Habington at nearby Hindlip. This friendship makes it likely that Pakington was known to the Jesuits, and still extant at Harvington is a cluster of hiding-places believed to be by Nicholas Owen.

The hides are situated around the massive Great Staircase, the design of which dates from about 1600. Given the upheaval to the household that its construction would have caused, it makes sense to suppose that it and its surrounding hides were built of a piece (the former providing cover for the latter), some time after this date.

Climb the Great Staircase at Harvington today, to the top landing, and before you is a set of five steps leading up to what is known as the Nine Worthies Passage.* Place your fingers under the top two treads of these

* The existing staircase is a replica: the original was dismantled and moved to Coughton Court in 1910.

steps and they hinge back to reveal a small triangular hide, suitable for books and massing equipment. In the far wall of this hide is a gap, once covered by a secret door (probably camouflaged to look like brickwork), through which you can climb to a larger, man-sized hide beyond.

Carry on along the passage and you come to the Marble Room, with a triangular fireplace built into the far corner. This fireplace has no chimney-stack beneath it and extends only as far as the ceiling: it is entirely false. If you were to climb up it, through its carefully blackened surrounds, you would enter a bewildering maze of attics above, with a second hide built into the end garret.

Below the Marble Room is Dr Dodd's Library. At the far end of this library, opposite the window, is a small raised stage, once used as a book-cupboard. The back wall of it is panelled; the sides would once have been panelled too, but now they consist of bare brick and upright timber beams. In the darkest recesses of this stage is a timber beam, the end of which can be raised. It pivots open to reveal a hide beyond, eight foot long, three foot wide and five foot high – luxury in terms of hiding-places – with a small wooden joint-stool as furniture. This stool is too big to pass through the ten-inch-wide entrance to the hide, so it must have been built, or assembled, *in situ.* The hide was discovered by accident in 1897 by boys playing in the, then, derelict hall. It and its companion pieces are some of the very finest hiding-places ever to have been built. They provide enduring evidence of the genius of Nicholas Owen.

Endnotes

CHAPTER ONE

1 Neale, II, pp. 108–9; Mattingley, p. 159; Somerset, p. 451.
2 Mattingley, pp. 159–60.
3 *Ibid.*, p. 161.
4 *Ibid.*
5 *Ibid.*, pp. 166–7; Somerset, pp. 451–2; Neale, II, p. 193.
6 Milne-Tyte, pp. 10, 14, 23; Somerset, p. 446.
7 Milne-Tyte, pp. 17, 18, 29; Wernham, p. 342.
8 Milne-Tyte, pp. 45, 129–30; Mattingley, p. 166.
9 Milne-Tyte, p. 10; Somerset, p. 458.
10 Milne-Tyte, pp. 32–33.
11 Somerset, p. 458; Milne-Tyte, pp. 18, 41; C.S.P. Spanish, IV, pp. 373, 479–83.
12 *Harleian Miscellany*, I, pp. 119ff.; Somerset, pp. 458–9; Milne-Tyte, p. 84.
13 Milne-Tyte, pp. 84, 90, 101–2, 151.
14 Milne-Tyte, pp. 112–13, 145; C.S.P. Spanish, IV, pp. 419, 479–83.
15 Milne-Tyte, pp. 107–8.
16 C.S.P. Spanish, IV, p. 493; Caraman, *Garnet*, pp. 82–3; *Harleian Miscellany*, I, pp. 132–3.
17 *Harleian Miscellany*, I, p. 115; Davies, *Europe*, p. 520; MacCulloch, pp. xix–xx.
18 Somerset, p. 467; Caraman, *Garnet*, p. 83.
19 Gerard, p. 9 and note; Morris, *Condition*, p. 280; Jessopp, p. 161; Elton, p. 370.
20 Gerard, pp. 9ff.; Morris, *Condition*, pp. 280ff.
21 Gerard, p. 1; Morris, *Condition*, pp. ix–xi.
22 Gerard, p. 2.
23 Davies, *Europe*, p. 496; Gerard, pp. 2–3.
24 Gerard, p. 3; S.P. Domestic 12/clxviii/35.
25 Gerard, pp. 3–4.
26 Gerard, pp. 4–5.
27 Gerard, p. 6.
28 Gerard, pp. 7–8, 213 note; S.P. Domestic 12/cxcix/95–6, 12/ccxvii/3, 12/ccxvii/81; Devlin, *Life*, p. 248.
29 Gerard, p. 8.
30 Neale, II, pp. 37ff.; Tierney, III, pp. xxxiii–xxxvii; Aydelotte, pp. 54–5.
31 Gerard, pp. 10–11.
32 Gerard, pp. 11–12; Jessopp, p. 162.
33 Gerard, pp. 12–13.
34 Gerard, p. 13; Jessopp, p. 162.
35 Gerard, pp. 13–14; Jessopp, pp. 162, 225.
36 Gerard, p. 14.
37 *Ibid.*, pp. 14–15.
38 *Ibid.*, p. 15; Jessopp, p. 164.
39 Strype, III, pp. 87–92; Devlin, *Life*, p. 175.
40 Devlin, *Life*, p. 172.

CHAPTER TWO

1 V.C.H. Oxford, IV, p. 74; Parish Registers of St Peter-le-Bailey; Salter, *Properties*, p. 203; Salter, *Survey*, II, pp. 120–22; Hodgetts, *Owens*, p. 427; Hogge, pp. 291ff.
2 Hogge, pp. 291–2; V.C.H. Oxford, IV, pp. 86, 401; Salter, *Survey*, II, p. 123; Salter, *Names*, pp. 11–12.
3 R.C.H.M. Oxford, pp. 156–8; V.C.H. Oxford, IV, pp. 296–300.
4 V.C.H. Oxford, IV, pp. 85, 365; R.C.H.M. Oxford, p. 155.

5 McCoog, p. 40; McConica, p. 364.
6 McCoog, p. 40; McConica, p. 125.
7 McCoog, p. 40; McConica, p. 125.
8 McCoog, p. 40; McConica, p. 128.
9 McConica, p. 365; Rashdall, pp. 106–7.
10 V.C.H. Oxford, IV, p. 12.
11 McConica, pp. 368–72, 404; McCoog, pp. 43–4.
12 McCoog, p. 45; Fowler, p. 97.
13 McCoog, pp. 47–9; V.C.H. Oxford, III, pp. 20, 238, 251.
14 Fowler, pp. 97, 357; McCoog, p. 49; McConica, pp. 375, 377.
15 Wernham, pp. 99, 181, 208; S.P. Domestic 11/I/7.
16 Plowden, p. 25.
17 Tierney, II, pp. ccxxix, 120.
18 Neale, I, p. 44; Plowden, p. 18.
19 Elton, p. 264; Plowden, p. 18; Tierney, I, pp. 176, 186–7; Neale, I, p. 57.
20 Neale, I, pp. 35, 83; C.S.P. Spanish, I, p. 64; Sander, p. 282 note; Plowden, p. 36.
21 Sander, p. 244.
22 Neale, I, pp. 78–9; Sander, pp. 268, 283–5; Duffy, pp. 567–8.
23 McCoog, p. 251; Plowden, p. 48.
24 Gerard, pp. 18–19 and note, 221 note.
25 C.S.P. Spanish, I, pp. 217–18; V.C.H. Oxford, IV, p. 412; McCoog, p. 54; Rowlands, p. 157.
26 McConica, p. 405; Tierney, II, p. 143.
27 C.S.P. Domestic 1547–1580, p. 186; Rashdall, pp. 111–15; McConica, pp. 381, 408; McCoog, pp. 50–51; V.C.H. Oxford, III, p. 159.
28 Somerset, pp. 122–3.
29 C.S.P. Spanish, I, pp. 156, 217–18; McCoog, p. 51.
30 C.S.P. Spanish, I, p. 303.
31 McConica, pp. 397ff.; Duncan-Jones, pp. 35ff.; Wallace, pp. 60–64; A. Wood, *History*, pp. 154ff.; Plummer, pp. 197ff.
32 C.S.P. Spanish, I, pp. 178–9.
33 Simpson, *Campion*, pp. 12–14; McCoog, p. 87; Watkins, pp. 83, 89.
34 Jessopp, p. 102.
35 Haile, pp. 2ff, 77; Clancy, p. 8.

CHAPTER THREE

1 Sander, p. 261; Tierney, III, pp. 7–8.
2 McCoog, p. 251 note.
3 Plowden, pp. 89–91; Haile, p. 108.
4 Neale, I, pp. 185–6.
5 Davies, *Europe*, pp. 496–502; Clancy, p. 7.
6 Tierney, III, p. 87.
7 Challoner, p. 5; Allen, pp. 108–9; Haile, p. 140.
8 Haile, p. 119.
9 *Ibid.*, pp. 34–5, 47.
10 Plowden, pp. 59–60; Haile, p. 58; Knox, *Letters*, p. 21.
11 Haile, pp. 71–5.
12 *Ibid.*, pp. 77–80.
13 Plowden, p. 63.
14 Neale, I, pp. 116–17.
15 Magee, pp. 31–2.
16 S.P. Domestic 12/xxi/10, 12/xxiii/9; *Richard II*, I.iii.176.
17 Gerard, p. 1; Cook, p. 42; Devlin, *Life*, p. 17.
18 S.P. Domestic 12/xxiii/9; Devlin, *Life*, p. 10.
19 Devlin, *Life*, p. 16.
20 Knox, *Diaries*, p. lxxvi; Haile, pp. 80ff.
21 Haile, pp. 127, 128, 185.
22 *Ibid.*, pp. 134–7, 144.
23 Knox, *Diaries*, pp. lvii–lviii; Munday, *Life*, p. 12.
24 Munday, *Life*, pp. 12ff.
25 *Ibid.*, pp. 15, 28; Porter, p. 13.
26 Munday, *Life*, pp. 21–7; Haile, p. 82.
27 Haile, pp. 149–50, 151, 154; Knox, *Letters*, pp. 178–9, 247–8, 264–5.
28 Haile, p. 146; Knox, *Diaries*, p. xlvi; C.S.P. Spanish IV, p. 37.
29 Haile, p. 140.
30 Allen, p. 105.
31 *Ibid.*
32 Haile, p. 140; Challoner, p. 5; Allen, pp. 105–6.
33 Haile, p. 141; Challoner, p. 5; Plowden, p. 121.
34 Allen, pp. 107–108; Haile, p. 141; Challoner, p. 6.
35 Haile, p. 141.

36 Clancy, p. 140; Simpson, *Campion*, pp. 69–70; Knox, *Diaries*, p. lxxxii; McCoog, p. 293 and note.
37 Simpson, *Campion*, p. 7.

CHAPTER FOUR

1 Simpson, *Campion*, pp. 378–9; Haile, pp. 85–6.
2 Simpson, *Campion*, pp. 1ff., 28–9; Haile, p. 86.
3 Simpson, *Campion*, pp. 73–4.
4 Davies, *Europe*, p. 496.
5 E. Waugh, *Campion*, p. 69.
6 Plowden, pp. 145–6; Harmsen, p. 9.
7 Plowden, pp. 152–3; E. Waugh, *Campion*, pp. 70–71.
8 C.R.S. 39, p. 4; E. Waugh, *Campion*, p. 102.
9 C.R.S. 39, pp. xii–xiii, 319–21; E. Waugh, *Campion*, pp. 72–3.
10 Simpson, *Campion*, pp. 151, 154–6.
11 E. Waugh, *Campion*, pp. 75–9.
12 C.R.S. 39, p. xiii; Simpson, *Campion*, pp. 143, 146; E. Waugh, *Campion*, p. 80; Meyer, p. 271.
13 E. Waugh, *Campion*, pp. 80–81; Simpson, *Campion*, pp. 148–9.
14 Somerset, pp. 321, 325.
15 Wernham, pp. 333, 358; Somerset, p. 326.
16 C.R.S. 39, p. xiii.
17 Simpson, *Campion*, p. 170; Plowden, p. 149.
18 McCoog, p. 68; E. Waugh, *Campion*, p. 82.
19 Simpson, *Campion*, pp. 174–6; Plowden, p. 155.
20 C.R.S. 39, p. xv; E. Waugh, *Campion*, p. 99.
21 C.R.S. 39, p. xv; Foley, III, p. 661.
22 Simpson, *Campion*, pp. 173–4, 223, 297; Simpson, *Associations*, p. 28.
23 Simpson, *Campion*, p. 223; Simpson, *Associations*, p. 28; C.R.S. 39, p. 84; Foley, III, pp. 658–704.
24 C.R.S. 39, p. 44.
25 E. Waugh, *Campion*, p. 82.
26 Knox, *Diaries*, pp. lxviii–lxix; E. Waugh, *Campion*, p. 92.
27 Simpson, *Campion*, p. 177.
28 Holinshed.
29 Holinshed; Hat. Cal. IV, pp. 335–6.
30 Wernham, p. 357.
31 C.S.P. Spanish II, pp. 709–10; Haile, p. 178; Somerset, p. 385; S.P. Domestic 12/cxiv/22; C.R.S. 22, pp. 2ff.
32 Haigh, *Continuity*, p. 51; C.R.S. 60, p. 10; Neale, II, p. 29.
33 E. Waugh, *Campion*, pp. 93–4; Plowden, pp. 157–8.
34 Plowden, pp. 157–8.
35 McCoog, p. 263.
36 Simpson, *Campion*, pp. 179–80.
37 C.R.S. 39, p. 44; E. Waugh, *Campion*, pp. 99–100, 102.
38 C.R.S. 39, p. 57.
39 E. Waugh, *Campion*, pp. 100–102.
40 Simpson, *Campion*, p. 254; C.R.S. 9, pp. 63–9; Knox, *Diaries*, p. lxx; C.R.S. 39, p. 108; V.C.H. Oxford, IV, p. 413.
41 C.R.S. 39, p. xxiii; E. Waugh, *Campion*, pp. 105–106.
42 E. Waugh, *Campion*, p. 110; C.R.S. 39, p. 45.
43 Plowden, p. 166; C.R.S. 39, p.xxx; Allen, p. 39.
44 Neale, I, pp. 383–4.
45 Tierney, III, p. xxiv; Neale, I, pp. 386–9.
46 Simpson, *Campion*, pp. 310–11.
47 *Ibid.*, p. 312.
48 Munday, *Answer*, pp. 1ff.; Pollard, pp. 458, 461.
49 Pollard, p. 461.
50 Pollard, pp. 462–3 and note; McCoog, pp. 124–5; Hodgetts, *Recusant History*, p. 151.
51 Pollard, pp. 454, 459.
52 E. Waugh, *Campion*, pp. 130–31.
53 Tierney, III, p. 151.
54 E. Waugh, *Campion*, pp. 149–50.
55 Pollard, pp. 454–5; E. Waugh, *Campion*, pp. 157–8.
56 Simpson, *Campion*, p. 135.
57 Plowden, p. 185.
58 Allen, p. 14; C.S.P. Spanish III, p. 231.
59 Simpson, *Campion*, pp. 378–9; E. Waugh, *Campion*, p. 147.

CHAPTER FIVE

1 Jones, p. 24; Devlin, *Life*, p. 240.
2 Caraman, *Garnet*, pp. 68–9, 123–6.
3 Caraman, *Garnet*, p. 81; Gerard, p. 16–17.
4 Caraman, *Garnet*, pp. 1ff.
5 Caraman, *Garnet*, p. 7; Sander, p. 369.
6 Caraman, *Garnet*, pp. 52–3, 68, 78, 80, 82; Devlin, *Life*, p. 175.
7 Caraman, *Garnet*, p. 86.
8 *Ibid.*; Devlin, *Life*, pp. 3ff.
9 Devlin, *Life*, pp. 23ff.
10 Caraman, *Garnet*, p. 65.
11 Jessopp, p. 179; C.S.P. Spanish III, p. 471; A.P.C. XV, p. 333; S.P. Domestic 12/ccviii/37.
12 Gerard, p. 17.
13 V.C.H. Norfolk, pp. 473, 486, 488.
14 *Ibid.*, pp. 253–4, 493, 500.
15 *Ibid.*, pp. 498–9; Magee, p. 27; Somerset, pp. 44–6; *Oxburgh*, p. 38.
16 Jessopp, pp. 164, 179; Gerard, p. 18; Simpson, *Associations*, p. 26.
17 Gerard, pp. 15, 17.
18 Ross, p. 51; Caraman, *Garnet*, p. 123 note 3.
19 *Shrew*, 2.i.79–81; Jessopp, p. 166; Hat. Cal. XI, p. 365; Gerard, pp. 17–18.
20 C.R.S. 39, pp. 331ff.; Simpson, *Campion*, pp. 297–8.
21 *Ibid.*; Gerard, p. 18.
22 Gerard, pp. 18–19; Jessopp, p. 151.
23 Devlin, *Life*, p. 3.
24 Jessopp, pp. 223–4, 238.
25 Nichols, *Elizabeth*, II, pp.115–16, 215–19; V.C.H. Norfolk, p. 269; Jessopp, p. 95.
26 S.P. Domestic 12/cxiv/22; Read, *Walsingham*, II, p. 283 note.
27 Read, *Walsingham*, II, p. 273 note.
28 S.P. Domestic 12/clxxxv/83; Caraman, *Garnet*, p. 51; Southwell, p. 43.
29 A.P.C. XIV, pp. 87–8; S.P. Domestic 12/clxxxiii/25, 12/clxxxiv/46, 12/clxxxv/3, 12/clxxxvii/48; Jessopp, pp. 226ff., 245.
30 *Harleian Miscellany*, III, pp. 93ff.; Neale, II, pp. 56–7, 176; Somerset, p. 489.
31 Magee, p. 24.
32 Pritchard, p. 51.
33 Caraman, *Garnet*, p. 21; Gerard, p. 33; Clancy, p. 132; C.R.S. 39, pp. xii–xiii; Basset, pp. 113–14.
34 Girouard, p. 82; Haile, pp. 98–9; Morris, *Condition*, p. 155.
35 Misc. Vol., A.5.3., *Inrolment of Apprentices 1514–1591*; *Shakespeare's England*, p. 119.
36 Morris, *Condition*, p. 182.
37 Tierney, III, pp. xxxiii–xxxvii; Neale, II, p. 38; Rogers, p. 4.
38 S.P. Domestic 12/clxvii/3, 12/clxxviii/39, 12/cxcv/21, 12/ccix/19; Weston, pp. 22–3, 79; Morris, *Two Missionaries*, pp. 6ff, 196.
39 Weston, pp. 66ff.; Caraman, *Garnet*, pp. 32–3.
40 Weston, p. 72; Hodgetts, *Secret Hiding-Places*, p. 9; Morris, *Two Missionaries*, p. 142.
41 Weston, pp. 44–6; Foley, 1, pp. 379–82.
42 Caraman, *Garnet*, p. 45; Simpson, *Associations*, p. 27.
43 C.R.S. 39, p. 332.
44 Morris, *Condition*, pp. 182, 184.
45 *Ibid.*, pp. 183–4, 187.
46 Hodgetts, *Recusant History*, XII, pp. 171–97.
47 S.P. Domestic 12/ccxxx/76.
48 E. Waugh, *Campion*, pp. 101–102.

CHAPTER SIX

1 *Harleian Miscellany*, III, pp. 93ff.; Neale, II, p. 176; Hasler, pp. 513–15.
2 Tierney, III, pp. 156–8; C.R.S. 5, pp. 211–12; Caraman, *Garnet*, p. 107.
3 Nichols, *Elizabeth*, II, p. 219; Jessopp, p. 111.
4 C.S.P. Domestic 1547–1580, p. 575; Caraman, *Garnet*, p. 107.
5 Neale, I, pp. 121, 391; Read, *Walsingham*, II, pp. 283–4.
6 Devlin, *Life*, p. 211; Neale, I, pp. 191–2; Porter, pp. 13, 18, 24–5.
7 Simpson, *Campion*, p. 440; Devlin, *Life*, pp. 212–13.
8 Devlin, *Life*, pp. 213–14.

9 Jessopp, pp. 127ff.; E. Waugh, *Campion*, pp. 166–7.
10 Jessopp, pp. 191ff.; Gerard, pp. 21–2, 27.
11 Jessopp, p. 211; Rogers, p. 11; Anstruther, *Seminary Priests*, I, p. 264.
12 Jessopp, pp. 211–18.
13 *Ibid.*, pp. 218, 255–6.
14 *Ibid.*, pp. 257, 262, 264; S.P. Domestic 12/ccxlvii/21.
15 Jessopp, pp. 277ff.; S.P. Domestic 12/ccxlviii/78, 112, 91, 12/ccxlix/12, 44.
16 Jessopp, pp. 297ff.
17 McCoog, pp. 146–7.
18 Simpson, *Campion*, p. 462; Somers, p. 164; C.R.S. 39, pp. 133–4; Knox, *Diaries*, pp. 197, 362–3; Hodgetts, *Owens*, p. 418; Hogge, pp. 291ff.
19 Ortiz, pp. 19, 61; Carvajal y Mendoza, pp. 15ff.
20 Carvajal y Mendoza, pp. 29–32.
21 Haile, pp. 119–20.
22 C.R.S. 39, pp. 132–3; Anstruther, *Seminary Priests*, I, pp. 36–8; Basset, pp. 69–70.
23 Simpson, *Campion*, pp. 413–14.
24 Morris, *Troubles*, I, pp. 172–3; Tierney, III, pp. 136–7; Anstruther, *Seminary Priests*, I, pp. 56–7.
25 Caraman, *Siege*, pp. 80–81; Caraman, *Garnet*, pp. 53, 222.
26 Caraman, *Garnet*, pp. 26–7, 53, 268–9, 275.
27 *Ibid.*, pp. 204–5.
28 *Ibid.*, pp. 223, 233–4, 296.
29 E. Waugh, *Campion*, p. 161.
30 Abbott, pp. 225ff.
31 Simpson, *Campion*, p. 443.
32 Allen, pp. xiv–xviii.
33 *Ibid.*, pp. 57–62.
34 *Martyrs of Chichester*, pp. 279–84; Caraman, *Garnet*, pp. 159–60, 217–18; Hodgetts, *Owens*, p. 420.
35 Devlin, *Life*, pp. 122–3, 225–6, 274; C.S.P. Domestic 1591–1594, pp. 255, 279, 304, 315, 357, 449, 469; Morris, *Two Missionaries*, pp. 208–9.
36 Basset, pp. 70–71; C.R.S. 39, p. 227; Weston, pp. 1–5; Caraman, *Garnet*, pp. 24–5, 118; Devlin, *Life*, pp. 107–108.
37 Caraman, *Garnet*, pp. 97–8, 118 and note, 119 and note, 183–4.
38 Morris, *Two Missionaries*, pp. 45, 67; Caraman, *Garnet*, pp. 28, 244; Devlin, *Life*, pp. 106, 108; Gerard, p. 17; Jessopp, p. 265; Morris, *Troubles*, I, p. 177, ; S.P. Domestic 12/ccxlix/12, 44.
39 Caraman, *Garnet*, p. 60.
40 *Ibid.*, p. 129.
41 *Ibid.*, pp. 127–8.
42 Gerard, pp. 40–41.
43 Caraman, *Garnet*, pp. 39, 128–9; Hodgetts, *Secret Hiding-Places*, pp. 67ff.
44 Caraman, *Garnet*, p. 129; Gerard, p. 41.
45 Gerard, p. 41; Caraman, *Garnet*, p. 131.
46 Gerard, p. 41–2.
47 *Ibid.*, p. 42.
48 *Ibid.*; Caraman, *Garnet*, pp. 132–3.
49 Gerard, p. 42; Caraman, *Garnet*, pp. 133–4.
50 Gerard, p. 42; Hodgetts, *Secret Hiding-Places*, pp. 68–70.
51 Gerard, p. 42; Caraman, *Garnet*, p. 134.
52 Caraman, *Garnet*, pp. 129–30, 135–6.
53 Caraman, *Garnet*, p. 130; Gerard, p. 42.

CHAPTER SEVEN

1 *King James Bible*, John I.i; S.P. Domestic 12/ccxxxviii/160–63.
2 *Harleian Miscellany*, III, pp. 93–7.
3 Somerset, p. 482.
4 Southwell, p. 1.
5 Anderson, p. 37.
6 *Shakespeare's England*, II, pp. 217–18; Anderson, p. 40.
7 Sander, pp. 307–308 and note.
8 C.R.S. 39, pp. xxxii–xxxiii; E. Waugh, *Campion*, p. 111.
9 C.R.S. 39, p. xxxii.
10 C.R.S. 39, p. xxxiii; Haile, p. 206; E. Waugh, *Campion*, pp. 117–18.
11 Devlin, *Life*, p. 140; Somerset, p. 430.
12 Devlin, *Life*, pp. 134–5, 141.
13 Pritchard, pp. 41–3.
14 Gerard, p. 26; Pollard Brown, p. 123.
15 Devlin, *Life*, pp. 145–7.
16 Devlin, *Life*, p. 147.
17 *Ibid.*, pp. 72, 85, 99.

18 *Ibid.*, pp. 3ff.
19 C.R.S. 5, pp. 321ff.
20 Devlin, *Life*, pp. 3, 6.
21 C.R.S. 5, pp. 308ff., 313; Devlin, *Life*, pp. 123–4; Garnet, *Caraman*, p. 44.
22 Gerard, p. 15.
23 Janelle, pp. 218, 255ff.; Hayward, p. 30.
24 Devlin, *Life*, p. 179.
25 Southwell, pp. 1ff.; Peck, pp. 3ff.; Somerset, p. 113.
26 Southwell, pp. 35, 39; Somerset, pp. 218–19.
27 Southwell, pp. xi–xii, 27; Devlin, *Life*, p. 247; Clancy, pp. 148, 150–51, 158.
28 Caraman, *Garnet*, pp. 141–3; Devlin, *Life*, p. 256.
29 Strype, IV, p. 186; Smith, p. 433; Southwell, p. xii.
30 Caraman, *Garnet*, pp. 294–5.
31 Tierney, III, pp. cxcviiiff.
32 Anstruther, *Seminary Priests*, I, pp. 52–3, 220–21; V.C.H. Middlesex, p. 260.
33 Tierney, III, pp. cxcvii–cxcviii; *Rambler*, VII, pp. 111–15.
34 Tierney, III, pp. cxcvii–cxcviii; *Rambler*, VII, pp. 111–15.
35 *Ibid.*
36 Tierney, III, p. cxcviii; C.R.S. 5, pp. 211–12; Caraman, *Garnet*, pp. 148–9; S.P. Domestic 12/ccxxxviii/179.
37 Caraman, *Garnet*, pp. 148–9.
38 *Ibid.*
39 Strype, IV, pp. 185–6; *Rambler*, VII, p. 111.
40 Strype, IV, pp. 185–6.
41 Devlin, *Life*, p. 287.
42 Devlin, *Life*, pp. 287–8.
43 C.R.S. 5, pp. 211–12; Devlin, *Life*, p. 289.
44 A.P.C. XXIII, p. 71.
45 *Rambler*, VII, p. 111; Tierney, III, pp. cxcviiiff.
46 *Rambler*, VII, pp. 111ff.; S.P. Domestic 12/ccxliii/26.
47 S.P. Domestic 12/ccxlvi/81, 12/ccxlix/31.
48 A.P.C. XXVI, p. 362; *Rambler*, VII, p. 111.
49 Devlin, *Life*, p. 303; Pendry, p. 107.
50 Caraman, *Garnet*, p. 134.
51 Devlin, *Life*, pp. 333–5.
52 Caraman, *Garnet*, pp. 55, 110; Devlin, *Life*, p. 334.
53 Caraman, *Garnet*, pp. 110, 134; Anstruther, *Seminary Priests*, I, pp. 90–91; Allen, pp. 84–5.
54 *Ibid.*, p. 195; C.R.S. 5, pp. 333ff.; Foley, I, pp. 364ff.
55 C.R.S. 5, p. 334.
56 Foley, I, pp. 369–70.
57 Foley, I, pp. 371–2; C.R.S. 5, p. 335.
58 Foley, I, pp. 372–3; C.R.S. 5, p. 336.
59 Simpson, *Campion*, p. 283; Foley, I, p. 373 and note.
60 C.R.S. 5, pp. 313, 318; Foley, I, p. 374.
61 Foley, I, p. 375.

CHAPTER EIGHT

1 *Lansdowne MSS*, 72/48; Jessopp, p. 194; Gerard, pp. 37, 40.
2 Gerard, pp. 164–5, 170.
3 *Ibid.*, pp. 170–71.
4 V.C.H. Suffolk, pp. 24ff.; Magee, p. 27.
5 S.P. Domestic 12/clxxv/75, 12/clxxxviii/38; Gerard, p. 23.
6 Gerard, pp. 24, 222 note.
7 Caraman, *Siege*, p. 38.
8 Caraman, *Garnet*, p. 215; Devlin, *Life*, p. 140.
9 C.R.S. 39, pp. 139–40.
10 Thomas, p. 62.
11 Thomas, pp. 59, 63.
12 Caraman, *Garnet*, p. 113.
13 *Ibid.*, p. 90; Thomas, pp. 42–3.
14 Caraman, *Siege*, p. 28; Gerard, p. 87.
15 Anstruther, *Vaux*, p. 321.
16 Parish registers for St Peter-le-Bailey, Oxford, 2 and 10 October 1585, 3 May 1590; C.R.S. 60, p. 222; Hodgetts, *Owens*, pp. 418–19.
17 Rowlands, pp. 157, 158, 161.
18 Somerset, p. 59; C.S.P. Scottish, pp. 602–603; Simpson, *Campion*, p. 20.
19 Pritchard, pp. 44ff.; Milne-Tyte, p. 98.
20 Pritchard, pp. 45–6; A.P.C. XXIV, pp. 328–9.
21 Devlin, *Life*, p. 219; Hodgetts, *Owens*, p. 418; Birt, p. 521.

22 Gerard, p. 169 and note.
23 *Ibid.*, pp. 161–3; V.C.H. Buckinghamshire, pp. 401–402.
24 Rowlands, p. 151.
25 A.P.C. XV, p. 368; Rowlands, p. 153; S.P. Domestic 12/clxxxv/17.
26 S.P. Domestic 12/ccviii/19, 58, 66; *Lansdowne MSS*, 72/48; S.P. Domestic 12/ccxlviii/9.
27 S.P. Domestic 12/clxxxiv/8, 12/cxcv/116; Rowlands, p. 152.
28 Neale, II, pp. 280ff.
29 S.P. Domestic 12/ccxliv/75; Neale, II, pp. 282, 283, 293–5; Tierney, III, p. xxxviii.
30 Jessopp, p. 254.
31 Morris, *Troubles*, III, pp. 413, 417; Abbott, pp. 313–16.
32 Morris, *Troubles*, pp. 430, 432.
33 Gerard, p. 28.
34 *Ibid.*, pp. 25, 27, 224 note.
35 *Ibid.*, pp. 28–9.
36 *Ibid.*, pp. 29, 224 note.
37 V.C.H. Essex, pp. 20, 21, 31ff.; Gerard, pp. 33–6, 226 note.
38 S.P. Domestic 12/ccxliii/95, 12/ccxliv/7.
39 S.P. Domestic 12/ccxlviii/103.
40 Caraman, *Garnet*, p. 171.
41 S.P. Domestic 12/ccxlvii/3; Gerard, pp. 51–2.
42 Caraman, *Garnet*, p. 185.
43 Gerard, pp. 54, 233 note; Caraman, *Garnet*, p. 187; S.P. Domestic 12/ccxlviii/31, 37–40.
44 S.P. Domestic 12/ccxlviii/36; Gerard, pp. 55–7.
45 Gerard, pp. 58ff.
46 Gerard, pp. 51, 63; 7th Report, R.C.H.M., p. 540; Simpson, *Campion*, p. 440.
47 Gerard, pp. 64–5.
48 S.P. Domestic 12/ccxlviii/31.

CHAPTER NINE

1 Mynshall, p. 49; Gerard, pp. 66–7 and note.
2 Gerard, pp. 65–6.
3 *Ibid.*, pp. 66ff.
4 Stow, *London*, p. 258; Pendry, pp. 4, 85; Gerard, p. 68; S.P. Domestic 12/ccxliv/40.
5 Pendry, p. 7; Gerard, pp. 68, 71 note 1.
6 Gerard, pp. 68–9.
7 *Ibid.*, p. 69.
8 *Ibid.*, p. 72.
9 Stow, *London*, p. 363; Pendry, p. 41; Gerard, p. 72.
10 S.P. Domestic 12/ccxlviii/103; Gerard, p. 74.
11 Gerard, pp. 102–103.
12 *Ibid.*, pp. 77–8.
13 C.R.S. 39, p. 180.
14 Gerard, pp. 78ff.
15 Pendry, pp. 7ff., 260; Hat. Cal. VI, p. 562; Gerard, p. 79.
16 Gerard, pp. 81, 82, 90; Knox, *Diaries*, p. lxix; Pendry, p. 249; Caraman, *Garnet*, p. 204.
17 Gerard, pp. 81–2.
18 *Ibid.*, pp. 82–6.
19 *Ibid.*, p. 98.
20 *Ibid.*
21 Cecil, p. 28.
22 S.T., I, pp. 1078, 1080, 1081; Tierney, III, pp. iv–vi.
23 Devlin, *Life*, pp. 164–6.
24 C.R.S. 5, pp. 322ff.
25 C.R.S. 5, p. 151; S.P. Domestic 12/ccxii/70; Devlin, *Life*, p. 170.
26 Gerard, p. 99.
27 A.P.C. XXVII, p. 38.
28 S.P. Domestic 12/ccxlvi/49.
29 S.P. Domestic 12/ccxlvii/33, 53.
30 C.S.P. Domestic 1591–1594, p. 398.
31 S.P. Domestic 12/ccxlvii/79.
32 S.P. Domestic 12/ccxlvii/66, 12/ccxli/2.
33 S.P. Domestic 12/ccxlvii/78, 79.
34 S.P. Domestic 12/ccxlviii/78, 12/ccxlix/12.
35 S.P. Domestic 12/ccxlvii/78, 91.
36 A.P.C. XXVII, p. 21; Hat. Cal. VI, p. 313; S.P. Domestic 12/cclii/14.
37 Hat. Cal. VI, p. 413.
38 Gerard, pp. 102–103; Hat. Cals. IV, p. 498, V, pp. 24–5, VI, pp. 311–13; S.P. Domestic 12/ccxlviii/41, 103.
39 Gerard, pp. 104–105.
40 Gerard, pp. 105–106.

41 S.P. Domestic 12/cclxii/123; Gerard, pp. 106–108; C.R.S. 39, pp. 221–4.
42 Gerard, p. 108.
43 Gerard, pp. 109ff. and note.
44 Gerard, p. 114.
45 Langbein, pp. 4–9, 73, 75, 81ff.; *Harleian Miscellany*, III, pp. 539, 583.
46 Hat. Cal VII, p. 260.
47 Gerard, p. 114.
48 *Ibid.*, pp. 116ff.
49 *Ibid.*, pp. 123–5.
50 *Ibid.*, pp. 125–7.
51 *Ibid.*, pp. 128ff.
52 *Ibid.*, p. 130.
53 Caraman, *Garnet*, pp. 67, 181; Weston, p. 119.
54 Caraman, *Garnet*, pp. 225–6; Gerard, p. 131.
55 Gerard, pp. 131ff.
56 *Southwark*, p. 13.
57 Gerard, pp. 133ff.
58 *Ibid.*, pp. 134, 138.
59 Hat. Cal. VII, p. 417.
60 Gerard, pp. 138–9; Caraman, *Garnet*, p. 238.

CHAPTER TEN

1 Hicks, *Father Persons*, p. 114; *Harleian Miscellany*, III, pp. 515–18.
2 Bruce, *Manningham*, p. 159.
3 Read, *Walsingham*, II, p. 3; Somerset, p. 89.
4 C.S.P. Spanish I, pp. 589, 591; Somerset, p. 189.
5 Hurstfield, p. 373; Somerset, p. 560.
6 Ashdown, p. 197.
7 S.P. Domestic 12/ccxlii/53, 12/ccxlii/116, 12/ccxlv/26, 12/ccxlix/12.
8 S.P. Domestic 12/ccxxxviii/163.
9 Hat. Cal. IV, pp. 461–3.
10 Ashdown, p. 199; Clifford, pp. 20–21.
11 *Conference . . .*, p. 194; C.S.P. Spanish IV, pp. 184–6; S.P. Domestic 12/ccxlix/92, 12/ccxlviii/53; Hat. Cal. V, pp. 251–4.
12 *Conference . . .*, pp. 1ff.
13 Caraman, *Garnet*, p. 253.
14 Smith, p. 433; Knox, *Letters*, pp. 348–51.
15 Weston, pp. 31, 39 note; Somerset, p. 520.
16 Gerard, p. 92; Hat. Cal. V, pp. 24–5.
17 S.P. Domestic 12/clxxiii/79, 80; C.S.P. Domestic 1581–1590, p. 223; Simpson, *Religious Assocs.*, p. 26; Morris, *Troubles*, II, pp. 25–6; Jessopp, pp. 97–8; C.R.S. 52, p. 232 note.
18 Jessopp, p. 98.
19 Hasler, pp. 513–15; Jessopp, p. 98.
20 C.R.S. 5, pp. 210–11.
21 C.R.S. 5, p. 209.
22 Gerard, p. 148.
23 *Ibid.*, pp. 144, 147.
24 *Ibid.*, pp. 164, 168.
25 Harmsen, p. 234; McConica, p. 430; C.R.S. 2, p. 285; Hat. Cal. X, p. 135; McKerrow, p. 209.
26 Hat. Cals. XI, p. 365, XII, p. 229, XIV, p. 194, XV, p. 25; S.P. Domestic 12/clxv/21.
27 Aydelotte, p. 67; Jessopp, p. 242; S.P. Domestic 12/xxxiv/38.
28 Gerard, p. 193.
29 Law, *Controversy*, I, pp. 26–7; Read, *Walsingham*, II, p. 430; Meyer, pp. 105, 391.
30 Read, *Walsingham*, II, p. 282 note; Caraman, *Garnet*, pp. 209–210; Law, *Controversy*, I, p. xv.
31 Clancy, pp. 162, 164, 167.
32 C.R.S. 39, pp. 303–309.
33 Hurstfield, p. 387; Pritchard, p. 52; Haile, p. 288.
34 Meyer, p. 396; Caraman, *Garnet*, p. 219.
35 Caraman, *Garnet*, pp. 207, 219.
36 Caraman, *Garnet*, pp. 240–41; Law, *Controversy*, II, p. xvii; Law, *Sketch*, p. lxvii; Meyer, p. 421.
37 S.P. Domestic 12/cclxxxvi/57; Law, *Controversy*, II, pp. 226–41.
38 Tierney, III, pp. cxxxiiiff.
39 Meyer, p. 432; Tierney, III, pp. cxxxiii–cxxxv; Caraman, *Garnet*, p. 268.
40 Tierney, III, p. cxxxv.
41 Caraman, *Garnet*, p. 272; Law, *Controversy*, I, pp. 206–208.
42 Law, *Sketch*, pp. xxii, lxxxvi, xcii;

Caraman, *Garnet*, p. 54; Devlin, *Life*, p. 257.
43 Law, *Sketch*, p. xciii; Caraman, *Garnet*, p. 289; Law, *Controversy*, I, p. xxi; Meyer, p. 431.
44 S.P. Domestic 12/cclxxxiii/70.
45 Law, *Sketch*, pp. xcvi–ciii; A.P.C. XXXII, pp. 205, 300, 316.
46 Law, *Sketch*, pp. ci note, cvi.
47 Tierney, III, p. clvii; Knox, *Diaries*, I, pp. 362–3; Law, *Controversy*, II, p. 15; S.P. Domestic 12/ccxxxviii/160–63, 168; C.R.S. 5, p. 269; Anstruther, *Seminary Priests*, I, pp. 63–8.
48 Hat. Cal. XII, p. 205.
49 Tierney, III, p. clvi; Southwell, p. xiii; Caraman, *Garnet*, p. 294; Devlin, *Patriotism*, p. 351.
50 Law, *Sketch*, p. cxix; Tierney, III, p. clxxxvi.
51 Southwell, p. xiv; Harmsen, pp. 201–202; C.R.S. 34, p. 5; S.P. Domestic 14/viii/21–5.
52 Somerset, p. 558; Hat. Cal. IX, pp. 345, 391, 440; Stafford, p. 215; Hicks, *Sir Robert Cecil*, p. 135.
53 Davies, *Europe*, pp. 497, 502; Law, *Controversy*, II, p. 247; Bossy, *Henri IV*, p. 89; Anstruther, *Seminary Priests*, I, pp. 36–8.
54 Somerset, p. 553; Foley, I, pp. 47, 50.
55 Law, *Controversy*, II, pp. 227–9.
56 Caraman, p. 296; Jardine, *Gunpowder Plot*, pp. 207–208; Clancy, p. 78.
57 Caraman, pp. 298–9; Gerard, p. 46.
58 Somerset, p. 566; Caraman, *Garnet*, p. 302.
59 C.S.P. Venetian IX, pp. 554, 557.
60 S.P. Domestic 12/cclxxxvii/49, 12/cclxxxvii/53; Foley, I, p. 54; Fraser, p. xxvi; C.S.P. Domestic 1601–1603, p. 303.
61 Weston, p. 222; Bruce, *Manningham*, p. 146.
62 Somerset, p. 568; Fraser, p. xxvi.
63 Law, *Controversy*, II, p. 241; C.S.P. Venetian IX, p. 565; Somerset, p. 569; Bruce, *Manningham*, pp. 147–8; T. Wright, II, p. 495.

CHAPTER ELEVEN

1 Wilson, p. 663; Patterson, pp. 38–9.
2 Patterson, pp. 41, 42, 54; Loomie, *Spain*, pp. 42, 56.
3 C.S.P. Venetian IX, pp. 566, 570; C.S.P. Venetian X, pp. 2, 15; Bruce, *Manningham*, p. 154; Tierney, IV, pp. lxiv–lxvi.
4 Loomie, *Toleration*, p. 9.
5 Caraman, *Garnet*, p. 306.
6 Fraser, p. xxxi; Nichols, *James*, pp. 75–6 and note; C.S.P. Venetian X, p. 9.
7 Gerard, p. 213 note; Morris, *Condition*, pp. 20, 23, 27; A.P.C. XXXII, pp. 495, 496; C.S.P. Venetian X, p. 17; Nichols, *James*, p. 52.
8 Patterson, pp. 18–19; Hicks, *Standen*, V, p. 205; Fraser, p. 15; C.S.P. Venetian X, pp. 77, 513; Jardine, *Remarks*, p. 33.
9 C.S.P. Spanish IV, pp. 682, 724; Loomie, *Spain*, pp. 10, 73, Hicks, *Standen*, V, pp. 195, 206.
10 Bruce, *Correspondence*, pp. 75–6; Tesimond, pp. 58–9.
11 Fraser, p. 41; Morris, *Condition*, p. 24.
12 Anstruther, *Priests*, I, pp. 372–4; Watson, pp. 89–90; Harmsen, p. 11; Bossy, *Henri IV*, p. 90; Tierney, IV, pp. xix–xx.
13 Nichols, *James*, pp. 70, 81, 83; Hat. Cal. XV, p. 232; Loomie, *Toleration*, p.14; Bruce, *Correspondence*, pp. 31, 37.
14 Morris, *Condition*, pp. 21, 22; Hat. Cal. XV, p. 73.
15 Stow, *Annals*, p. 823; Nichols, *James*, p. 113; Tanner, pp. 24–5; C.S.P. Venetian X, pp. 74–5; Fraser, pp. 62–3; Camden, *Annals*, p. 641.
16 Tierney, IV, p. i; Gardiner, *History*, p. 100.
17 Hicks, *Father Robert Persons*, p. 114.
18 Tierney, IV, pp. i, iii, xxxi; Gardiner, *History*, p. 101.
19 Tierney, IV, pp. xiii, xv, xxix, xxxvii.
20 Morris, *Condition*, p. 74; Tierney, IV, pp. v, xiii; Caraman, *Garnet*, p. 334; Hat. Cal. XV, pp. 154, 206.
21 Morris, *Condition*, p. 74.

22 Hat. Cal. XV, pp. 193, 194, 230; C.S.P. Domestic 1603–1610, pp. 20, 23, 27, 28, 30.
23 Tierney, IV, pp. xxvi, xxx, xxxii–xxxiii; Stow, *Annals*, pp. 829, 831; Birch, pp. 27, 31.
24 Gardiner, *History*, p. 115; Loomie, *Toleration*, p. 15.
25 Hat. Cal. XV, pp. 202, 219; Tierney, IV, pp. xv–xvi.
26 Loomie, *Toleration*, p. 20; C.R.S. 2, pp. 216–17.
27 Hat. Cal. XV, p. 2; Loomie, *Toleration*, pp. 6, 12, 15.
28 Loomie, *Toleration*, p. 21.
29 *Ibid.*, pp. 13, 15, 16.
30 *Ibid.*, p. 17; Loomie, *Spain*, p. 11.
31 Loomie, *Spain*, p. 13; Loomie, *Guy Fawkes*, pp. 17, 18, 22.
32 Hat. Cal. XVII, p. 512; Loomie, *Guy Fawkes*, p. 11.
33 Loomie, *Guy Fawkes*, pp. 18, 21, 22–3, 63.
34 *Ibid.*, pp. 28–9, 31–2; Loomie, *Spain*, p. 16.
35 Loomie, *Spain*, p. 31; Loomie, *Guy Fawkes*, p. 30; Loomie, *Toleration*, pp. 23, 27; C.S.P. Venetian X, p. 514.
36 Hat. Cal. XV, pp. 264, 278–9; C.S.P. Domestic 1603–1610, pp. 60, 111; Bruce, *Correspondence*, p. 36.
37 Morris, *Condition*, p. 41; Patterson, pp. 25–6, 49; Tanner, pp. 50, 83; Fraser, pp. 84–5; Stow, *Annals*, pp. 834–5; Hicks, *Standen*, V, p. 196.
38 Tanner, pp. 24ff., 83–5; Patterson, p. 52; Gardiner, *History*, pp. 202, 203; Fraser, p. 90.
39 Patterson, p. 53; Gardiner, *History*, p. 203.
40 Loomie, *Toleration*, p. 29; Foley, I, p. 61.
41 Loomie, *Toleration*, pp. 30–31; Hicks, *Standen*, V, p. 219.
42 Fraser, p. 85; Loomie, *Toleration*, p. 34.
43 Loomie, *Toleration*, pp. 21–2, 34, 40; Fraser, p. 79; Tesimond, p. 74.
44 Loomie, *Toleration*, pp. 35, 36; Loomie, *Spain*, p. 37.
45 Tierney, IV, pp. xxi, xxii and note, xxvi, xxxv.
46 Loomie, *Toleration*, pp. 25, 32, 39; S.P. Domestic, 14/xix/40.
47 Caraman, *Garnet*, pp. 310, 313–4.
48 S.P. Domestic 14/xix/41; S.T. II, p. 288; JH, *A True and Perfect* . . .

CHAPER TWELVE

1 Jardine, *Trials*, p. 127; Somerset, pp. 262–3; King's Book, p. 247.
2 Hat. Cal. XVII, pp. 512–15; Caraman, *Garnet*, pp. 242, 264.
3 Hodgetts, *Secret Hiding-Places*, p. 76; Loomie, *Guy Fawkes*, p. 2 note.
4 S.P. Domestic 14/xviii/87.
5 S.P. Domestic 14/ccxvi/205; Hat. Cal. XVIII, p. 96; S.P. Domestic 14/xix/11.
6 Jardine, *Trials*, pp. 100, 274, 307; Fraser, pp. 203–204; S.P. Domestic 14/ccxvi/117, 124.
7 Hat. Cal. XVIII, p. 96; S.P. Domestic 14/xix/41.
8 King's Book, pp. 247ff.
9 King's Book, p. 248.
10 *Ibid.*; Fraser, p. 97; Tanner, p. 29; Gerard, pp. 180–81.
11 S.P. Domestic 14/xix/44.
12 S.P. Domestic 14/xix/41, 42.
13 Caraman, *Garnet*, p. 317; S.P. Domestic 14/xx/45; Fraser, p. 46; Loomie, *Toleration*, p. 40; Morris, *Condition*, pp. 72–3.
14 Tesimond, pp. 81, 82, 87, 90; Jardine, *Trials*, p. 308; Loomie, *Toleration*, p. 41; Loomie, *Spain*, pp. 46–7.
15 Loomie, *Toleration*, pp. 38, 39, 55–6; *Rymer*, pp. 597–9; Gardiner, *History*, p. 224.
16 Hat. Cal. XVII, p. 611; Hicks, *Standen*, V, p. 202.
17 Hat. Cal. XVII, pp. 23–5; *King Lear*, p. xviii, I.ii.100.
18 C.S.P. Domestic 1603–1610, pp. 240, 256, 259; Hat. Cal. XVII, pp. 569–70; Fraser, p. 126; Mathew, p.16.
19 McClure Thomson, p. 36.
20 Caraman, *Garnet*, p. 319; Stow, *London*, p. 107; S.P. Domestic

14/xix/40; Gardiner, *Declarations*, pp. 510–11, 517.

21 S.P. Domestic 14/xix/41; Loomie, *Spain*, p. 66; Gardiner, *Declarations*, pp. 510ff.

22 Gardiner, *Declarations*, pp. 510ff.

23 Foley, IV, pp. 141–2; Caraman, p. 320; S.P. Domestic 14/xix/40.

24 Gardiner, *Declarations*, pp. 513–15.

25 S.P. Domestic 14/xiv/41; Morris, *Condition*, pp. 76–7; Foley, IV, pp. 61–2 note; Tierney, IV, pp. cix–cxi and note.

26 Stow, *Annals*, p. 844; Fraser, pp. 102, 120, 141.

27 Fraser, pp. 133–4; Nichols, *James*, pp. 537ff.; Tyacke, p. 7.

28 Morris, *Condition*, pp. 78–9.

29 Jardine, *Gunpowder Plot*, p. 179; Hodgetts, *Secret Hiding-Places*, pp. 163–4; Fraser, pp. 143–7; Jardine, *Trials*, pp. 24–5.

30 Morris, *Condition*, pp. 79–80; Anstruther, *Vaux*, p. 278; Gardiner, *Declarations*, p. 517.

31 S.P. Domestic 14/xix/40; Gardiner, *Declarations*, pp. 515, 516.

32 Fraser, pp. 116–17, 158; Gardiner, *Declarations*, p. 516; S.P. Domestic 14/xviii/87 and 14/ccxvi/166.

33 S.P. Domestic 14/ccxvi/2; Fraser, p. 151; King's Book, pp. 235ff.

34 Fraser, p. 151.

35 Loomie, *Spain*, pp. 70–71; Loomie, *Toleration*, p. 42.

36 Hat. Cal. XVII, pp. 475–6, 477; S.P. Domestic 14/ccxvi/16, 17.

37 C.S.P. Venetian X, pp. 290–92; McClure Thomson, pp. 57–9; Hat. Cal. XVII, pp. 484–6.

38 Hat. Cal. XVII, p. 479; S.P. Domestic 14/ccxvi/53, 54, 70, 97, 106, 107.

39 S.P. Domestic 14/ccxvi/92; Gerard, pp. 197–8.

40 Hat. Cal. XVII, pp. 490–91.

41 Hat. Cal. XVII, pp. 477–8; C.S.P. Venetian X, pp. 290–92; King's Book, p. 236; JH, *A True and Perfect . . .*; Clancy, p. 85.

42 C.S.P. Domestic 1603–1610, pp. 254, 259; Hat. Cal. XVII, pp. 500–501; S.P. Domestic 14/xvi/113, 14/xvii/31, 14/ccxvi/97.

43 *B.M. Add. MSS*, 6178, 98.

44 S.P. Domestic 14/ccxvi/145.

45 S.P. Domestic 14/xvii/18; Tesimond, p. 9; Hat. Cal. XVII, pp. 528, 567; Fraser, p. 211; S.P. Domestic 14/ccxvi/166; Jardine, *Trials*, pp. 281–2;

46 *Rymer*, pp. 639–40.

47 Tesimond, pp. 162–3; S.P. Domestic 14/xviii/21.

48 Tesimond, pp. 162–3.

49 Hat. Cals. XV, p. 219, XVI, pp. 399–401, XVIII, p. 41; Anstruther, *Seminary Priests*, II, pp. 138–9; Foley, I, pp. 466–8; Morris, *Condition*, pp. 210–11, 220; S.P. Domestic 14/ccxvi/197, 198.

50 Gerard, p. 203; S.P. Domestic 14/xviii/35.

51 Hat. Cal. XVIII, pp. 99, 105; C.S.P. Domestic 1603–1610, pp. 284, 310; Gerard, pp. 204–7.

52 Foley, IV, pp. 202ff.; S.P. Domestic 14/ccxvi/187, 197.

53 Gerard, p. 45; Tesimond, p. 165; Jardine, *Gunpowder Plot*, p. 182.

54 S.P. Domestic 14/ccxvi/194; *MSS Anglia* 3, 58; Foley, IV, pp. 66–9.

55 Hodgetts, *Secret Hiding-Places*, pp. 170–71; Fraser, p. 213.

56 Gilbert, p. 417.

57 S.P. Domestic 14/xviii/29.

58 Gilbert, p. 417; S.P. Domestic 14/xviii/38; Nash, I, pp. 585ff.; *Harleian MSS*, 360, 58; Morris, *Condition*, pp. 152–3.

59 Nash, I, pp. 585ff.; *Harleian MSS*, 360, 58.

60 Caraman, pp. 316, 343; S.P. Domestic 14/xviii/52.

61 Jardine, *Trials*, pp. 120–21 note; JH, *A True and Perfect . . .*

62 Fraser, pp. 218, 229–34; Tesimond, pp. 205, 217, 229.

63 Jardine, *Trials*, pp. 194–5; King's Book, p. 237; S.P. Domestic 14/xviii.19.

64 S.P. Domestic 14/xviii/64, 14/xix/11; Nash, I, pp. 585ff.; *Harleian MSS*, 360, 58; Caraman, *Garnet*, pp. 244–5, 227.
65 S.P. Domestic 14/xviii/87, 14/xix/11; Caraman, *Garnet*, pp. 253–5.
66 Anstruther, *Vaux*, p. 341; S.P. Domestic 14/ccxvi/192.
67 S.P. Domestic 14/ccxvi/194; Star Chamber Proceedings, James I, bundle 5, file 16.
68 Hepworth Dixon, pp. 393–4.
69 C.S.P. Venetian X, pp. 327–8; Morris, *Condition*, pp. 185ff.
70 Tesimond, pp. 197–201; Morris, *Condition*, pp. 185ff.
71 Kilburn, p. 17; Caraman, *Owen*, p. 14.
72 Hat. Cal. XVIII, pp. 60–61; S.P. Domestic 14/ccxvi/242; C.S.P. Ireland 1603–1606, pp. 412–13.
73 Morris, *Condition*, pp. 173–5; Hat. Cal. XVIII, p. 108.
74 Hat. Cal. XVIII, p. 74; S.P. Domestic 14/xix/27.
75 Hat. Cal. XVII, pp. 594–5.
76 S.P Domestic 14/ccxvi/216; C.S.P Venetian X, pp. 337–8.
77 S.T., II, pp. 217ff.; Morris, *Condition*, pp. 224ff.
78 S.P. Domestic 14/xix/15, 14/ccxvi/210, 212; Fraser, p. 207.
79 Morris, *Narrative*, p. 175; Hat. Cal. XVIII, p. 98; McClure Thomson, p. 37; S.P. Domestic 14/xix/59.
80 McClure Thomson, p. 37; C.S.P. Venetian X, pp. 337–8; Hat. Cal. XVIII, pp. 95–6, 107–111; Jardine, *Trials*, pp. 322–3, 326–7; S.P. Domestic 14/xx/11.
81 King James Bible, John 15:13.
82 S.P. Domestic 14/xxi/4; C.S.P. Venetian X, pp. 321, 350–51.
83 S.T., II, pp. 355–8; Morris, *Condition*, pp. 288–97.
84 Caraman, *Garnet*, pp. 236, 244.
85 Loomie, *Spain*, pp. 81–4; Jardine, *Trials*, pp. 346–7; Pollen, *Father Garnet*, p. 49; Hat. Cal. XVIII, p. 455.
86 Gerard, pp. 207–9.
87 Gerard, pp. 181 note, 269 note; Morris, *Condition*, pp. cxciii, cxcv, ccviii.
88 Morris, *Condition*, pp. ccxiii–ccxxv, 201.
89 S.P. Domestic 14/xxiv/38, 14/xxiv/40; Hat. Cal. XVIII, pp. 47, 49.

EPILOGUE

1 Reynolds, p. 210; Patterson, p. 79; Statutes, 3 Jac. I. c. 4.
2 Notestein, p. 145; Statutes, 3 Jac. I. c. 1, 4, & 5; Kilburn, p. 14; Hat. Cal. XVIII, p. 68.
3 Dures, p. 46; Statutes, 3 Jac. I. c. 1; Patterson, pp. 78, 92; Lunn, p. 241.
4 Donne, pp. xxxviii, xxxix, 152, 165, 228; Patterson, pp. 81–2; C.S.P. Domestic 1603–1610, pp. 363, 370, 397; Lunn, pp. 241–2.
5 Dures, p. 46; Donne, pp. 167–8; Hat. Cal. XVIII, pp. 419–20; C.S.P. Domestic 1603–1610, pp. 330, 425.
6 Hat. Cal. XVIII, pp. 173, 234; Tierney, IV, pp. cxxxvii–cxxxviii; Davies, *Isles*, p. 403.
7 Rowlands, p. 156; Fraser, p. 283; Southwell, p. 41; Spratt, pp. 26–8.
8 Brock, p. 56; Stow, *London*, pp. 319–20; Stow, *Annals*, p. 912; Parmiter, pp. 96, 99.
9 Stewart, pp. 244, 344, 346; Hodgetts, *Secret Hiding-Places*, pp. 196–215; Fraser, p. 278.
10 Ollard, p. 142; Davies, *Isles*, pp. 516–17, 522; *Twelfth Night*, 5.i.375; I am grateful to Fr. Thomas M. McCoog S.J. for the information about the Jesuit descendants of Cecil and Walsingham.
11 Fraser, pp. 267–8, 269–70.
12 Foley, IV, pp. 202–44; Fraser, p. 275.
13 Jardine, *Trials*, p. 380; *Macbeth*, pp. xx, 59 note 9, 2.iii.4–12; Greenblatt, pp. 336–8.
14 Southwell, p. 8 note; McCoog, p. 265; Goldsmith, 5.200–202; J. Wright, pp. 132, 135, 147.
15 Weston, p. 28 note.
16 McCoog, pp. 62–3; Patterson, p. 93 note; Tyacke, pp. 22, 223, 733, 866, 888.
17 Brock, pp. 1, 49, 54–5, 237; Hastings, pp. 97, 225, 227.

18 Sewell, pp. 170–75; *The Daily Telegraph*, 5 March 1998; Philanglus, p. 851; *Curse*, pp. i, iv, 37.
19 *The Times*, 18 January 2005; Downing Street Press Office, 15 October 2004; *The Herald*, 4 June 2001; *The Guardian*, 15 January 2005.
20 Sewell, pp. 2–3; E. Waugh, *Brideshead*, p. 105.
21 *Los Angeles Times*, 8 November 2001; Bond, pp. 43–53.
22 C.R.S. 60, pp. 160–601.
23 Cecil, p. 36.
24 Tesimond, p. 154; S.P. Domestic 14/xx/11.
25 Tierney, III, p. clxi note.
26 Hat. Cal. XVIII, p. 108; S.P. Domestic 14/xix/41.

Bibliography

Unless otherwise specified, place of publication is London.

7th Report of the Royal Commission on Historical Manuscripts, Vol. 6.I, 1879

Abbott, Geoffrey, *The Book of Execution: An Encyclopaedia of Methods of Judicial Execution*, 1995

Ackroyd, Peter, *London: The Biography*, 2001

Allen, Cardinal William, *A Brief History of the Glorious Martyrdom of Twelve Reverend Priests*, ed. J.H. Pollen, S.J., 1908

Allison, A.F., 'John Gerard and the Gunpowder Plot', *Recusant History*, Vol. V, No. 2, 1959–1960

An Advertisement written to a Secretary of My L. Treasurer of England, by an English Intelligencer . . ., 1592

Anderson, Benedict, *Imagined Communities: Reflections on the Origin and Spread of Nationalism*, 2nd edn, 11th imp., 2002

Anstruther, Godfrey, O.P., *Vaux of Harrowden: A Recusant Family*, Newport, 1953

——, *Seminary Priests*, Vols. I and II, Great Wakering, Essex, 1975

(A.P.C.) Acts of the Privy Council, Vols. X–XXXII, 1895–1907

Ashdown, Dulcie M., *Tudor Cousins: Rivals for the Throne*, Stroud, Glos., 2000

Aydelotte, Frank, *Elizabethan Rogues and Vagabonds*, 1967

Baddesley Clinton, Warwickshire, The National Trust, 1998

Basset, Bernard, S.J., *The English Jesuits, from Campion to Martindale*, 1967

Birch, Thomas, ed., *The Court and Times of James the First*, 1848

Birt, H.N., O.S.B., *The Elizabethan Religious Settlement: A Study of Contemporary Documents*, 1907

B.M. Add. MSS, 6178, 98
B.M. Add. MSS, 21, 203
B.M. Add. MSS, 34, 218
Bond, Michael, Gisli Gudjonsson, and Ian Robbins, 'Interrogation', *New Scientist*, Vol. 184, No. 2474, 20 November 2004
Bossy, John, *The English Catholic Community 1570–1850*, 1975
——, 'Henri IV, the Appellants and the Jesuits', *Recusant History*, Vol. VIII, No. 2, 1966
Briggs, Robin, *Early Modern France 1560–1715*, Oxford, 1977
Brock, M.G., and M.C. Curthoys, eds., 'The Nineteenth Century', *The History of the University of Oxford*, Vol. VI, Oxford, 1997
Bruce, John, ed., *Correspondence of King James VI of Scotland*, printed for the Camden Society, 1861
——, ed., *Diary of John Manningham, of the Middle Temple, 1602–1603*, printed for the Camden Society, 1868

Camden, William, *The Annals . . . of King James I*, reprinted 1719
——, *The History of the Most Renowned and Victorious Princess Elizabeth*, Book 4, reprinted 1630
Camm, Dom. Bede, O.S.B., *Forgotten Shrines*, reprinted 1936
Caraman, Philip, S.J., *Henry Garnet 1555–1606 and the Gunpowder Plot*, 1964
——, *Saint Nicholas Owen, Maker of Hiding Holes*, 1980
——, ed., *The Other Face: Catholic Life under Elizabeth I*, 1960
——, ed., *The Years of Siege: Catholic Life from James I to Cromwell*, 1966
Carvajal y Mendoza, Luisa de, *A Spanish Heroine in England: Doña Luisa de Carvajal*, 1905
(C.C.H.) Classical County Histories, Essex, II, republished Wakefield, Yorks., 1978
Cecil, William, *The Execution of Justice in England*, ed. Robert M. Kingdom, Cornell, 1965
Challoner, Bishop Richard, *Memoirs of the Missionary Priests*, reprinted 1924
Character of a Popish Successor and what England may expect from Such a One . . ., 1680
Clancy, Thomas H., S.J., *Papish Pamphleteers: The Allen-Persons Party*

and the Political Thought of the Counter-Reformation in England, 1572–1615, Chicago, 1964

Clark, Andrew, ed., *Register of the University of Oxford*, Vol. II, Part I, Oxford, 1887

——, ed., *Register of the University of Oxford*, Vol. II, Part III, Oxford, 1888

Clifford, Arthur, ed., *The State Papers and Letters of Sir Ralph Sadler*, Edinburgh, 1809

Conference About the Next Succession to the Crown of England . . ., reprinted 1681

Cook, A.H., Prisoners of the Tower, Vol. I, 1959

(C.R.S.) Catholic Record Society, Miscellanea, Vol. 2, 1905

(C.R.S.) Catholic Record Society, Vol. 5, 'Unpublished Documents relating to the English Martyrs, Vol. I, 1584–1603', ed. J.H. Pollen, 1908

(C.R.S.) Catholic Record Society, Miscellanea VII, Vol. 9, 1911

(C.R.S.) Catholic Record Society, Miscellanea XII, Vol. 22, 1921

(C.R.S.) Catholic Record Society, Vol. 34, 'London Sessions Records 1605–1685', ed. Hugh Bowler, 1934

(C.R.S.) Catholic Record Society, Vol. 39, 'Letters and Memorials of Father Robert Persons S.J., 1578–1588', ed. L. Hicks, S.J., 1942

(C.R.S.) Catholic Record Society, Vol. 52, 'Verstegan Papers', ed. Anthony G. Petti, 1959

(C.R.S.) Catholic Record Society, Vol. 60, 'Recusant Documents from the Ellesmere Manuscripts 1577–1715', ed. Anthony G. Petti, 1968

(C.S.P.) Calendar of State Papers, Domestic, 1553–1558, 1998

(C.S.P.) Calendar of State Papers, Domestic, 1547–1580, 1856

(C.S.P.) Calendar of State Papers, Domestic, 1581–1590, Nendeln, Liechtenstein, reprinted 1967

(C.S.P.) Calendar of State Papers, Domestic, 1591–1594, 1867

(C.S.P.) Calendar of State Papers, Domestic, 1595–1597, 1869

(C.S.P.) Calendar of State Papers, Domestic, 1598–1601, 1869

(C.S.P.) Calendar of State Papers, Domestic, 1601–1603, 1870

(C.S.P.) Calendar of State Papers, Domestic, 1603–1610, 1857

(C.S.P.) Calendar of State Papers, Domestic, Addenda, 1566–1579, 1871

(C.S.P.) Calendar of State Papers, Domestic, Addenda, 1580–1625, 1872

(C.S.P.) Calendar of State Papers, Ireland, 1603–06, 1872
(C.S.P.) Calendar of State Papers, Scottish, II, 1563–1569, Edinburgh, 1900
(C.S.P.) Calendar of State Papers, Spanish, I, 1558–1567, 1892
(C.S.P.) Calendar of State Papers, Spanish, II, 1568–1579, 1894
(C.S.P.) Calendar of State Papers, Spanish, III, 1580–1586, 1896
(C.S.P.) Calendar of State Papers, Spanish, IV, 1587–1603, 1899
(C.S.P.) Calendar of State Papers, Venetian, VIII, 1581–1591, 1894
(C.S.P.) Calendar of State Papers, Venetian, IX, 1592–1603, 1897
(C.S.P.) Calendar of State Papers, Venetian, X, 1603–1607, 1900
Curse of Popery, and Popish Princes, to the Civil Government and Protestant Church of England. Demonstrated from the Debates of Parliament in 1680 . . ., 1716

Davies, Norman, *Europe: A History*, reprinted 1997
——, *The Isles: A History*, 1999
Devlin, Christopher, *The Life of Robert Southwell, Poet and Martyr*, 1956
——, 'The Patriotism of Robert Southwell', *The Month*, December 1953
Donne, John, *Pseudo-Martyr*, ed. and with an Introduction by Anthony Raspa, Montreal, 1993
Duffy, Eamon, *The Stripping of the Altars: Traditional Religion in England, c.1400–c.1580*, Yale, 1992
Duncan-Jones, Katherine, *Sir Philip Sidney: Courtier Poet*, 1991
Dunn, Jane, *Elizabeth and Mary: Cousins, Rivals, Queens*, 2003
Dures, Alan, *English Catholicism 1558–1642: Continuity and Change*, Harlow, Essex, 3rd imp., 1988

Edwards, Francis, S.J., *Guy Fawkes: The Real Story of the Gunpowder Plot?*, 1969
Elton, G.R., *England Under the Tudors*, 2nd edn, 1974

'Father Southwell and his Capture', *The Rambler*, Vol. VII, 1857
Fea, Allan, *Secret Chambers and Hiding-Places*, 3rd edn, 1908
Foley, Henry, S.J., *Records of the English Province of the Society of Jesus*, 7 vols., 1st Series, 1877–83

Foster, Joseph, *Alumni Oxonienses 1500–1714*, Wiesbaden, 1968
Fowler, Thomas, *The History of Corpus Christi College, with Lists of its Members*, Oxford, 1893
Fraser, Antonia, *The Gunpowder Plot: Terror and Faith in 1605*, 1997

Gardiner, Samuel R., *History of England, Vol. I, 1603–1642*, 1883
——, 'Two Declarations of Garnet relating to the Gunpowder Plot', *English Historical Review*, Vol. III, July 1888
Gerard, John, S.J., *The Autobiography of an Elizabethan*, tr. from the Latin by Philip Caraman, S.J., 2nd edn, 1956
Gilbert, C. Don, 'Thomas Habington's Account of the 1606 Search at Hindlip', *Recusant History*, Vol. XXV, 2000–1
Girouard, Mark, *Life in the English Country House: A Social and Architectural History*, 1979
Goldsmith, Oliver, *The Good Natured Man*, ed. A.S. Collins, 1936
Greenblatt, Stephen, *Will in the World: How Shakespeare became Shakespeare*, 2004
Gunpowder Plot Book, P.R.O., SP 14/216

Haigh, Christopher, *English Reformations: Religion, Politics, and Society under the Tudors*, Oxford, 1993
——, 'The Continuity of Catholicism in the English Reformation', *Past & Present*, No. 93, November 1981
Haile, Martin, *An Elizabethan Cardinal: William Allen*, 1914
Harleian Miscellany, Vol. I, reprinted, New York, 1965
Harleian Miscellany, Vol. III, 1745
Harleian MSS, 360, 58
Harmsen, T.H.B.M., *John Gee's Foot out of the Snare (1624)*, Netherlands, 1992
Hasler, P.W., 'The House of Commons, 1558–1603', in *The History of Parliament*, Vol. 3, 1981
Hastings, Selina, *Evelyn Waugh: A Biography*, 1994
(Hat. Cal.) Historical Manuscripts Commission, Calendar of the Manuscripts of the Most Hon. the Marquis of Salisbury, Vols. IV–XIX, 1892–1965
Hayward, John, ed., *The Penguin Book of English Verse*, reprinted 1984
Hepworth Dixon, William, *Her Majesty's Tower*, 7th edn, 1885

Hicks, Leo, S.J., 'Sir Robert Cecil, Father Persons and the Succession', *Archivum Historicum Societatis Iesu*, Vol. 24, Fasc. 47, Rome, 1955
——, 'Father Robert Persons S.J. and the Book of Succession', *Recusant History*, Vol. IV, No. 3, October 1957
——, 'The Embassy of Sir Anthony Standen in 1603', *Recusant History*, Vols. V–VII, 1959–1964
Hodgetts, Michael, *Secret Hiding-Places*, Dublin, 1989
——, *Harvington Hall*, Archdiocese of Birmingham Historical Commission, Stafford, 1991
——, 'In Search of Nicholas Owen', *The Month*, October 1961
——, 'Nicholas Owen in East Anglia', *The Month*, August 1962
——, 'Elizabethan Priest-Holes: I–VI', *Recusant History*, Vols. XI (1971–2); XII (1973–4); XIII (1975–6)
——, 'Tanner on Nicholas Owen: A Note', *Recusant History*, Vol. XV, No. 6, October 1981
——, 'A Topographical Index of Hiding-Places', *Recusant History*, Vol. XVI, No. 2, October 1982
——, 'The Owens of Oxford', *Recusant History*, Vol. XXIV, October 1999
Hogge, Alice, 'Closing the Circle: Nicholas Owen and Walter Owen of Oxford', Recusant History, Vol. XXVI, October 2002
Holinshed, Raphael, *The Third Volume of Chronicles . . . continued to the year 1586*, 1587
Hurstfield, Joel, 'The Succession Struggle in Late Elizabethan England', in S.T. Bindoff, J. Hurstfield, and C.H. Williams, eds., *Elizabethan Government and Society: Essays presented to Sir John Neale*, 1961

Janelle, Pierre, *Robert Southwell the Writer: A Study in Religious Inspiration*, 1935
Jardine, David, *A Narrative of the Gunpowder Plot*, 1857
——, ed., *Criminal Trials*, Vol. II, 1832
——, 'Remarks upon Letters of Thomas Winter and Lord Mounteagle, lately discovered by John Bruce, Esq., F.S.A.', *Archaeologia*, XXIX, 1844
Jessopp, Augustus, *One Generation of a Norfolk House*, 3rd edn, 1913
J.H., *A True and Perfect Relation . . . of the Gunpowder Treason*, reprinted 1662

Jones, Norman L., *Faith by Statute: Parliament and the Settlement of Religion, 1559*, 1982

Judges, A.V., ed., *The Elizabethan Underworld: A Collection of Tudor and Early Stuart Tracts*, 1930

Kilburn, E.E., *Blessed Nicholas Owen, Maker of Hiding Holes*, 1931

(King's Book) 'His Majesty's Speech in this Last Session of Parliament . . . Together with a discourse of the manner of the discovery of the late intended Treason . . .', *Harleian Miscellany*, Vol. IV, 1745

Knox, T.F., ed., 'The Letters and Memorials of William, Cardinal Allen', in *Records of the English Catholics Under the Penal Laws*, II, 1882

——, ed., *The First and Second Diaries of the English College, Douai*, 1878

Langbein, John H., *Torture and the Law of Proof*, Chicago, 1977

Lansdowne MSS, 72/48

Law, T.G., *A Historical Sketch of the Conflicts between Jesuits and Seculars in the Reign of Queen Elizabeth*, 1889

——, ed., *The Archpriest Controversy: Documents Relating to the Dissensions of the Roman Catholic Clergy, 1595–1602*, 2 vols., printed for the Camden Society, 1896 and 1898

Loomie, Albert J., S.J., *Spain and the Jacobean Catholics*, Vol. I, 1603–1612, Catholic Record Society, 1973

——, 'Toleration and Diplomacy, the Religious Issue in Anglo-Spanish Relations, 1603–1605', *Transactions of the American Philosophical Society*, New Series Vol. 55, Part 6, Philadelphia, 1963

——, 'Sir Robert Cecil and the Spanish Embassy', *Bulletin of the Institute of Historical Research*, 42, 1969

——, 'Guy Fawkes in Spain: the "Spanish Treason" in Spanish Documents', *Bulletin of the Institute of Historical Research*, Special Supplement No. 9, November 1971

Lunn, Maurus, 'The Anglo-Gallicanism of Dom. Thomas Preston, 1567–1647', in Derek Baker, ed., *Schism, Heresy and Religious Protest*, Cambridge, 1972

MacCulloch, Diarmaid, *Reformation: Europe's House Divided, 1490–1700*, 2003

Magee, Brian, *The English Recusants: A Study of the Post-Reformation Catholic Survival and the Operation of the Recusancy Laws*, 1938

'Martyrs of Chichester, The', *The Rambler*, Vol. VII, April 1857

Mathew, David, *Catholicism in England, 1535–1935*, 1936

Mattingley, Garrett, *The Defeat of the Spanish Armada*, 1959

McClure Thomson, Elizabeth, ed., *The Chamberlain Letters: A Selection of the Letters of Sir John Chamberlain . . . from 1597 to 1626*, 1966

McConica, James, ed., 'The Collegiate University', in *The History of the University of Oxford*, Vol. III, Oxford, 1986

McCoog, Thomas M., S.J., ed., *The Reckoned Expense: Edmund Campion and the Early English Jesuits*, Woodbridge, Suffolk, 1996

McKerrow, R.B., ed., *Dictionary of Printers and Booksellers 1557–1640*, 1910

——, ed., *Dictionary of Printers and Booksellers 1557–1640*, 1968

Meyer, A.O., *England and the Catholic Church under Queen Elizabeth*, tr. Revd J.R. McKee, reprinted 1967

Middleton, Thomas, *A Game at Chess*, ed. Richard Dutton, Oxford, 1999

Milne-Tyte, Robert, *Armada! The Planning, the Battle and After*, Wordsworth Editions, 2nd edn, 1998

Miscellaneous Volume, A.5.3., *Inrolment of Apprentices 1514–1591*. From the Oxford City Archives, held in the Oxfordshire Record Office

Monod, Paul Kléber, *The Power of Kings: Monarchy and Religion in Europe, 1589–1715*, Yale, 1999

More, Henry, S.J., *Historia Missionis Anglicanae Societatis Jesu (1660)*, ed. and tr. from the Latin by Francis Edwards, S.J., 1981

Morris, John, S.J., *The Condition of Catholics under James I: Father Gerard's Narrative of the Gunpowder Plot*, 1871

——, *The Troubles of Our Catholic Forefathers*, 1872

——, ed., *Two Missionaries under Elizabeth: A Confessor and an Apostate*, 1891

MSS Anglia, MSS Grene Collectanea. From the Jesuit Archives, 114 Mount Street, London

Munday, Anthony, *A Brief Answer made unto Two Seditious Pamphlets . . .*, 1582

——, *The English Roman Life*, Amsterdam, 1972

Mynshall, Geoffrey, *Essays and Characters of a Prison and Prisoners*, Edinburgh, reprinted 1821

Nash, Thomas, *The History and Antiquities of Worcestershire*, 2 vols., 1781
Neale, J.E., *Elizabeth I and Her Parliaments, 1559–1581*, I, 1953
——, *Elizabeth I and Her Parliaments, 1584–1601*, II, 1957
Nichols, John, *Progresses of Queen Elizabeth*, Vol. 2, 1823
——, *The Progresses of King James the First*, Vol. 1, 1828
Notestein, Wallace, *The House of Commons, 1604–1610*, Yale, 1971

Ollard, Richard, *The Escape of Charles II after the Battle of Worcester*, reprinted 2002
Ortiz, Don Antonio, *A Relation of the Solemnity wherewith the Catholic Princes King Philip III and Queen Margaret were received in the English College of Valladolid . . .*, tr. Francis Rivers, 1601
Oxburgh Hall, Norfolk, The National Trust, reprinted 2002

Parish registers for St Peter-le-Bailey, Oxford, held in the Oxfordshire Record Office
Parmiter, Geoffrey de C., *The King's Great Matter: A Study of Anglo-Papal Relations, 1527–1534*, 1967
Patterson, W.B., *King James VI & I and the Reunion of Christendom*, Cambridge, 1997
Peck, D.C., ed., *Leicester's Commonwealth: The Copy of a Letter written by a Master of Art of Cambridge*, Ohio, 1985
Pendry, E.D., 'Elizabethan Prisons and Prison Scenes', in Dr James Hogg, ed., *Elizabethan and Renaissance Studies*, Salzburg, 1974
Phelips MSS, P.R.O. SP 31/6/1
'Philanglus', *England's Great Interest in the Choice of this New Parliament, Dedicated to all her Free-holders and Electors*, 1680
Plowden, Alison, *Danger to Elizabeth: The Catholics under Elizabeth I*, Stroud, Glos., 2nd edn, 1999
Plummer, Charles, ed., *Elizabethan Oxford. Reprints of Rare Tracts*, Oxford, 1887
Pollard, A.F., *Tudor Tracts*, 1903
Pollard Brown, Nancy, 'Paperchase: The Dissemination of Catholic Texts in Elizabethan England', in Peter Beal and Jeremy

Griffiths, eds., *English Manuscript Studies 1100–1700*, Vol. I, Oxford, 1989
Pollen, J.H., S.J., *Father Garnet and the Gunpowder Plot*, Catholic Truth Society, 1888
Pollen, Revd J.H., 'An Error in Simpson's "Campion"', *The Month*, June 1905
Porter, Bernard, *Plots and Paranoia: A History of Political Espionage in Britain 1790–1988*, reprinted 1992
Pritchard, Arnold, *Catholic Loyalism in Elizabethan England*, 1979

Ransomer, The, XII, 5, 1948
Rashdall, H., and R.S. Rait, *New College*, 1901
Read, Conyers, *Mr Secretary Walsingham and the Policy of Queen Elizabeth*, 2 vols., Oxford, 1925
——, *Lord Burghley and Queen Elizabeth*, 1960
Reynolds, E.E., *Campion and Persons: The Jesuit Mission of 1580–1*, 1980
Rogers, D.M., ed., *English Recusant Literature 1558–1640*, Vol. 351, 1977
Ross, Stewart, *History in Hiding: The Story of Britain's Secret Passages and Hiding-Places*, 1991
Rowlands, Marie B., 'Recusant Women 1560–1640', in Mary Prior, ed., *Women in English Society 1500–1800*, 1985
Rowse, A.L., *The England of Elizabeth*, 1950
Royal Commission on Historical Monuments, 'City of Oxford', 1939
Royal Commission on Historical Monuments, 'London', Vol. V, 1930
Royal Commission on Historical Monuments, 'N-W Essex', Vol. I, 1916
Rymer Foedera, Vol. 16, 1715

Salter, H.E., *The Historic Names of the Streets and Lanes of Oxford*, Oxford, 1921
——, 'Oxford City Properties', in *Oxford Historical Society*, Oxford, 1926
——, 'Survey of Oxford, Vol. II', in *Oxford Historical Society, New Series*, Vol. XX, Oxford, 1969
Sander, Nicholas, *The Rise and Growth of the Anglican Schism, with a Continuation of the History by Edward Rishton*, tr. from the Latin by David Lewis, 1877
Sewell, Dennis, *Catholics: Britain's Largest Minority*, 2002

Shakespeare, William, *Complete Works of Shakespeare*, Alexander Text, reprinted 1991
——, *King Lear*, ed. Kenneth Muir, The Arden Shakespeare, reprinted 1980
——, *Macbeth*, ed. Kenneth Muir, The Arden Shakespeare, reprinted 2003
——, *Taming of the Shrew, The*, ed. H.J. Oliver, The Oxford Shakespeare, reprinted 1994
Shakespeare's England: An Account of the Life and Manners of his Age, 2 vols., Oxford, 1916
Simpson, Richard, *Edmund Campion. A Biography*, 2nd edn, 1896
——, 'Religious Associations in the Sixteenth Century', *The Rambler*, May 1889
Smith, Alan G.R., *The Emergence of a Nation State: The Commonwealth of England, 1529–1660*, 8th imp., 1991
Somers, John, Baron, *4th Collection of Scarce and Valuable Tracts*, Vol. I, 1752
Somerset, Anne, *Elizabeth I*, 3rd imp., 1999
Southwark: A History of Bankside, Bermondsey and 'The Borough', Southwark Heritage Association, reprinted 1997
Southwell, Robert, S.J., *An Humble Supplication to Her Majesty . . .*, ed. R.C. Bald, Cambridge, 1953
Spratt, T.A.B., ed., *The Autobiography of the Revd. Devereux Spratt*, 1886
Squiers, Granville, *Secret Hiding-Places*, 1934
(S.T.) *State Trials, Cobbett's Complete Collection of . . . From the Earliest Period to the Present Time*, Vols. I and II, 1809
Stafford, H.G., *James VI of Scotland and the Throne of England*, New York, 1940
Star Chamber Proceedings, James I, Vol. 8, Bundle 5, No. 16
Statutes of the Realm . . . from Original Records and Authentic Manuscripts, Vol. IV, Part I, 1819
Stewart, Alan, *The Cradle King: A Life of James VI & I*, 2003
Stow, John, *A Survey of London, written in the year 1598*, Stroud, Glos., reprinted 1997
——, *The Annals or General Chronicles of England*, 1615
Strype, John, *Annals*, Vols. III and IV, Oxford, 1824

Tanner, J.R., *Constitutional Documents of James I, 1603–1625*, Cambridge, 1930
Tawney, R.H., and E. Power, eds., *Tudor Economic Documents*, 1924
Tesimond, Oswald, S.J., *The Gunpowder Plot*, tr. from the Latin by Francis Edwards, S.J., 1973
Thomas, Keith, *Religion and the Decline of Magic*, reprinted 1991
Thomas, Werner, and Luc Duerloo, eds., *Albert and Isabella*, essays, Royal Museums of Art and History, Brussels, 1998
Tierney, Revd M.A., ed., *Dodd's Church History of England with Notes, Additions and a Continuation*, 4 vols., 1840
Torture and Punishment, Royal Armouries, 1997
Treasure, G.R.R., *Seventeenth Century France*, 2nd edn, 1981
Turner, W.H., *Selections from the Records of the City of Oxford, 1509–1583*, Oxford, 1880
Tyacke, Nicholas, ed., 'Seventeenth Century Oxford', in *The History of the University of Oxford*, Vol. IV, Oxford, 1997

Urquhart, Alastair, 'Nicholas Owen: The Master Builder of Secret Passages and Hiding Places', *International History Magazine*, No. 28, April 1975

(V.C.H.) Victoria County History, Buckinghamshire, III, 1925
(V.C.H.) Victoria County History, Essex, II, 1907
(V.C.H.) Victoria County History, Middlesex, IV, Oxford, 1971
(V.C.H.) Victoria County History, Norfolk, II, Folkestone, Kent, reprinted 1975
(V.C.H.) Victoria County History, Northampton, II, i, 1906
(V.C.H.) Victoria County History, Northampton, IV, 1937
(V.C.H.) Victoria County History, Oxford, III, Oxford, 1954
(V.C.H.) Victoria County History, Oxford, IV, Oxford, 1979
(V.C.H.) Victoria County History, Suffolk, II, 1907

Wallace, M.W., *The Life of Sir Philip Sidney*, Cambridge, 1915
Watkins, Susan, *In Public and in Private: Elizabeth I and her World*, 1998
Watson, William, 'A Decacordon of Ten Quodlibetical Questions', in D.M. Rogers, ed., *English Recusant Literature, 1558–1640*, Vol. 197, 1974

Waugh, Evelyn, *Edmund Campion*, 2nd edn, reprinted 1957
——, *Brideshead Revisited*, rev. edn, reprinted 1981
Waugh, Margaret, *Bl. Nicholas Owen*, 1961
Wernham, R.B., *Before the Armada: The Emergence of the English Nation, 1485–1588*, New York, 1966
Weston, William, S.J., *The Autobiography of an Elizabethan*, tr. from the Latin by Philip Caraman, S.J., 1955
W.I. 49/1/21, Walter Owen's will, held in the Oxfordshire Record Office
Wilson, Arthur, 'The Life and Reign of James . . . King of Great Britain', in *A Complete History of England*, 1719
Wood, Anthony à, *Athenæ Oxonienses*, reprinted 1813
——, *History of the University of Oxford*, Vol. II, Part I, Oxford, 1796
Wood, Michael, *In Search of Shakespeare*, BBC, 2003
Wright, Jonathan, *The Jesuits: Missions, Myths and Histories*, 2004
Wright, Thomas, ed., *Queen Elizabeth and Her Times*, 1838

Index